L. A. Utracki (Editor)

Two-Phase Polymer Systems

POLYMER PROCESSING SOCIETY

PROGRESS IN POLYMER PROCESSING

Series Editor: L. A. Utracki

A. I. Isayev
Modeling of Polymer Processing

L. A. Utracki
Two-Phase Polymer Systems

A. Singh/J. Silverman
Radiation Processing of Polymers

L. A. Utracki (Editor)

Two-Phase Polymer Systems

With contributions from

J.-F. Agassant · D. M. Bigg · P. J. Carreau · M. M. Dumoulin
J. J. Elmendorp · D. F. Hiscock · P. Hold · P. R. Hornsby · T. Inoue
J. P. Jog · M. R. Kamal · T. Kawaki · M. J. Kiesel · Y. Kijima · B. Ch. Kim
K. U. Kim · T. Kyu · T. Mutel · V. M. Nadkarni · J. M. Saldanha · D. B. Todd
L. A. Utracki · A. K. Van der Vegt · M. Vincent · S. F. Xavier

Hanser Publishers, Munich Vienna New York Barcelona

Distributed in the United States of America and in Canada by
Oxford University Press, New York

Editor:
Dr. Leszek A. Utracki, National Research Council of Canada, Industrial Materials Research Institute, Boucherville, Quebec, Canada

Distributed in USA and in Canada by
Oxford University Press
200 Madison Avenue, New York, N. Y. 10016

Distributed in all other countries by
Carl Hanser Verlag
Kolbergerstr. 22
D-8000 München 80

The use of general descriptive namens, trademarks, etc, in this publication, even if the former are not especially identified, is not to be taken as a sign that such names, as understood by the Trade Marks and Merchandise Marks Act, may accordingly be used freely by anyone.

While the advice and information in this book are believed to be true and accurate at the date of going to press, neither the authors nor the editors nor the publisher can accept any legal responsibility for any errors or omissions that may be made. The publisher makes no warranty, express or implied, with respect to the material contained herein.

Library of Congress Cataloging-in-Publication Data

Two-phase polymer systems / L. A. Utracki, editor.
    p. cm. –– (Progress in polymer processing series ; v. 2)
    Includes bibliographical references and index.
    ISBN 0-19-520915-X
    1. Polymers. I. Utracki, L. A., 1931-. II. Series.
    TP 1087. T89 1991   91-73604
    668.9 –– dc20   CIP

Die Deutsche Bibliothek – CIP-Einheitsaufnahme

Two-phase polymer systems / L. A. Utracki (vol. and series ed.). With contributions from J.-F. Agassant ... – Munich ; Vienna ; New York ; Barcelona : Hanser ; New York : Oxford Univ. Press, 1991
(Progress in polymer processing series ; Vol. 2)
ISBN 3-446-15576-7
NE: Utracki, Leszek A. [Hrsg.]; Agassant, Jean-François; GT

ISBN 3-446-15576-7 Carl Hanser Verlag, Munich, Vienna, New York, Barcelona
ISBN 0-19-520915-X Oxford University Press, New York

The cover picture is described on page 264.

PROGRESS IN POLYMER PROCESSING SERIES,

Leszek A. Utracki, *Series Editor*

ADVISORY BOARD

FOREWORD

The term "polymer processing" encompasses the whole spectrum of polymer science and engineering domains, from polymerization, compounding, forming, decorating, and assembling to properties. The polymers in question can be either natural or man-made, elastomeric, thermoset or thermoplastic. Formation method may be by extrusion (including profiles, films or fibers), molding in solid or molten state, casting, etc.

In polymer processing there is an uneasy balance between the technology and science. Most frequently the technology has led the way in opening new areas, such as in development of rubbers or thermosets, in extrusion or molding. In this case the science followed, providing understanding and tools for optimization. However, in the area of thermoplastics frequently the reverse was true; here the scientific curiosity and directed research led to synthesis of these market dominating polymers. Nowadays development of new materials or products first and foremost requires solid understanding of basic principles, followed by good engineering design of the forming process and sound knowledge of the market. The successful polymer process researcher or engineer must have a broad knowledge of fundamental principles and engineering solutions.

There are several polymer engineering societies, based on the national membership, promoting interests of the industry and disseminating pertinent technological information. There are also chemical and physical societies with polymer divisions taking a more basic approach to polymer science. However, searching for information to solve a specific polymer processing problem still leads to difficulties. There are hundreds of thousands of articles published annually. They provide a fragmented body of information requiring, to start with, either familiarity within the narrow domain or an encyclopedic knowledge of the field. On the other hand, with the exponential expansion of information the preparation of monographic single-authored books is a tedious and frustrating task, particularly when trying to catch up with recent and most pertinent developments.

There is also another element - the globalization of science and technology. As the development of polymer blend technology demonstrates, the center of activity within a domain may rapidly shift from one country to another, from one continent to the next. The recognition of these globalization tendencies led to the formation in 1985 of the Polymer Processing Society, PPS, the first fully international professional organization dedicated to the promotion, growth and development of scientific understanding, and innovation in polymer processing. It provides a forum for the world-wide community of engineers and scientists and publishes the Society journal, *International Polymer Processing*, as well as the book series, *Progress in Polymer Processing*.

Progress in Polymer Processing was initiated by PPS in 1986 and formally established in 1988. Its aim is to provide complete, in-depth, up to date information on various aspects of polymer science and engineering. The Series Editor, with the internationally based Advisory Board, is responsible for selecting the volume topics and Volume Editor(s), as well as for general supervision of content, quality, and form of the publication.

The series aims to provide multi authored monographic books on the subject of current interest to the international polymer processing community. Depending on the breadth of the selected topic, the volume may have the character of either an exhaustive monograph, providing an actual, complete picture of the selected field, or of an in-depth progress report of its dominant aspects. As time progresses we hope to shift more and more toward the former character. Using the multi authored format we expect on the one hand to provide a more complete picture of the selected topic viewed from different perspectives, and on the other to shorten the production time, keeping the published information up to date. We expect to produce two to three volumes a year. Current list of titles has a dozen positions.

To accomplish these goals, we shall need the help and cooperation of the international community, as well as serious effort by everyone involved in the process, the authors, editors, members of the Advisory Board, and the publisher. It is hoped that in time we will become more accustomed to our tasks and hence more efficient, always guided by the needs of our colleagues within the polymer processing community.

L.A. Utracki
Series Editor

CONTENTS

CONTRIBUTORS

J. F. Agassant, École Nationale Supérieure des Mines de Paris, Centre de Mise en Forme des Matériaux, Unité associée au CNRS no. D13740, Sophia Antipolis, VALBONNE 06560, FRANCE

D. M. Bigg, BATELLE, 505 King Avenue, Columbus, OH 43201-2693, U.S.A.

P. J. Carreau, Ecole Polytechnique, Chemical Engineering Department, Montreal (Quebec) H3C 3A7, CANADA

M. M. Dumoulin, National Research Council of Canada, Industrial Materials Institute, 75 de Mortagne, Boucherville, QC, J4B 6Y4, CANADA

J. J. Elmendorp, Koninklijke/Shell Laboratorium, Amsterdam, Badhuisweg 3, 1031 CM Amsterdam, THE NETHERLANDS

D. F. Hiscock, BATELLE, 505 King Avenue, Columbus, OH 43201-2693, U.S.A.

Peter Hold, Polymer Processing Research Center, 32 Gulf View Court, Milford, CT 06460, U.S.A.

Peter R. Hornsby, Brunel University, Department of Materials Technology, Uxbridge, Middlesex, UB8 3PH, UNITED KINGDOM

Takashi Inoue, Tokyo Institute of Technology, Department of Organic and Polymeric Materials, Ookayama, Meguro-ku, Tokyo 152, JAPAN

J. P. Jog, National Chemical Laboratory, Chemical Engineering Division, Pune, 411008, INDIA

M. R. Kamal, McGill University, Department of Chemical Engineering, Montreal (Quebec) H3A 2A7, CANADA

Takao Kawaki, Mitsubishi Gas Chemical Co., Inc., Tokyo Central Research Laboratories, Niijuku, Katsushika-ku, Tokyo 125, JAPAN

Mark J. Kiesel, Caterpillar Co. Ltd., Peoria, IL 61614, U.S.A.

Yasuhiko Kijima, Mitsubishi Gas Chemical Co., Inc., Tokyo Central Research Laboratories, Niijuku, Katsushika-ku, Tokyo 125, JAPAN

Byoung Chul Kim, Korea Institute of Science and Technology, Polymer Processing Laboratory, P.O. Box 131, Cheongryang, Seoul, KOREA

Kwang Ung Kim, Korea Institute of Science and Technology, Polymer Processing Laboratory, P.O. Box 131, Cheongryang, Seoul, KOREA

Thein Kyu, University of Akron, Center for Polymer Engineering, Akron, OH 44325, U.S.A.

A. T. Mutel, DuPont Canada Inc., Research Center, P.O. Box 5000, Kingston (Ontario) K7L 5A1, CANADA

V. M. Nadkarni, National Chemical Laboratory, Chemical Engineering Division, Pune 411008, INDIA

Jeanne M. Saldanha, Lord Corporation, Materials and Process Division, Erie, PA 16514, U.S.A.

David B. Todd, APV Chemical Machinery Inc., 901 Durham Avenue, S. Plainfield, NJ 07080, U.S.A.

L. A. Utracki, National Research Council of Canada, Industrial Materials Institute, 75 de Mortagne, Boucherville, QC, J4B 6Y4, CANADA

A.K. Van der Vegt, University of Technology, Delft, Laboratory of Polymer Technology, Julianalaan 136, 2628 BL, Delft, THE NETHERLANDS

M. Vincent, Ecole Nationale Superieure des Mines de Paris, Centre de Mise en Forme des Materiaux, Unite associee au CNRS no. D13740, Sophia Antipolis, VALBONNE 06560, FRANCE

S. F. Xavier, Indian Petrochemicals Corporation Ltd., Research Centre, Baroda 391 346, Gujarat, INDIA

PREFACE

This second volume in the **Progress in Polymer Processing** series took nearly three years to prepare. During this period, the publication procedure of the series was established, necessitating some modification of the texts. However, the long preparation time was caused primarily by expansion of the chapters in terms of depth of subject treatment and length.

There is no need to stress the importance of two-phase polymeric systems both to science and technology. Mixtures of polymer melts with gas (*foams*), with another molten polymer (*alloys* and *blends*) and with solid particles (*filled systems* and *composites*) constitute the majority of modern polymeric materials. This volume aims to stress the common denominators of these materials, methods of combining the ingredients (mixing or compounding), the need for care in structure development during processing as well as the effects of the two-phase nature on properties of finished products. Note that such a volume program resulted in the elimination of an important subclass of two-phase systems, the coprocesssed materials, but it still left such a vast area of science and technology to be covered as to make a complete presentation of the topic impossible. For this reason, from the point of view of **Progress in Polymer Processing** goals this volume lies somewhere between the "progress on" and a monographic book.

The fourteen chapters were written by prominent, competent and internationally known experts in the field. With the Editor's intervention limited to rules on the mechanics of text preparation and a general outline of topics, the chapter development was left to each author's good judgment. An attempt has been made to preserve uniform nomenclature in the volume. However, in order to avoid any possibility of confusion, a list of symbols and abbreviations is provided after each chapter. The common subject index illustrates the generality of behavior of the two-phase systems.

The volume begins with an overview "*On Processing Two-Phase Polymer Systems.*" This chapter attempts to provide a general outlook on the problems associated with mixing, extruding, and molding two-phase systems, as well as information on the most recent developments in the field. Chapter 2 by *Hold* and Chapter 3 by *Todd* deal with complementing aspects of mixing - a general introduction is provided by the first author followed by a treatise on the use of the ever popular twin-screw extruder prepared by the second.

The next two chapters, Chapter 4 by *Hornsby* and Chapter 5 by *Kim* and *Kim*, discuss processing and properties of rigid, structural polymeric foams. *Hornsby* provides a general overview on processing and properties of thermosets and thermoplastics based systems with and without reinforcement, while *Kim* and *Kim* concentrate on the flow behavior of polyvinylchloride during the chemical foaming process.

Chapters 6 to 10 deal with polymer blend science and technology. In Chapter 6, *Elmendorp* and *van der Vegt* discuss the fundamental aspects of microrheology during flow and processing. Since control of the blend structure is essential for optimization of performance, the chapter's importance exceeds the goals modestly stated by the authors. In Chapter 7, *Dumoulin* et al. review the flow behavior of polymer blends with special emphasis on the polyethylene/polypropylene two-phase mixtures. The crystallization of polymer blends is comprehensively examined in Chapter 8 by *Nadkarni* and *Jog*. The authors stress the diversity of effects observed upon blending as well as the importance of crystalline morphology on properties of finished products. A new use of blend technology for controlling birefringence in injection molded optical disks is discussed in Chapter 9 by *Kijima* et al. In Chapter 10, *Kyu* et al. provide an important insight into the blend fracture. The authors observed that two-phase morphology (generated via controlled spinodal decomposition) leads to enhancement of mechanical properties when compared to that of single-phase specimens having identical composition.

The remaining Chapters 11 to 14 concentrate on processing and process-related behavior of reinforced polymer composites. *Vincent* and *Agassant* in Chapter 11 discuss principles of flow orientation in the fiber-filled composites. The authors developed a predictive model and evaluated its validity by comparing it with experimental data on injection molded simple geometry specimens. The work by *Mutel* and *Kamal* on rheology of fiber-reinforced polymer melts is presented in Chapter 12. It complements the preceding chapter well. The authors provide insight into the orientational effect on flow behavior especially within the transient zone. It is well-known that properties of the fiber-filled systems greatly depend on the fiber length-to-diameter (aspect) ratio. Chapter 13, by *Hiscock* and *Bigg*, discusses new technology for manufacturing the long fiber composites in which the large aspect ratio is retained. Finally, in Chapter 14, *Xavier* reviews the crystallization phenomenon in polymer composites with particular attention to transcrystallinity and modification of performance it introduces.

Although the volume provides but a glimpse of the vast domain of two-phase polymeric materials, the selected topics illustrate their importance and congruency. The mixture of academic and technological approaches stems from our basic philosophy that progress in polymer processing depends on the integration of both. As Editor, I wish to thank all the authors for the efforts and cooperation.

L. A. Utracki
Montreal, 1990

CHAPTER 1

PROCESSING TWO-PHASE POLYMER SYSTEMS

by L.A. Utracki

National Research Council Canada
Industrial Materials Institute
75 de Mortagne
Boucherville, QC, J4B 6Y4
CANADA

The term "two-phase polymer systems" encompasses the polymer-based foams, blends, and particulate-filled materials, i.e., filled polymers and composites. There are significant differences as far as rheology and thermal properties are concerned between these systems and the more traditional single-phase polymers. As a result, processing of two-phase materials usually requires special care. This chapter provides a broad outline of the flow and thermal behavior, followed by discussion on compounding, extrusion, and injection molding. There were two goals in preparing this chapter: 1) to provide an overview of the field, and 2) to indicate how topics treated in the following chapters fit into the general picture. In the overview, the differences in behavior and processing between single-phase and two-phase systems were stressed taking into account the most recent developments in the field.

1.1   INTRODUCTION

World production of plastics in 1995 is projected at 76 million metric tons with the annual growth rate (AGR) throughout the decade of 3.7% [1]; this volume is sufficient to form a ring 1.6 m in diameter all around the earth. During the same period the expected AGR of polymer alloys and blends is 12%, and that of composites 16%. In 1987, already 21% of polymers were used in blends and 29% in composites and filled plastics [2]. It is not difficult to calculate that if these tendencies persist, by 1995 all manufactured resins will be used in multi-phase systems. It should be noted that this trend is particularly strong for the fastest growing segment of the polymer industry, the engineering resins. There are two factors that will moderate this tendency: 1) a need for single-phase polymers in some applications (e.g., PTFE, UHMWPE, light-sensitive polymers), and 2) use of polymer alloys and blends as matrices for composites, filled materials, or foams.

Presence of the second phase necessitates changes in processing. Great care must be taken to distribute uniformly the minor component in the matrix, to generate and preserve the optimum morphology, guaranteeing the desired properties of the final product. Furthermore, with the growing concern for the environment, the recycling of the two-phase polymeric systems is becoming increasingly important for the plastics industry.

This chapter starts with an outline of flow behavior in the two-phase systems, followed by a short discussion of their thermal properties. Compounding methods and equipment are then discussed, followed by two sections dealing with the special aspects of extrusion and injection molding of polymeric foams, blends, and composites. Throughout the text the reader is referred to topical chapters of the volume addressing specific aspects of the two-phase polymeric system processing or properties.

1.2   FLOW BEHAVIOR IN TWO-PHASE POLYMERIC SYSTEMS

The rheology of two-phase systems was recently reviewed [2,3]. Here only a brief outline will be made.

1.2.1   Newtonian Flows

The Newtonian flow is limited to low concentrations and/or low deformation rates. It is useful to consider the two-phase polymeric systems as either suspensions or emulsions based on viscous matrix liquid. In this region the flow is described by the relative viscosity equation:

$$\eta_r = \eta/\eta_0 = f(\phi, p) \tag{1.1}$$

where η and η_0 is viscosity of the system and the matrix, respectively, ϕ is the volume fraction of the dispersed phase and p the aspect ratio of particles. Eq (1.1) can also be considered as a two-parameter equation of the maximum packing volume fraction, ϕ_m, and the intrinsic viscosity, $[\eta] = \lim_{\phi \to 0} (\eta_r - 1)/\phi$, e.g., [4]:

$$\eta_r = 1 + [\eta] \, \tilde{\eta} \, \phi$$

$$\tilde{\eta} \equiv 4 (1 - Y^7) / [4 (1 + Y^{10}) - 25 Y^3 (1 + Y^4) + 42 Y^5] \tag{1.2}$$

$$Y \equiv [2 (\phi_m/\phi)^{1/3} - 1]^{-1}$$

For hard spheres, monodispersed suspensions $[\eta]$ take on the Einstein value of 2.5, whereas for emulsions [5]:

$$[\eta] = 2.5 \, (\eta_d + 2\eta_0/5 + \eta_i/5d) / (\eta_d + \eta_0 + \eta_i/5d) \tag{1.3}$$

where η_d and η_i represent the viscosity of the dispersed liquid and the liquid–liquid interphase, respectively, with d being the droplet diameter. $[\eta]$ strongly depends on shape and the aspect ratio, p. For example, for fibers with $p \leq 50$ experimentally [3]:

$$[\eta] = 2.34 + 0.1636 \, p \tag{1.4}$$

On the other hand, ϕ_m depends on the shape, aspect ratio, spatial orientation (and sometimes on ϕ). For example, suspensions of polydisperse hard spheres can show variation of ϕ_m from 0.52 to 0.94, whereas for uniform fibers with $p \leq 150$, experimentally ϕ_m varies from 0.70 to 0.001 [3]:

$$1/\phi_m = 1.38 + 0.0376 \, p^{1.4} \tag{1.5}$$

For mineral suspensions of irregular shape both $[\eta]$ and ϕ_m can be determined experimentally and then η_r calculated from Eq (1.2).

The anisometric particles in Newtonian liquid rotate with the period:

$$t_p = 2\pi \, (p + 1/p)/\dot{\gamma} \tag{1.6}$$

where $\dot{\gamma}$ is the rate of shear. It is worth noting that t_p is symmetrical, numerically identical for a prolate, $p > 1$, and oblate, $p < 1$, ellipsoid of rotation: $t_p(p) = t_p(1/p)$. The $[\eta]$ is a measure of the rotational encompassed volume. As the concentration $\phi \to \phi_m$ the three-dimensional rotation of anisometric particles changes to a two-dimensional one which causes both $[\eta]$ and ϕ_m to change [6].

In flow the particle is subjected not only to convective but also to Brownian forces. Their relative intensity is expressed by the Peclet number:

$$Pe = \dot{\gamma}/D_r \tag{1.7}$$

where D_r is the rotational diffusion coefficient. Presence of Brownian motion is responsible for randomization of particle orientation imposed by the flow, which in turn leads to appearance of the elastic, normal forces. For $Pe \ll 1$ and $p = (1+e) \to 1$ (e is particle eccentricity) theoretically [7]:

$$\sigma_{12} = \sigma_{12}^0 \{5/2 + e^2 [78/441 + 109/(180 + 5Pe^2)] + 0\,(e^3)\} \tag{1.8}$$

$$N_1 = \sigma_{11} - \sigma_{22} = \phi\sigma_{12}^0 \{216e^2 Pe/35\,(36 + Pe^2) + 0\,(e^3)\} \tag{1.9}$$

$$N_2 = \sigma_{22} - \sigma_{33} = \phi\sigma_{12}^0 \{-36e^2 Pe/35\,(36 + Pe^2) + 0\,(e^3)\} \tag{1.10}$$

Thus, suspensions of anisometric particles in Newtonian liquid may exhibit the first, N_1, and the second, N_2, normal stress difference proportional to the matrix shear stress, σ_{12}^0, and to concentration. Note that for spheres, $e = 0$, $N_1 = N_2 = 0$, and $[\eta] = 5/2$; for $e \neq 0$, $N_1 = -6N_2$.

The extensional relative viscosity of anisometric particles suspended in Newtonian liquid can be expressed as [8]:

$$\eta_{r,E} = 3 + 4\phi p^2/3\ln\,(\pi/\phi) \tag{1.11}$$

Note that for $\phi \to 0$ the Troutonian limit $\eta_E/\eta = 3$ is recovered. Eq (1.11) shows a strong dependence on the aspect ratio, p^2. However, it is interesting to note that for spheres $(p = 1)$ $\eta_{r,E}$ slowly increases with ϕ. Experimentally, owing to the irrotational nature of the extensional flow, the relation was found valid up to relatively high concentration.

From the processing point of view the flow of two-phase systems is crucial in controlling the morphology of the finished product. Even in the Newtonian region of small stresses and small deformation rates the orientational effect can be important. In the irrotational flow (e.g., in uni- or biaxial elongation) there is a transient part of the initial stress increase related to the alignment of anisometric particles and/or extension of deformable drops. In homogeneous steady-state shear flow the vorticity counteracts the orientational effects of strain. However, due to periodic accelerated rotation of the anisometric particles (rapid flip then almost stable particle position), they retain preferred orientation controlled by the total strain.

Flow also imposes another morphological feature - the particle exclusion near the wall, leading to a skin-core structure and/or the particle size fractionation within the extrudate. These effects are enhanced at higher concentrations and deformation stresses, i.e., within the non-Newtonian flow region.

To summarize this part, the Newtonian flow of a two-phase system is characterized by: (a) an increase of shear viscosity dependent on concentration of the dispersed phase, polydispersity of particle size, aspect ratio, and geometrical arrangements of the particles; (b) suspensions of anisometric solid particles or elongated drops show presence of the first, N_1, and the second, N_2, normal stress difference, strongly dependent on the aspect ratio, p; (c) similarly, a strong dependence on p is observed for the elongational viscosity, η_E, but by contrast with N_1 and N_2, the η_E does not vanish for spherical suspensions.

1.2.2 Non–Newtonian Flows

The proportionality between the stress and deformation rate requires that the liquid structure remains stable in the full range of explored variables. In the non–Newtonian region this fundamental condition is no longer valid. Most multiphase systems exhibit the apparent yield stress, σ_y, defined as:

$$\sigma_{12} - \sigma_y = \eta(\dot{\gamma})\dot{\gamma} \tag{1.12}$$

where $\eta(\dot{\gamma})$ is shear stress, σ_{12}, or rate of shear, $\dot{\gamma}$, dependent coefficient of viscosity. It can be seen that for $\sigma_y = 0$ and $\eta(\dot{\gamma}) = \text{const} = \eta_0$ the Newtonian relation is recovered.

In rheology it is useful to relate the time of experiment, t, to the characteristic time factor of the material, τ. The ratio is known as the Deborah number:

$$De = \tau/t \tag{1.13}$$

Frequently τ is identified with the longest relaxation time, τ_1, which allows classification of all materials:

Material	De
Newtonian liquid	0 (by definition)
Water	10^{-12}
Lubricating oil	10^{-6}
Polymer melts	10^0 to 10^1
Glasses	10^9 to 10^{11}
Hookean solid	∞ (by definition)

The yield stress indicates presence of a three-dimensional, solidlike structure with interactions able to support stresses $\sigma_{12} < \sigma_y$ but being broken at $\sigma_{12} \geq \sigma_y$. In terms of De all substances (with exception of the ideal Newtonian liquid) must show solidlike behavior, i.e., yield stress, provided that the test time $t \ll \tau$. Thus, it is useful to express σ_y as a function of De_y [2]:

$$\sigma_y = \sigma_y^0 [1 - \exp\{-De_y\}]^u \tag{1.14}$$

where $De_y = \tau_y \dot{\gamma} = \tau_y \omega = \tau_y \dot{\varepsilon} = \ldots$ is the Deborah number in steady–state shear, dynamic or extensional flow, respectively, σ_y^0 is the instantaneous (Hookean) yield stress, τ_y is the longest relaxation time of the three-dimensional structure, and the exponent $u = 0.2$ to 1.0 provides a measure of the structure uniformity.

The presence of σ_y has a profound influence on the type of flow and the resulting morphology of any two-phase system. In the nonuniform stress field present in a tube or a Couette-type flow (e.g., in extruders, and compounding or injection-molding machines), the yield stress leads to a plug flow where only a skin layer undergoes deformation and orientation. This in turn may result in the absence of extrudate swell, wall slip, lack of mixing, jetting, and multiple weld lines in injection molding.

Experimentally, σ_y increases with the strength of interparticle interactions, with the aspect ratio, and with concentration of the dispersed phase, the latter being solid (suspensions), liquid (emulsions), or gaseous (foams). Since the strongest interparticle interactions are observed for the first systems, the suspensions are known to possess yield stress even at relatively low concentration.

Wildemuth and Williams [9] derived:

$$\sigma_y = M \left[\phi - \phi_{m,0}\right] / \left[\phi_{m,\infty} - \phi_{m,0}\right]^{1/m} \tag{1.15}$$

where m and M are material parameters, while $\phi_{m,0} = \phi_m$ and $\phi_{m,\infty}$ represent the maximum packing volume fraction in the absence of stress and at $\sigma_{12} \to \infty$, respectively.

The function $\eta(\dot{\gamma})$ in Eq (1.12) can locally either increase or decrease with the deformation rate. Accordingly, these systems are classified respectively as shear thickening (dilatant) and shear thinning (pseudoplastic). The dilatant systems are relatively rare, observed primarily for suspensions of uniform particles, e.g., blood cells, PS or PVC lattices, mineral powders, etc. [2,3]. By contrast, the pseudoplasticity is nearly universal for polymeric systems. In general:

$$\eta(\dot{\gamma}) = \eta_\infty + (\eta_0 - \eta_\infty)/f(\dot{\gamma}) \tag{1.16}$$

with

$$f(\dot{\gamma}) = [1 + (\dot{\gamma}\tau)^{m_1}]^{m_2} / [1 + b_0 \exp\{(\dot{\gamma}\tau - b_1)^{2b_2}\}] \tag{1.17}$$

where b_i and m_i are parameters. Eqs (1.16) and (1.17) describe both the pseudoplastic ($b_0 = 0$) and dilatant flow.

This volume is dedicated to two-phase systems. During the processing of the immiscible polymer blends one may, by means of composition, pressure, or temperature, move into a single phase region of miscibility. Crossing the boundary condition (binodal or spinodal) may generate a sharp increase of viscosity and elasticity of the system [2]. Two mechanisms of phase separation are known: the nucleation and growth (leading to droplet dispersions) and the spinodal decomposition (responsible for creating a three-dimensional cocontinuous system). Control of processing to ascertain that the second mechanism is responsible for the blends morphology may lead to significant improvement of the performance. Toughening by means of the controllable spinodal decomposition is discussed in Chapter 10.

At constant stress, the viscosity of a particulate-filled melt vs. concentration (curve *a* in Fig. 1.1) shows a monotonic increase with an asymptote at $\phi = \phi_m$. On the other hand, this dependence for blends may show either a positive (over 60% of blends) or negative (about 30%) deviation from the log additivity rule (curves *b* and *c* in Fig. 1.1). For about 10% blends the η vs ϕ dependence passes through a local minimum and maximum (curve *d* in Fig. 1.1). Blends with concentrations corresponding to these extrema are of special interest - they are easier to process and may show more interesting properties [2].

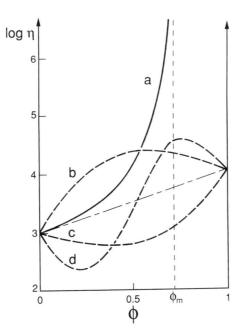

Figure 1.1 *Concentration dependence of shear viscosity for: (a) solid particle–filled liquid, and (b–d) polymer blends.*

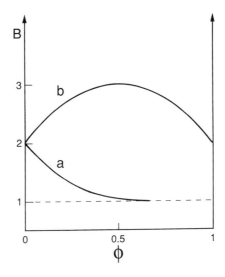

Figure 1.2 *Concentration dependence of extrudate swell for: (a) solid particle–filled melt, and (b) polymer blend.*

Another important distinction between the particulate-filled system and blends originates in the extrudate swell B. For the former B decreases with composition (curve *a* in Fig. 1.2). This lack of swelling is responsible for jetting filled polymers into mold cavity. By contrast, blends usually show enhanced swelling caused by a form recovery (curve *b* in Fig. 1.2) [2].

1.2.3   Flow–Induced Morphology

Flow plays a dominant role in the structure of the processed material. Compounding or mixing must break the dispersed phase (aggregates) and randomly distribute the fragments within the matrix. Thus, the compounding equipment is designed to provide dispersing and distributive stress/strain fields. It is well known that the same compound may process well on one processing line but poorly on another of the same design. Apparently, a small variation in screw and/or die wear, a seemingly insignificant variation in the runners may lead to variation in flow-induced morphology (FIM) and property changes in the finished product. It is useful to consider FIM in relation to an individual particle and then to the whole system.

1.2.3.1   FIM–particle ought be discussed in terms of the orientation, deformation, and migration.

1.2.3.1a   Orientation is limited to anisometric (fiber or platelet) particles. The effect of irrotational (extensional) field is particularly pronounced. On starting, the particle aligns in the direction of the extensional force and then preserves this orientation during the flow. Experimentally, the orientation angle may show $\pm 5°$ variations. In the shear field all particles rotate with the period given by Eq (1.6). The rotation is periodic: a rapid change followed by a nearly stationary flow [10]. This may lead to orientation in the steady-state pipe or annular flows.

1.2.3.1b   Deformation affects both liquid drops and solid particles (especially those with high aspect ratio). Again, the irrotational field is more powerful in deforming liquid drops [2]:

$$D_E/D_S \cong 3.3 \left[1 + (1.9\lambda\kappa)^2\right]^{1/2} / [1 + 1.5\kappa] \qquad (1.18)$$

where (D_E and D_S) are, respectively, deformability in uniaxial extension and in shear, λ is the viscosity ratio, and κ the capillarity parameter:

$$\lambda = \eta_d/\eta_m; \qquad\qquad \kappa = \sigma_{ij}\, d/\nu \qquad (1.19)$$

with subscripts d and m indicating dispersed and matrix phase, respectively, σ_{ij} being the stress, d the equilibrium drop diameter, and ν the interfacial tension coefficient. Thus, fibrillation is predicted to occur preferentially in extensional field, especially in systems with $\eta_d > \eta_m$. Deformation of solid particles, e.g., fibers, requires the vorticity component of the stress tensor. Breaking of glass fibers (GF) in shear field has frequently been observed [11].

1.2.3.1c   Migration of solid particles toward the low stress in the capillary center was reported [10]. Such a hydrodynamic effect differs from the wall exclusion one, in which the solid particles are expelled away from the wall zone. The shear field is also known to be able to fractionate a polydisperse mixture of particles, placing the smaller ones in the higher stress field [12]. These observations are in accord with known "laziness of nature," otherwise known as the principle of energy minimization. Thus, it is no surprise that during flow of a two-phase system the low-viscosity component migrates toward the high-stress domain, e.g., lubricating the flow through pipes, runners, dies, and capillaries [13]. The migration is a kinetic process whose rate depends on the stress gradient, as well as the relative viscosity and normal stress of the components. Figure 1.3 shows schematically the morphology of HDPE/PA blend extruded through a long capillary at 150 and 250°C. The flow induced encapsulation is evident. Another example is the addition of glass fiber, GF, to polypropylene, PP, which frequently leads to apparent lack of change in capillary viscosity up to a concentration exceeding ϕ_m. Similarly, foaming may generate an insignificant effect on the steady-state shear viscosity. In the flow of immiscible blends the capillary viscosity may show a drastic drop upon addition of only a small amount of the second ingredient [14]. The drop is related to the magnitude of the polymer-polymer interaction coefficient, χ_{23}; the larger it is, the deeper the viscosity drop. The mechanism may be a "roll bearing" effect or a preferential adsorption of the lubricating ingredient on the capillary wall and subsequent slip.

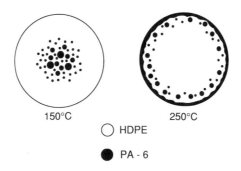

150°C   250°C

○ HDPE

● PA - 6

Figure 1.3 *Schematic representation of skin-core morphology of HDPE blended with 15 wt% PA-6, extruded at 150° C (left) and 250° C (right).*

1.2.3.2 FIM–system strongly depends on concentration. Already at $\phi \gg 0.02$ the solid particles preferentially combine into doublets, while the liquid drops undergo a continuous process of dispersion and coalescence. With an increase of concentration, invariably the yield stress appears leading to the plug flow. This in turn may be responsible for several morphological features: skin-core, absence of extrudate swell (in systems containing solid particles or in blends with large values of λ), jetting into the mold, and poor weld lines, etc. In polymer blends the morphology strongly depends on concentration and stress field. While at $\phi \leq 0.12$ generally only drops are observed, at the higher range, say up to $\phi \cong 0.35$, fibers are apparent. In the middle concentrations, near the (dependent on λ) phase inversion, the two phases are cocontinuous. Accordingly, performance of the material is expected to undergo a nonlinear variation at concentrations of the morphological change. The value of this transitional concentration is stress dependent. In Chapter 7 blend rheology and FIM are discussed in more detail. The effect of blending on crystallinity is reviewed in Chapter 8, and that on the optical properties in Chapter 10.

The above discussion is pertinent to a steady-state flow. However, during the processing the multiphase material is subject to different types of stress field, some generating transient effects, some superimposing different a flow deformation. The multiplicity of the deformation modes may lead to complex morphologies giving evidence of intricate flow history. The simplest example is the variable orientation of fibers across the length and thickness of the injection-molded plates [15,16].

To sum up this part, one observes that: 1) there is great similarity of flow behavior for all two-phase systems: emulsions, suspensions, foams, blends, and composites; 2) the flow-induced morphology depends on the flow field, aspect ratio, and concentration; for blends the thermodynamic properties (χ, ν) must be taken into account; 3) in the presence of the yield stress, the morphology preserves memory of deformation history; and 4) flow controls the morphology which in turn determines the final properties of the material.

1.2.4   Thermal Properties

Every processing operation involves heating, forming, and cooling. Thus the specific influence of the two-phase nature systems on thermal properties needs to be considered.

The energy balance equation per unit volume is usually written equaling the rate of total accumulated energy $(\dot{H})$ with the rate of thermal energy "in," $\dot{H}_H$, minus the rate of thermal energy "out," $\dot{H}_K$, plus the rate of viscous heating, $\dot{H}_\eta$. The rate of heat conduction can be expressed as:

$$\dot{H}_{cond} = k\nabla^2 T \tag{1.20}$$

where k is the thermal conductivity coefficient and $\nabla^2 T$ is the Leplacian of the temperature, whereas the rate of heat convection:

$$\dot{H}_{conv} = \vec{u}\,\rho\,c_p\,\nabla T \tag{1.21}$$

where $\vec{u}$ is the velocity, ρ the density, c_p the heat capacity and ∇T the temperature gradient. Since, during processing, the heat is being conducted and convected both in and out of the elementary volume:

$$\dot{H}_H - \dot{H}_K = \dot{H}_{cond} - \dot{H}_{conv} \tag{22}$$

In multicomponent systems mixing rules are needed to express an average heat conductivity and capacity.

Assumption of a multilayer structure with each layer characterized by its density, ρ_i, conductivity, k_i, and weight fraction, w_i, the average conductivity can be expressed as:

$$\bar{k} = (\sum_i w_i/\rho_i k_i)^{-1} \tag{1.23}$$

The inverse additivity rule is also valid for $\bar{\rho}$. On the other hand, the heat capacity is a weight average quantity:

$$\bar{c}_p = \sum_i w_i c_{pi} \tag{1.24}$$

The inverse additivity rule means that the layer characterized by the lowest value of $\rho_i k_i$ is primarily responsible for $\bar{k}$. Since relative heat conductivity of gas, polymer, and metal are respectively 1:10:10,000, the presence of gas (e.g., in foam) strongly affects $\bar{k}$, but not that of solids.

The relative heat capacity (per gram) of gas, polymer, and metal are 2:4:1, respectively. Thus, the heat capacity of polymer being relatively large, it acts as a heat sink for spurious variations of heating or cooling during the process. Addition of gases or inorganic solids lowers this capacity. For this reason processing foams or filled systems requires finer temperature control than that for neat polymers.

The energy generated in flow is given by:

$$\dot{H}_\eta = \sum_i w_i \sigma_i \dot{\gamma}_i \qquad (1.25)$$

where index i refers to the local stress, σ_i, and rate of straining, $\dot{\gamma}_i$. For uniformly dispersed ingredients in a uniform stress field the relation predicts an average response. However, since most processing equipment has a nonuniform stress field causing ingredient segregation, the flow invariably leads to minimization of $\dot{H}_\eta$.

Thus the total energy balance per unit volume can be written as:

$$\bar{\rho}\bar{c}\dot{T} = \bar{k}\nabla^2 T + \overline{\sigma\dot{\gamma}} \qquad (1.26)$$

with the net convection term on the left hand side and the net conduction and shear heating terms on the right.

1.3   COMPOUNDING

Compounding is a four stage process involving 1) preparation of ingredients (drying, sizing, heating, etc.), 2) premixing (dry blending, homogenization, breakage of agglomerates, fluxing, etc.), 3) melt-mixing (usually with degasing), and 4) chopping, e.g., granulation, pelletization, or dicing.

During the preparation stage the ingredients are analyzed (to verify specifications), and dried. The solid filler or reinforcing particles undergo surface treatment, they are weighed and sometimes preheated. In some formulations (especially those of PVC) up to 40 ingredients are to be mixed together, with proportions usually adjusted to suit the customer processing capability. The maximum permissible moisture level depends on the type of material and process. In general, the radical polymers tolerate higher moisture levels (0.1 to 0.2%) than polycondensates: polyamide (≤ 0.1) or polyester (0.01 to 0.02%). Similarly, extrusion with usual venting/devolatilization does not require as strict drying as injection molding; e.g., maximum moisture level for Bayblend extrusion is 0.1 to 0.2% while for injection molding it is 0.02% [17]. In most cases the air-circulation drier is sufficient. However, for the moisture-sensitive resins the vacuum drying cabinets may be necessary. This is the case for polyamides which must be dried at low temperature to a low moisture level ($< 0.1\%$). Before compounding, care must be taken to dry all ingredients to the level recommended for the most sensitive component. Improperly dried material may depolymerize during the compounding/processing, and form bubbles or blisters causing deterioration of the ultimate properties.

Premixing is accomplished in several types of devices, starting with rotating cone tumblers, ribbon, propeller, kinetic energy, screw, or planetary mixers, ball mills, and fluodized beds. Some mixers (e.g., propeller and kinetic energy type) preblend and flux the material. Equipment selection depends on the type of operation (batch vs. continuous), properties of the material, and cost.

1.3.1 Compounding Equipment

The melt compounders are either batch or continuous type. The former requires lower investment cost, is more labor-intensive, has low output and poor batch-to-batch reproducibility. Recent developments in process control and automatization eliminated some of these traditional disadvantages.

Batch mixers include the two-roll mills, and internal and kinetic energy mixers. The two-roll mill has an open construction allowing for observation of the mixing process. Its primary function is dispersion in the high shear and extensional stress field generated in the gap. Since the gap can be adjusted during mixing, the stresses can be tailored to the variable processing conditions. The distributive action is provided by continuous cutting and folding of the mixed slab. Modern mills can be programmed to perform automatically some or all these operations. Like two-roll mills the internal mixers were originally developed for the rubber industry. Today they find use in mixing not only rubbers but a diversity of plastics formulations, e.g., PVC, ABS, EVAc, PP, PE, engineering resins, master batches, blends, and filled and reinforced systems. Some of the internal mixers are fully automatic with capacity from 0.5 to 300 liters [18]. There is a wide variety of rotor designs. Mixing is carried out under controlled ram pressure.

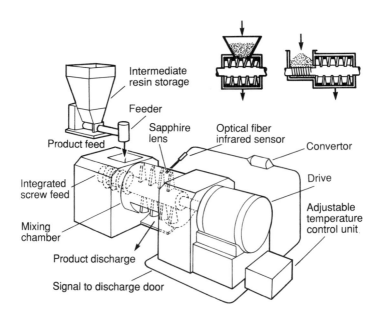

Figure 1.4 *Kinetic energy mixer (Gelimat) [19].*

The kinetic energy, or the high intensity mixer, (see Fig. 1.4) operates on the principle of high-speed rotation of a central shaft with a few sturdy blades. The tip speed of the blades can be adjusted within the range of 15 to 48 m/s to provide residence time between 10 and 20 sec. The apparatus is automatic with the sapphire-tipped IR sensor causing the mixing chamber doors to open when the preset temperature is reached. The batch-to-batch reproducibility is high with the maximum spread of temperature within ± 1.5°C [19].

All the batch mixers require additional equipment either to pelletize the compound or process it into the final product. With the exception of the kinetic energy mixer (which scales at constant tip speed) compounders are difficult to scale-up. In most cases the delivery of a larger unit necessitates reoptimization of mixing parameters for each formulation. Some theoretically based guidance for internal mixer scaling is available [20].

The continuous mixers require high capital investment, but are easily robotizable, have high output, and can be run with a statistical quality loop control (QC) providing a high quality, reproducible compound. To this group belong: 1) twin-shaft continuous mixers, 2) single screw extruders, 3) twin-screw extruders, and 4) speciality mixers.

The twin-shaft continuous mixer (CM), or continuous internal mixer, (see Fig. 1.5) combines the high intensity mixing with continuous operation. The first nonintermeshing, counter-rotating mixer was developed nearly 30 years ago. Since that time the machine underwent several modifications. The unit provides solid conveying, melting, and mixing functions. Like the batch mixers, CMs also require a separate forming unit, usually an in-line melt extruder. There are several barrel and rotor designs available, including the split-barrel models for ready access to the solidified mixed material making it possible to analyze the mixing process. The rotors (single or double stage) are designed with two principal elements: the solid conveying screw and two mixing wings twisted in opposite directions. An adjustable outlet gate is used to control the pressure and residence time. The relative angular positions of the wings are adjustable providing an added control element to the process. The material is continuously pumped from one rotor chamber to another. Thus, the maximum stress, randomization, and total mixing time can be independently adjusted providing a wide variety of mixing conditions. The distribution of residence time is relatively narrow with a tail at longer times [21-23].

The single screw extruder is an inexpensive, difficult to scale up, poor mixer with broad residence time distribution and relatively long residence time [24]. There are two exceptions to this rule: 1) speciality mixing extruders and 2) modified mixing extruders. To the first category belong Buss Kneader or Barmag extruders (neither cheap nor poor), to the second the standard single screw extruders equipped with special mixing screws and mixing add-on devices.

The Buss Kneader (Fig. 1.6) has been in operation for 50 years and it developed a large body of enthusiastic admirers. Here, the constant channel depth screw consists of interrupted flights and three rows of stationary pins in the barrel. During operation the screw rotates and axially oscillates providing good self-wiping of the blades. There is a good deal of distributive mixing during the high-stress dispersive process. The residence time distribution is relatively broad, with only a short tail at longer times. The clam-shell barrel design makes it easy to clean, service, analyze, and optimize the process. The machine is used for blends, PVC formulations, engineering resins (with and without reinforcement), foodstuffs, carbon electrode pastes, etc. [25,26].

Other compounding extruders are: the planetary gear extruders (originally designed for PVC compounding) and the FN-Plastifier (developed for polymer blending and recycling). The planetary gear extruder is a single screw unit with the standard screw stretching about half-way into the barrel with the second (front) part made into a central gear

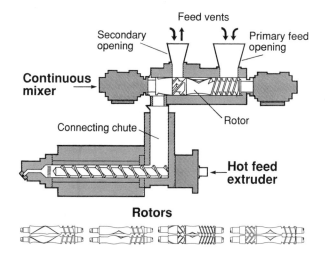

Figure 1.5 *Twin–shaft continuous mixer compounding line. Four standard rotors for (from left) temperature–sensitive materials, filled systems, devolatilization, and short–residence-time materials (courtesy Farrel Co.) [22].*

Figure 1.6 *Open processing section of Buss Kneader, Model MDK/P–140 (courtesy Buss A.G.) [26].*

spindle, rotating several planetary spindles engaged in the barrel. The planetary spindles provide both the dispersive and distributive mixing. Uniform stress and temperature combined with good self-wiping and short residence time make it a popular compounder for PVC formulations [27]. A planetary extruder for high performance resins and blends was recently introduced [28]. It continuously rolls and folds the melt into 200 to 300 μm thin layers. There is good temperature control of both the central screw-spindle and the barrel. Degasing and addition of fillers or reinforcements into melt are possible.

The FN-Plastifier (Fig. 1.7) is a short (L/D = 5) single-screw extruder with a three-start screw extended from the feed zone 2/3 over the screw length. The frontal part of the screw is smooth ending with a flat disk. The material is transported and partially fluxed by the grooved part of the screw. Melting and mixing is performed in the small gap between the smooth part of the screw and the barrel. The pumping is assured by the normal stresses between the flat part of the screw-end and the die (Maxwell-Scalora normal stress extruder principle). The short screw assures short (and narrow distribution) residence time. Large thrust bearing allows for adjustment of the die gap, controlling the magnitude of the normal stresses, i.e., the morphology and the output [29-31].

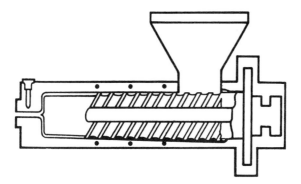

Figure 1.7 *FN short-screw plastifier with a three-start screw, L/D = 5 screws, and a Maxwell-Scalora normal stress extruder front end [29].*

There is a wide variety of modifications of a single-screw extruder to improve its mixing capabilities. Diverse screw designs with mixing elements and/or mixing heads (Fig. 1.8) are available from most screw manufacturing companies. Some of the mixing heads can be provided as add-ons either to be affixed to the end of the screw or operated independently with a separate drive [32-36]. Both the Barmag and the Cavity Transfer Mixer (CTM) require machining of both the screw and the barrel add-ons. By contrast, the newest in this family of mixers, the University of Twente (UT) mixing head uses a smooth barrel [37]. All three mixers are primarily distributive with very little dispersive shear stress provided. Here, the melt flows continuously from dents in the rotor to those in the barrel. Since from one dent it is directed to two others, a continuous splitting of melt streams and folding is achieved. In case of UT, a sleeve (with holes) placed between the barrel and the screw plays the role of the dented barrel. The sleeve slowly rotates under influence of the drag force.

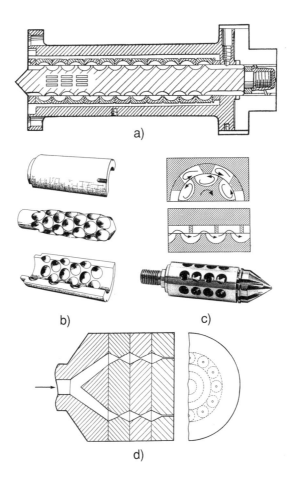

Figure 1.8 *Single–screw extruder mixing heads: (a) Barmag, (b) CTM, (c) University of Twente, and (d) Showa Denko convergent/divergent homogenizer.*

The motionless or static mixers (MM) are efficient in distributing the dispersed melts, blending in two fluids and homogenizing the molten materials (Fig. 1.9). There are more than thirty types of MM available [38]. The principle is to split the flow stream in each element into h channels. Thus the striation number (reflecting the degree of mixedness or homogeneity) for n^* element MM is:

$$S = h^{n^*} \tag{1.27}$$

and the striation thickness $D^* \leq D/S$ with D being the element diameter. The pressure drop of the mixer is proportional to that of a tube of the same length and diameter multiplied by a factor of $K = K(Re)$. For the Reynolds numbers, Re, of interest $K = 3$-5 for Kenics ($h = 2$) and $K = 160$ for Ross ($h = 4$) mixer [39].

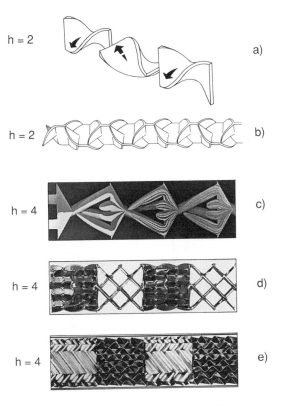

h = 2   a)

h = 2   b)

h = 4   c)

h = 4   d)

h = 4   e)

Figure 1.9 *Motionless mixers: (a) Kenics Corp., h = 2; (b) Komax, h = 2; (c) Ross, h = 4; (d) Sulzer, h = 4; and (e) Koch, h = 8.*

A special add–on motionless, dispersive mixer for homogenization of liquid mixtures, blends, and alloys was developed by Suzaka in Showa Denko (see Fig. 1.8d). The mixer operates on the principle of a flow through consecutive convergent and divergent channels where the drops are elongated and broken into smaller ones. Great improvement of mechanical properties was reported for PE/PP blends homogenized by the mixer [40].

At the present time the most popular compounders are twin-screw extruders. An excellent review of their use is given by Todd in Chapter 2 of this volume. There is a wide variety of twin-screw extruders available on the market (Fig. 1.10): co- and counter-rotating; fully, partially, or not at all, intermeshing (lengthwise or crosswise); with parallel or conical screws; and with solid, segmental, and clam-shell barrels [41]. In addition, there is a possibility of creating hundreds screw designs and configurations making selection of equipment difficult. Several existing rules simplify this task.

In general, the lower the degree of intermeshing, the broader the distribution of the relaxation times and the lower dispersive mixing is achieved. At the limit, the nonintermeshing twin-screw extruders behave as a single screw one with enhanced distributive mixing. Thus, the main application of low intermeshing twin-screw extruders is for dispersing reinforcements, and mixing blends with low ($\lambda \simeq 1$) viscosity ingredients.

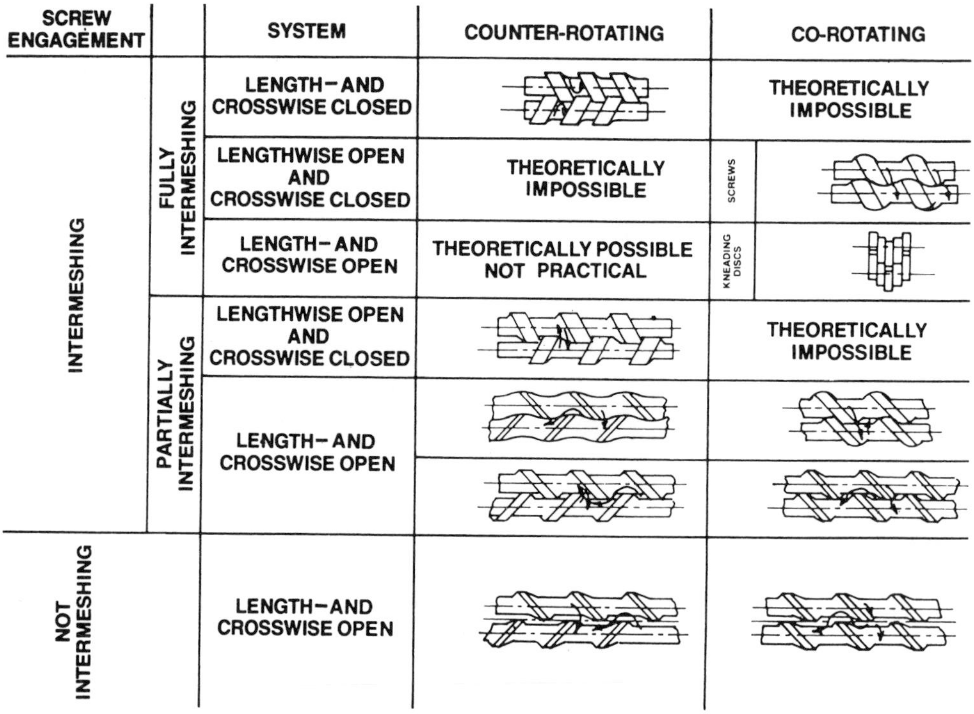

SCREW ENGAGEMENT		SYSTEM	COUNTER-ROTATING	CO-ROTATING
INTERMESHING	FULLY INTERMESHING	LENGTH–AND CROSSWISE CLOSED	*(diagram)*	THEORETICALLY IMPOSSIBLE
		LENGTHWISE OPEN AND CROSSWISE CLOSED	THEORETICALLY IMPOSSIBLE	*(diagram — SCREWS)*
		LENGTH–AND CROSSWISE OPEN	THEORETICALLY POSSIBLE NOT PRACTICAL	*(diagram — KNEADING DISCS)*
	PARTIALLY INTERMESHING	LENGTHWISE OPEN AND CROSSWISE CLOSED	*(diagram)*	THEORETICALLY IMPOSSIBLE
		LENGTH–AND CROSSWISE OPEN	*(diagram)*	*(diagram)*
			(diagram)	*(diagram)*
NOT INTERMESHING		LENGTH–AND CROSSWISE OPEN	*(diagram)*	*(diagram)*

Figure 1.10 *Classification of twin-screw extruders [41].*

By contrast, the intermeshing twin-screw machines are characterized by relatively narrow residence time distribution. They provide a high and controllable degree of dispersive and distributive mixing leading to homogeneous melt. In addition, here the melting is largely independent of the metal/melt friction coefficients (controlling the process in single-screw extruders); there is a low pressure, the positive and controllable conveying of material, good venting, self-wiping action, surge-free discharge, and possibility of multiple feeding.

The intermeshing counter-rotating extruders generate higher shear and extensional stresses in the gap than the co-rotating ones. In addition, there is a different flow pattern in these two types with a higher flux of radial flow in the former. Thus the co-rotating intermeshing twin-screw machine provides more uniform and controllable stress field. These units are used for reactive processing, compounding of alloys and blends, for color masterbatches, and fiber reinforced composition. Co-rotating extruders with a low degree of intermeshing are also used for the latter application. The quality of the product critically depends on the selection of screws and their configuration. However, optimization of screws and operating condition is still an art [42–46].

Recent developments in process control place emphasis on use of the critical product parameters to control the process (e.g., rheological properties, chemical composition, degree of dispersion, color) instead of guarding the processing parameters constant as the older controllers used to do [47–49]. For large volume production ($\dot{Q} > 20\,\text{t/h}$) the twin-screw compounders outperform the single-screw extruders.

The last group of continuous compounders is made of different types of machines which cannot be classified as either shaft or screw processor. To this group belong several rotating disk extruders. The first in this series was the Maxwell and Scarola normal stress extruder, schematically presented in Fig. 1.11 [50]. The machine is made of two discs, one (conical) rotating and the other stationary, with a die hole in the middle. As the material moves toward the center it melts and develops the normal stress which pulls it out through the die. The flow is dispersive laminar, generating co–continuous blend structures which result in good properties [51]. Neither this design nor its modifications [29–31,52] found general acceptance in the industry.

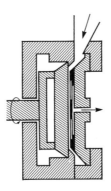

Figure 1.11 *Maxwell and Scalora normal stress extruder [50].*

The second design in this category is the Diskpack (Fig. 1.12) [53]. The machine has the inherent capability of performing all the elementary steps of plastics processing by combination of differently shaped rotating disks in a drumlike housing. The wiping action is provided by stationary channel blocks causing the material to transfer from one disk–gap to another. Melting, laminar mixing, venting, and pumping functions are clearly separated. Diskpack has been used for reactive processing, blending, compounding, mixing, and devolatilizing [54–55].

When selecting and optimizing the equipment the following two parameters may be found useful; 1) goodness of dispersion:

$$M^* \propto d_0/\phi\gamma \qquad (1.28)$$

where d_0 is initial size of the dispersed phase, ϕ its volume fraction, and γ the total strain, and 2) the specific power consumption:

$$E^* = gE/\rho N^3 D^5 \qquad (1.29)$$

with g, gravitational constant; E, power supplied, N, rotational speed; and D, diameter of the compounding machine.

The total strain in Eq (1.28) is usually taken as a weighted average:

$$\bar{\gamma} = \int \gamma f(t)\, dt \qquad (1.30)$$

where $\mathfrak{f}(t)$ is the residence time distribution. It is evident that M* strongly depends on material properties and compounder design - both affecting f(t). In plug flow, where within the plug $\gamma = 0$ and deformation is limited to the high stress annulus, the f(t) resembles a narrow, sharp peak. From Eqs (1.28) and (1.29) it follows that index M* in this case is large, indicating poor mixing.

1.3.2 Compounding Industry

The annual growth rate in the compounding industry has been significantly higher (16 to 18%) than that of the whole plastics industry (3-5%) [56,57]. The growth rate of compounded polymer alloys and color masterbatches has been the highest (18.5%), while that for reinforced and filled plastics lower but still impressive at 10%.

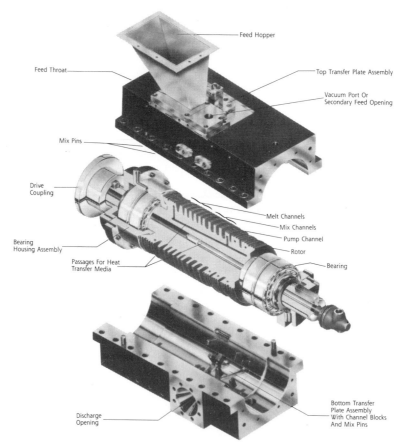

Figure 1.12 *Diskpack FDC/1–V 150 mm compounder (courtesy Farrel Co.) [55].*

In the U.S.A., nearly half of all plastics are being compounded by independent operators. Their market share was distributed as follows: those with sales below one million dollars, 22%; sales between 1 and 5 million, 28%; sales between 5 and 20 million, 25%; and sales over 20 million dollars, 25%. There is a great diversity of compounding equipment, with 44% being more than 15 years old. At the same time, there is a growing tendency for preferential purchase of twin-screw extruders and continuous mixers, with increasing emphasis on micro-processor control of the mixing and statistical quality control of the product. Growing competition in the industry and modernization is a sign of dynamism and strength of the compounding industry [58].

The economy of compounding depends on several parameters: 1) productivity of the compounding line, Pr, expressed as a product of the volumetric output, $\dot{Q}$, and equipment availability, A:

$$Pr = \dot{Q}A \qquad (1.31)$$

where A depends on reliability of the machine, lifetime of wearing parts, availability of spare parts, ease of maintenance, and speed of formulation change-over (dependent on the residence time distribution), 2) low amount of low-quality products and rejects, 3) low cost-to-throughput ratio, 4) low energy consumption, and 5) low personnel cost (automatization, quality control, self-cleaning capability, and robotization).

There are twelve criteria for economic, high-quality compounding [59]: 1) sound fundamental design principles of the compounding machine; 2) flexibility of operation; 3) high screw torque; 4) appropriate selection of the screw elements; 5) a balanced combination of the high-shear dispersive and low-shear distributive screw elements; 6) low pressure at the die; 7) reliability of components and parts; 8) long lifetime of parts; 9) reliable feeding/dosing equipment; 10) reliable granulation method; 11) automation of the process; 12) on-line quality inspection and closed-loop control.

1.3.3   Loop Control

The last two criteria, automation and closed-loop controls, require a comment. While the conventional processing line provides constant screw speed and automatic temperature adjustments along the barrel, an automated line additionally provides an automatic start-up and shut-off, making it possible to store and repeat formulation recipes and providing visual display of operating parameters as well as of data storage. The main advantage of automation comes at the next stage when on-line quality inspection and closed-loop control come into effect. The type of inspected parameters depends on the required product performance, e.g., composition, dispersion, rheology, color, electrical conductivity, optical properties, and mechanical behavior may need to be controlled. The ultimate aim is to integrate the quality parameters with automatic line control so a disturbance (for example, caused by screen-pack blockage, a feed change or equipment wear) will be compensated for by an automatic adjustment of specific process parameters. As shown in Fig. 1.13, starting with conventional line control there are two more steps toward total automatization.

Three basic parameters: melt temperature, pressure, and the screw torque are controlled either by dynamic valve or a throughput. On-line melt rheology is usually adjusted by varying the feed ratio of different streams of material. Degree of dispersion (inspected, for example, by ultrasonic or photo-optic transducers) can be controlled by screw speed and/or dynamic valve. The aim is to reduce the time span between a detected change in material parameter and the corrective actions. With the output rate of modern twin-screw extruders $\dot{Q} \leq 50$ t/h every second counts. Some parameters can be continuously monitored (e.g., dispersion by ultrasonic means), some require seconds to be measured (e.g., viscosity), some minutes (e.g., color; a measurement by side-stream injection molding, cooling, and colorimetry takes 12-15 min.). To these times one must add the intervention and an average dynamic response time effects [48,60-63].

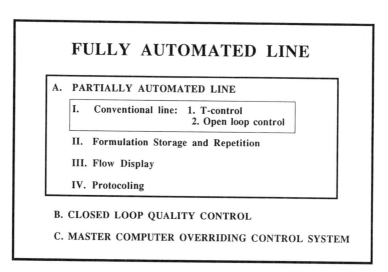

Figure 1.13 *Stages of automatization of a production line.*

Owing to the relatively rapid response and accessibility of commercial on-line rheometers, these are being used in closed-loop control. Two types of rheometers have been developed: dynamic and steady-state. In dynamic scan the frequency (ω) dependence of storage and loss moduli, G' and G", are determined. In principle they can be directly used to control both the molecular weight (MW) and the molecular weight distribution (MWD). It was observed that coordinates of the cross point [60,64]:

$$G_x \equiv G' = G''; \quad \omega_x = \omega\,(G' = G'') \tag{1.32}$$

empirically correlate with: $\omega_x \propto MW^{-a}$ and $G_x \propto MWD^{-b}$. On the other hand, the steady-state (capillary or slit) rheometer determines $\eta = \eta(\dot{\gamma})$. These data can be fed to simplified Eqs (1.16-1.17):

$$\eta\,(\dot{\gamma}) = \eta_0 \left[1 + (\dot{\gamma}\tau)^{m_1}\right]^{-m_2} \tag{1.33}$$

The parameters of this equation also depend on MW and MWD; τ and η_0 on the first, whereas m_1 and m_2 on the latter parameter, respectively [65]. However, as stated in Section 1.2, for multiphase systems the steady-state flow may introduce morphological changes and the measurements may reflect not the bulk material properties but rather those of the skin layer. Furthermore, since the steady-state shear rates are usually higher than frequencies in the dynamic tests, the steady-state data usually follow the power-law dependence:

$$\eta\,(\dot{\gamma}) = \eta_0\,(\dot{\gamma}\tau)^{m_1 m_2} = K\dot{\gamma}^{n-1} \tag{1.34}$$

(n is the power-law index) from which estimation of MW and MWD is problematic [2].

In the absence of flow segregation Eq (1.2) can be used to estimate the volume fraction of the dispersed phase. Since interphase thickness increases the apparent diameter of drops and solid particles, the calculated effective volume fraction can be correlated with the degree of dispersion. For good reliability, the η_r in Eq (1.2) should be calculated from low frequency $\eta' = G''/\omega$ values. Note that η_r depends on concentration, agglomeration, and particle-size distribution. If it is possible to ascertain the constancy of the remaining parameters the measurements of η_r can be used to control any one of these variables.

The importance of closed-loop line control is easily understood - it usually reduces the amount of recyclable products from 20-30% to 1-3%.

1.4   EXTRUSION

Mathematical description of extrusion of the two-phase systems starts with the set of equations describing conservation of mass, momentum, and energy [66]. The crucial step involves selection of the constitutive equation describing the stress-deformation dependencies. Computations provide pressure, stress, and velocity fields, which in turn can be used for calculation of bubble/drop deformation or solid particle motion. So far only severely simplified cases (e.g., dilute suspension or fibers in Newtonian or power-law liquid) have been treated. Two recent volumes on mixing and extruding are worth study [67,68]. The extruder provides the following functions: solid conveying, plasticating or melting, and melt conveying or pumping. The multiphase nature of the two-phase systems becomes important in the latter two.

Assuming a nonisothermal power-law flow in a corotating twin-screw extruder with heat flux through the barrel, one can compute the temperature, stress, pressure, and velocity profiles [69]. The numerical analysis for each screw section starts with assumption of Newtonian flow, from which the "length" of drag and pressure flow are calculated. Next, at the incremental step the local pressures, temperatures, leakage flow, and power consumption are determined, then the computation loops to correct for leakage flow and the absolute pressure. Once the correct value of the absolute pressure is obtained the next screw element is computed in an analogous manner. Excellent predictive ability of the method was reported; the melt temperatures in ZSK-300 under diverse processing conditions were predicted with a difference of ± 3°C from the measured values, whereas the specific energies were predicted with a difference of ± 3%. Surprisingly, as shown in Fig. 1.14, the simple schematic worked better for the kneading disks sections than for the conveying screw elements [69].

1.4.1   Foam Extrusion

There are two types of foam extrusion: 1) with a chemical blowing agent, and 2) with an expanding fluid injection. In the first case the chemical blowing agent decomposes at a well-defined temperature, in the second the injected fluid expands at the melt temperature once the melt pressure is reduced. In both cases homogenization of the blowing agent is the prerequisite of high quality foam [70,71].

For systems with a chemical blowing agent a standard extruder with L/D ≥ 28 and good mixing screw (selected for the matrix polymer) ought to be used. Since good mixing and temperature control are required, the size of the single-screw extruder is limited. For high throughput lines twin-screw machines are the must. In both cases it is important to select

equipment with relatively narrow distribution of the residence time devoid of the long
residence time tail. Since the chemical agent stability may depend on the purity of the
ingredients (e.g., possible catalytic decomposition in the presence of metals and ions) the line
automatization is strongly recommended.

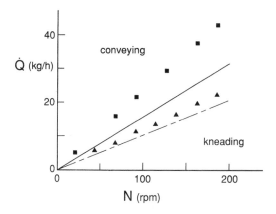

Figure 1.14 *Calculated and measured throughput vs. screw speed for conveying and kneading
elements of a twin-screw extruder [69].*

The extrusion lines with fluid injection are more complex. The fluid must be injected
under pressure at a location where polymer is already molten. It is preferable that the fluid be a
good solvent for the polymer in the whole range of temperatures and pressures experienced in
the extruder. Thus, addition of fluids drastically reduces the viscosity which may necessitate
cooling. Both actions: mixing in low-viscosity solvent and heat removal requires either a long
barrel extruder or two extruders in tandem. In both cases the barrier (or blister seal) screw is
required to isolate the solid and melting zone from the mixing and pumping one into which the
fluid is injected. The motionless mixers find good use on the foam extrusion lines. In Chapter 4
of this volume, Hornsby discusses processing and properties of foamed thermoplastics. In
Chapter 5 Kim and Kim provide an excellent review of the literature. The latter authors
concentrate on the flow rheology of foamed polymers.

1.4.2 Blends

It is difficult to generalize blend extrusion. Such blends as ABS, HIPS, Noryl, and Bayblend
can be processed as single-phase polymers provided that processing parameters specified by
the manufacturer are strictly adhered to. However, blends are generally more sensitive to
process variables (e.g., stress, temperature, pressure, and residence time) and need more
sophisticated line control than the single-phase polymers [2,72-75]. The key is stability of the
morphology, or in other words, a degree of dispersion and stabilization achieved during the
compounding stage.

All blends can be grouped into two types: well-stabilized and not. In the first type, it is
the intention of the blend manufacturer to generate stable morphology and processability not
very different from that of a single-phase polymer [2]. Within this type one can identify three

categories defined by the volume concentration of the second phase: (a) 0-12%, (b) 12-35% and (c) 35-50%. In (a) the dispersed phase exists in the form of drops, usually quite uniform in size, with drop diameter of the order of one micron. This category includes most of the toughened engineering polymers (usually not identified as blends). In (b) the toughening is accomplished by addition of a multifunctional copolymer. At 35% loading the morphology is complex, requiring intensive mixing during the compounding stage. In (c) the co-continuity of two polymers is usually achieved. Blends with this structure (e.g., PA/ABS or PE/PS) are known commercially. The interlocking generates relatively stable morphology. Thus, within the type of stabilized morphology blends, category (b) is the most sensitive to processing conditions. The manufacturer usually specifies a relatively narrow range of temperatures for each of the screw zones, as well as the screw speed (stress). Since across a channel of a single-screw extruder, temperature gradients of up to 40°C have been measured, twin-screw extruders with milder and more uniform process conditions are increasing in preference.

The second, nonstabilized morphology type includes blends designed to create particularly desirable structures during the final processing step. This includes blends used to manufacture blow-molded containers with superior barrier properties, or fibers reinforced by the presence of microfibrils of the dispersed phase. In both systems the concentration of the dispersed phase is below 15%, i.e., in the compounded blends there are spherical drops dispersed in a matrix. The key to the successful product is the size of the drops [2]. Since deformability is proportional to stress and the original drop diameter, the drops must be relatively large and uniform in size. During the processing step the limits of temperature and stress must be critically observed.

In Chapter 6 Elmendorp and van der Vegt review the fundamental principles of morphology formation during polymer blending. The discussion is pertinent not only for the compounding but for the processing stage as well.

1.4.3   Particulate-Filled Systems

Devolatilization during extrusion of particulate-filled polymer is required. Thus, the screws used for this application are of multi-stage type. Owing to abrasive wear the screw, the barrel and the die should be hardened and the wear systematically measured [67,76,77].

As mentioned before, flow of highly filled thermoplastics is characterized by the yield stress leading to plug flow in the extruder and through the die, which in turn results in small, if any, extrudate swell at the exit [78-82]. Processing these materials requires relatively low flight depth, which means either a small single-screw extruder, a mixing extruder, or a twin-screw extruder. Orientation of the anisometric particles is easily controlled by convergence or divergence. Since in most extruders the flow stream converges to the die, the usual orientation of fiber reinforcement is in the machine direction [83]. Diverging the flow at the exit from the die orients the fibers in the transverse direction [84].

Fiber breaking during processing of fiber-filled thermoplastics (attrition) originates in flow vorticity and occurs at any filler concentration [11,85]. By contrast, mica platelet breaking is primarily due to the mechanical platelet-platelet contact and occurs only above the dilute range of concentration $\phi \geq p^2$ [85]. The higher the stress and the longer the residence time, the higher the degree of attrition. In Fig. 1.15 the effect of concentration on particle attrition is illustrated. An interesting solution to both principal problems of particulate-filled (sheet) extrusion, namely the abrasive wear and particle attrition can be solved using a short, noncontacting cantilever screw and roller die [86].

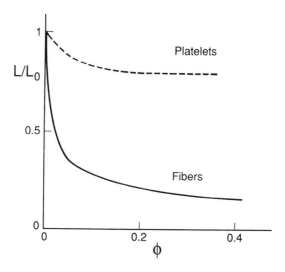

Figure 1.15 *Concentration dependence of glass fiber and mica flakes attrition during 2 min. compounding with PP in an internal mixer at 200° C [85].*

Increasingly, the compounding and forming of reinforced systems is combined into a single operation. The automatization, closed–loop control as well as new developments in side (or bottom) feeding of the reinforcements to molten thermoplastics are responsible for the trend. Several machine manufacturers developed the latter devices [87,88]. The side feed introduces the chopped glass fiber (GF) at a point where there is no melt pressure. It uses a short side, starve–fed twin–screw extruder and a vent. The main extruder screw is of the three–stage type, allowing for further devolatilization and a two–stage GF mixing. The advantage of the side feed is the rapid filler incorporation into the melt which reduces attrition and machine wear.

1.5   INJECTION MOLDING

Each year the injection–molding units are becoming larger. It is estimated that worldwide there are over 50,000 injection–molding plants, consuming annually about $4 billion worth of processing equipment (or 40% of total), manufactured by 400 companies. The injection–molding machines with 90,000 kN clamping force, shot size up to 90 liters from two injection units are available. Two giant multiprocessing machines are being built for injection and compression molding of thermoplastics–based composites for the automotive and housing industries [89]:

	Alpha 1	Quota 8000
Manufacturer	Kraus-Maffei	Sandretto
Lenght x height (m)	23 x 17	25 x 17
Clamping force (kN)	50,000	80,000
Compressor piston stroke (m)	2.1	2.5
Shot size (l)	11	50
Total weight (tons)	730	800

However, the biggest changes in the field are primarily due to robotization, automation, and control. Robots (e.g., 20 t, 7 axes unit from Battenfeld) perform multiple tasks at a speed of several m/s and payload capacity of several hundred kilograms with an estimated payback time of 6 to 12 months. Larger (up to 1000 t) robots are also available (e.g., from ATM Automation). These can handle up to 100 programs (from sprue removal to insert placement and mold change) and operate with 100 μm precision at speeds up to 1 m/s. Robotization and computerization leads to automated computer-integrated manufacturing (CIM) where flow of material (just-in-time delivery), control of the injection-molding machine, statistical process control, parts handling, and documentation are integrated for maximum efficiency and economy. The integration includes peripheral operations such as tool transport, handling, changing and preheating facilities, and transport of raw materials and finished goods. In the near future the system will also include the data base for technical and managerial information to cover the whole spectrum of operations from mold part design to the product sale. The heart of CIM is optimization and control of the injection molding stage.

Computer-aided design and manufacture (CAD/CAM) is reaching its maturity and universal acceptance. As indicated in Fig. 1.16, the process starts with the definition of requirements for the plastic part and selection of material. Next, the part is designed and the computer analyzes its performance in the intended environment as well as verifies the design in respect to the accepted criteria. The runners are then positioned and melt flow in the cavity computed with particular attention given to the weld lines, air entrapment, as well as pressure (P) and temperature (T) profile. In the last stage the holding pressure and mold shrinkage are computed. After each stage the computer program makes it possible to go back in order to optimize the product molding and performance. The final stage generates the set of suggested process parameters and TC type for mold cutting [75]. Once the mold is installed on the injection-molding machine these processing parameters can be used during the initial stage of process optimization. That stage is usually done by an operator, although automatic, self-optimization programs are being developed. For example, the PVT-holding pressure program was found to be successful [90].

Different manufacturers develop different control systems. These usually incorporate: user-friendly programming, integrated quality control (surveys basic molding parameters, makes statistical control, rejects defects), molding system (automatic adjustment of molding parameters to preserve molding quality), fill and feed control (automatic adjustment of shot volume and hold-on time), and PVT program (check the holding stage, as far as P, T, and time is concerned to minimize stresses and shrinkage in the finished product) [91].

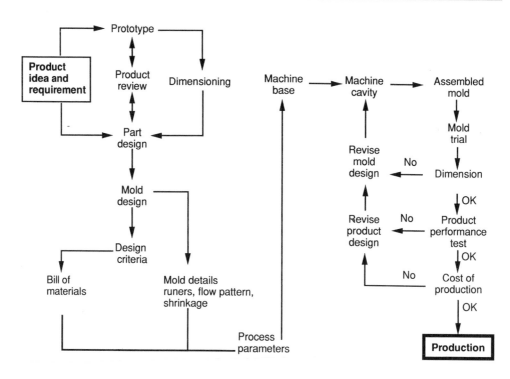

Figure 1.16 *Schematic representation of CAD/CAM mold and part design.*

The final part of automation of the injection-molding line is the mold change. The programmable quick mold changing (QMC) systems are available either from the injection-molding manufacturers or from independent houses [91-94].

1.5.1   Foams

Injection molding of foamed thermoplastics takes several forms. Using a standard method with a chemical foaming agent at relatively low pressure of 10 to 20 MPa a smaller volume of the thermoplastic is injected than that of the mold cavity. By contrast with normal operation, the material is injected using short and narrow runners into the thickest part. The mold ought to have air vents of 50 to 80 μm diameter. The resulting part has a skin-core structure, with 1 to 2 mm thick unfoamed skin. Thus, this method can be used for articles with thickness exceeding 5 mm. For further details please consult Chapters 4 and 5.

There are a few drawbacks to the standard method, namely a swirl-pattern, surface rugosity, and coalescence of air bubbles in the thickest part of the molded article. Numerous solutions to these problems have been proposed. These can be divided into modification of the mold, process, and the thermoplastic. Mold modification consists of redesigning the runners and mold thickness to eliminate jetting and large differences in gasing time at different locations. Process modification deals primarily with injection rate, counter pressure (during injection), mold cooling, and mold release. For example, to improve the surface finish

Variotherm uses a programmable temperature profile while Hunkar uses a variable speed of mold filling.

The most pronounced modifications are in the formulation, varying the method of gasing and prevention of gas-bubble coalescence. Injection of gases, N_2, Ar, CO_2, or addition of H_2O can improve foaming of specific polymers. Addition of fillers and reinforcement (e.g., glass fibers or mica flakes) also may help stabilizing the foam and generate more uniform structural products [70,95-97].

Molding of thermosets is not a subject of this book. However, the reactive injection molding (RIM) can be used for molding such thermoplastics or polyamide-6, polydicyclopentadiene [98]. The new technologies with macromers and cyclomers may broaden the use of this low molding pressure technology [99].

In a search for better surface finish of foamed articles a method of gas injection into the mold cavity was developed. Several modifications of the method exist including injection into unfoamed thermoplastics.

1.5.2 Blends

There is a strong correlation between the morphology and properties of polymer blends [2,100–105]. For materials which fail by the crazing mechanism, e.g., for PS, there is a critical particle size of the elastomeric component ($d_c \simeq 1$ to $2\,\mu$m) below which the blend toughness decreases. Optimized bimodal distribution of particle size can lower this limit without toughness loss [106]. On the other hand, for systems failing by the shear banding mechanism the critical d_c has not been found; the smaller the particle size, the lower the brittle-tough transition temperature [104]. In blends containing a higher concentration of the second phase the phase co-continuity may provide the desired behavior [2]. Thus the process must assure control of morphology [13,74,75]. As stressed in Section 1.4.2, it is critical that the processing conditions suggested by the manufacturer be strictly observed. The morphology is particularly sensitive to high temperature and stress. However, long dwell time can also severely affect some blends. Furthermore, the processor should avoid mixing two batches of the same blend - a small difference in composition or compounding may lead to batch-to-batch immiscibility and a dramatic drop of properties.

For blends, the preplasticating injection-molding machines are not recommended. Instead, a single, vented, in-line reciprocating screw unit with $L/D \simeq 20$ and compression ratio 2 to 2.5 should be used. The size of the machine should be such that 50 to 75% of the barrel capacity is used per shot. The unit must provide precise and reliable control of key molding parameters. Screw-tip nonreturn valve, shortest possible nozzle (with good temperature control), wide (non-jetting) runners, and slow to moderate injection rates are recommended [107-109].

In some aspects the blends are similar to solid-particle filled or reinforced polymers. Both these systems are the subject of flow orientation and suffer from the weld-line problems. The weld (joint) lines are formed in the mold where two melt fronts meet head on. Due to extensional flow field in the fountain-flow, the matrix macromolecules and the dispersed drops are strained. In the region where two fronts meet there is parallelism of orientation (Fig. 1.17) without interpenetration, which leads to weakness, a susceptibility to premature fracture [74,110,111]. There is no universal cure for the problem. Finer dispersion of well-stabilized drops, and optimization of the mold and molding parameters have been used for this purpose. Recently a new dynamic method of mold filling has been successful in erasing the weld line. The idea originated at Brunel University as an outcome of the development of technology for molding thick parts [112]. The "live-feed molding method" is based on pumping the material back-and-forth by means of two computer-controlled pistons (see Fig. 1.18a). The amplitude and frequency of oscillation can be precisely controlled. So far, the fiber-reinforced and LCP systems have been studied. The technology has been licensed to companies in Europe and Japan [113]. Another system developed around the oscillatory flow idea utilizes two injection units which alternately inject and suck back the material (Fig. 1.18b). The cycle time is longer but properties of molded LCP articles are superior [114].

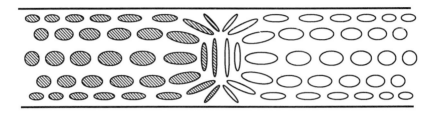

Figure 1.17 *Schematic representation of weld-line morphology in injection-molded immiscible blends, injected from the opposite gates.*

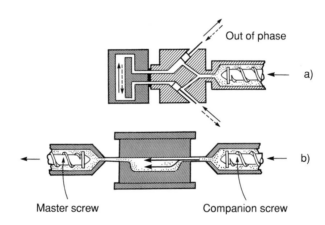

Figure 1.18 *Live-feed (a) and push-pull (b) methods for elimination of weld-lines in injection molding.*

     The other similarities between blends and filled system are in: flow behavior (in both cases the yield stress may require large submerged runners), reduction of mold stress (leading to a decrease of post-molding shrinkage), and problems with surface finish (swirls and rugosity) [115]. Thus, for both these multiphase systems large tabs or fan gates, short but large sprues and runners, generous tapers, round corners, precise temperature control of hot runners, designs without stagnation points, and high shot volume capacity are recommended [116].

For several reasons most of the engineering polymer blends are solid-filled or reinforced [2]. For automotive or household applications the reinforced blends are extruded, molded, and stamped. The standard processing methods produce parts unsuitable for more stringent requirements, i.e., for exterior automobile horizontal panels. There is a hope that the new method of injection and compression molding (see Section 1.5) using a large multipurpose unit (such as e.g., Alpha 1 or Quota 8000) will overcome these obstacles.

1.5.3   Particulate-Filled Melts

The heat capacity of mineral fillers and reinforcements is about half of that for the thermoplastics. At the same time the viscosity of filled systems is usually higher by a factor of 2 to 10. Since the polymer/metal friction coefficient is the same as for neat polymer the processing of particulate-filled melts necessitates lower screw speed, higher (by about 10-30°C) barrel temperature, and precise temperature control. As stated in the preceding part, short and wide runners and sprue, tapered molds with round corners, non-return valves, as well as abrasion-resistant injection units should be used [95,111,113]. The general recommendation for injection molding of reinforced polymers is to avoid diameters and thicknesses smaller than 6 mm. Since their shrinkage is 2 to 3 times smaller than that of unfilled resins, 2-3° taper and low back pressure should be used. Large vents, 0.1 by 0.5 mm, are advisable. Streamlining of the mold flow should be optimized to avoid excessive fiber breakage and lumping [117]. Warpage and surface roughness are the main problems. These are particularly acute for high loadings of high aspect-ratio fillers. The weld line defect is usually minimized by high-speed and high-temperature injections. Unfortunately, these conditions enhance attrition of particles and abrasion-wear of tooling. The screw rotation should be low.

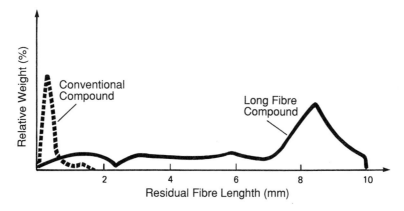

Figure 1.19 *The fiber length in an injection–molded part from short and long glass fiber filled PP; starting fiber lengths were, respectively, 0.25 and 6 mm (courtesy ICI).*

The mechanical performance of composites reinforced with anisometric particles increases with particle aspect ratio. The ultimate, continuous fiber composites suffer from costly and labor-intensive processing. An excellent alternative is the long-fiber composite technology [117]. Here the matrix polymer is one of the engineering resins (PA, PET, PBT,

POM, PU, PES, PEI, PEEK, or PPS) whereas the reinforcement is either glass, carbon, or Kevlar fiber with lengths up to 12 mm and loading 30 to 60 wt% [118]. Short runners and gates, with diameters above 6 mm as well as adequate venting (25 μm thick and 3 mm wide) are specifically recommended [119]. An example of the fiber length in injection-molding product is presented in Fig. 1.19.

For the long-fiber injection the attrition is of utmost importance. There are several mechanisms responsible for this effect: 1) mechanical damage of the compounded granules, 2) friction between fibers and metal, 3) stress field in the injection-molding unit, and 4) mutual interference of the fibers. Thus, the attrition depends on the properties of the pelletized material (type of matrix and the fiber, fiber length and length distribution, concentration, additives, and dimensions of the pellets), processing parameters (type of plasticating unit, screw speed, temperature profile, injection rate, mold material, and back pressure), as well as on the geometry (screw, nonreturn valve, runners, sprue, gate, and mold). There is a good correlation between areas of high abrasion and fiber attrition, i.e., reciprocating screw, small gates, thin-walled parts, sharp changes in flow pattern, and long flow paths.

A thorough investigation of the role of screw geometry, speed and length, injection rate, nonreturn valve design, and gate dimensions, as well as wall thickness and the flow length/wall thickness ratio on long-fiber attrition was recently reported [120]. In general the fiber attrition increased with the screw length at L/D > 14, especially within the metering zone. Doubling the screw speed increases the short fiber fragments by about 25%. Vented screws with short compression zones generated more damage than the standard ones. The ball-type nonreturn valve was found to be superior in performance to the standard ring-type. Wall thickness and injection rate also had a large influence on attrition. The fiber orientation has a strong bearing on flow properties of the reinforced polymer melt. On the other hand, the flow affects the orientation. Thus, during the processing there is an intimate relation between the processing parameters, flow behavior, and orientation. Orientation strongly affects the mechanical properties of the injection-molded part; the higher the fiber concentration, the stronger the effect. Deviations as small as by 5° may lower the strength by 40% [75].

Detailed discussions on the flow behavior of fiber-reinforced polymer melts is presented in Chapters 11 (by Vincent and Agassant) and 12 (by Mutel and Kamal). A new method of calculation of the fiber orientation in the injection-molded parts is also described in the former chapter.

Numerous authors studied fiber orientation in simple geometry molds, e.g., a center-gated disk or an edge-gated plate. Depending on the system, the mold geometry and the molding conditions the number of layers observed across the thickness varied from three to nine (Fig. 1.20) [121-128]. In most cases the five-layer structure was observed. Here there are two skins, two intermediary ones and one central layer each comprising about 3/16 of the specimen thickness [122].

There is a general consensus that the central layer has a transverse orientation, and the intermediate ones a radial one. However, for the skin layer there are reports of either transverse or random orientational arrangements. In the seven- or nine-layer structures, a thin fiber-free layer of polymer melt has been reported. For the three-layer structures only two radial (outer) and one transverse or random (central) layer were reported [125,127].

The diverse fiber orientations illustrate the multivariable nature of the problem. It is worth noting that theoretical computations in Chapter 11 for a five-layer structure predicted transverse orientation in the core, parallel to flow direction in the skin, and nearly random in the intermediate layers. The initial results are promising. One may expect that with refinements the model will be able to correctly predict variations of structure with the process-independent variables.

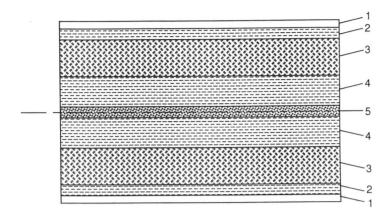

Figure 1.20 *Schematic representation of symmetrical nine-layer cross section in injection-molded specimen; layer 1 is PP virtually free of GF, layers 2 and 4 are oriented with the flow, whereas layers 3 and 5 are oriented perpendicularly to the flow direction.*

The orientation of fibers discussed above does not comprise the total orientation imposed by the flow on the reinforced thermoplastics. Studies of matrix orientation indicate that it depends on the specimen thickness and distance from the gate [125]. It has also been reported that transcrystallinity in injection-molded specimens depends on processing parameters and spatial location of the observed element within the injection-molded specimen [121]. The transcrystallinity phenomenon is reviewed in Chapter 14 by Xavier.

1.6   OTHER PROCESSING METHODS

The extrusion and injection molding hardly exhaust the processing methods used to form foams, blends, and composites. However, the aim of this chapter is not to provide a complete list of processing methods, but rather to touch on those considered more important.

Some blends have been specifically designed to be used in fiber spinning or blow molding. As mentioned in Section 1.2.3 the extensional flow deforms the dispersed phase drops into fibrillas or platelets, respectively, enhancing the processing and/or performance of the formed articles. In the first case better line stability, better surface finish, zero thermal shrinkage, and improved fiber strength have been observed [129]. In the second case improved barrier properties were achieved [130].

There is a large potential for use of metal-forming methods with blends and reinforced thermoplastics. In particularly, hydrostatic extrusion, rolling, and forging (see Fig. 1.21) have been successfully used for neat polymers [131-133]. The preliminary results on rolling of mica-filled polypropylene [134] and rotational extrusion of glass-fiber-filled polypropylene [135] are quite promising. Good control of orientation and improvement of mechanical properties were demonstrated. These methods have not been used for blend processing.

There is a growing trend for replacing the advanced composites based on thermosetting resins by those containing thermoplastics [136]. Formation of the thermoplastic matrix composites is reviewed by Bigg in Chapter 13. The author proposes a new method of prepreg preparation based on paper-making technology. Consolidation of the prepreg was done above the polymer melting point under compression. The compression molding of long-fiber composites was recently modeled for fiber orientation and melt separation [137]. The simulation agrees very well with the experiment, especially at lower fiber loading.

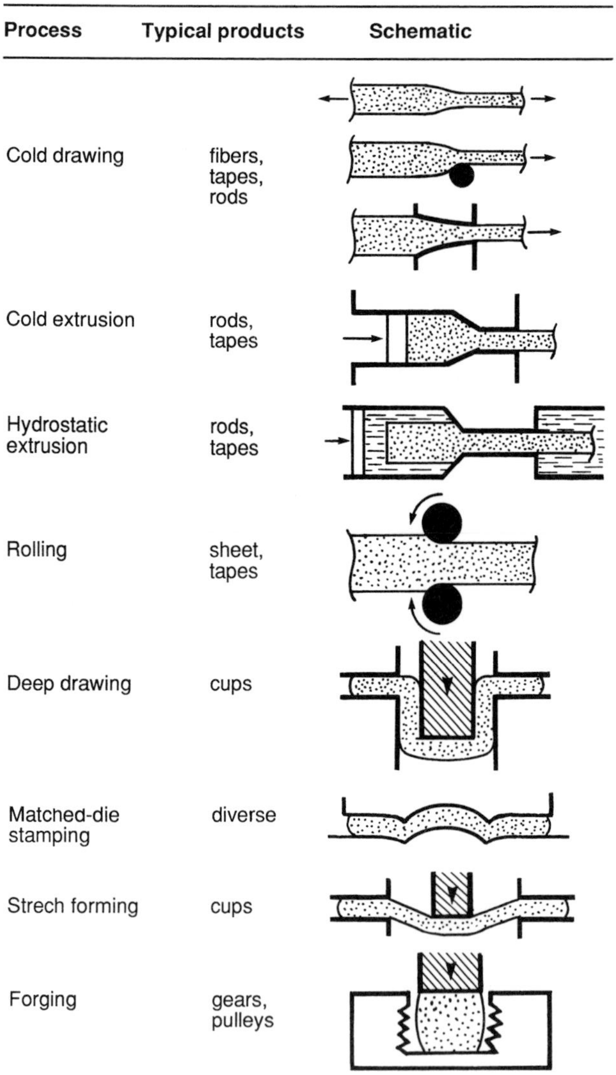

Process	Typical products	Schematic
Cold drawing	fibers, tapes, rods	
Cold extrusion	rods, tapes	
Hydrostatic extrusion	rods, tapes	
Rolling	sheet, tapes	
Deep drawing	cups	
Matched-die stamping	diverse	
Strech forming	cups	
Forging	gears, pulleys	

Figure 1.21 *Example of solid-state forming methods.*

The recently proposed shear-roll extruder, CMS, is an interesting development in compounding and processing a highly viscous system [138]. The machine combines functions of a two-roll mill with that of a nonintermeshing, counter-rotating twin-screw extruder (Fig. 1.22). It operates under ambient pressure at T = 20 to 280°C in two temperature control zones. The material (up to 25 kg) always stays on the front roll, from which it can be cut as a strip or granulated to feed the forming stage. Dispersion and distribution of ingredients can be controlled by independent adjustments of T, gap pressure, rotational speed of each roll, shear stress, and feed. The rolls, with $L/D = 5$ to 7.5, rotating with speeds $N = 5$ to 50 rpm, have shearing grooves tailored to user needs. The gap can be adjusted from 5 to 50 mm. Depending on the speed and size of the rolls, throughput $\dot{Q} = 3$ to 500 kg/h can be achieved. CMS is primarily used for compounding highly filled polymer systems, e.g., masterbatches, concentrates, ceramic or metal pastes, floor covering, blends, and rubbers. It can be operated enclosed, with a cover on, to prevent dust or fume contamination of the working environment.

Figure 1.22 *Color Metal CMS shear roller extruder [138].*

1.7   CONCLUSIONS

In this chapter the rheological and thermal properties of multiphase polymeric systems were outlined. Presence of the yield stress, extrudate swell, and the general increase of viscosity with concentration are the main rheological features. However, the flow–induced morphology is of prime importance, as it is responsible to a great extent for finished product performance. Thus, control of morphology through process design, toolings, and processing parameters is the most important goal for the processor.

From the energy balance point of view the presence of gas or solid particles in the polymeric matrix may have serious repercussions on frictional heat generation, thermal conductivity, and heat capacity. In general, these systems require more precise control of processing. Addition of gas lowers the wall stresses, the thermal conductivity, and the heat capacity. The foam processing involves injection of gas into homogeneous melt followed by nearly adiabatic distributive mixing. The addition of solid particles may enhance wall stress and plug flow as well as increase the heat conductivity and lower the heat capacity. Thus, in order to prevent overheating of the sheared layer adjacent to the wall, a good distributive mixing and precise process control are required.

For particulate-filled polymers and blends the compounding step is usually separated from that of forming. The compounding ought to generate uniform precursor morphology. For the particulate-filled systems, such as glass-fiber composites, this translates into uniform dispersion of GF without extensive attrition, whereas for blends it means uniform dispersion of stabilized minor phase. Nowadays increasing attention is being paid to closed-loop process control. In most cases continuous process is being used.

Owing to the yield stress the scale-up of two-phase system extrusion is difficult and for large volume lines it is possible only if special compounding machines with good distributive and self-wiping capabilities are used. The preferred extruders must also be streamlined, devoid of dead spaces in which the material can accumulate, phase-separate, and decompose, contaminating the production stream. The closed-loop control based on the key product properties is highly recommended.

The two-phase nature also imposes more stringent conditions for the mold design and processing conditions. Computer-integrated manufacture is becoming more and more common. Injection molding of particulate-filled melts, as well as that of blends, poses the inherent problem of poor weld-line strength. For many systems the newly developed oscillatory flow may provide a solution. However, in the case of long-fiber injection the oscillation was reported to cause extensive fiber-length reduction. Injection molding of blends requires strict control of the processing conditions recommended by the manufacturer. As in extrusion, in injection molding the flow-induced morphology is of prime importance.

Some blends have been specifically designed for use in processes other than extrusion or injection molding. Blow molding of partially compatibilized PE/PA blends is used for manufacturing containers with excellent barrier properties for gases and solvents. Fiber spinning of e.g., PET/PA has advantages in both the processing operation and the fiber properties. There is also slow progress in adapting the metal-forming methods to polymer processing. So far the compression molding and stamping of fiber-reinforced thermoplastics have been most successful.

NOMENCLATURE

Abbreviations

ABS	Acrylonitryle-butadiene-styrene terpolymer
AGR	Annual growth rate
CAD/CAM	Computer-aided design and manufacture
CIM	Computer-integrated manufacture
CM	Continuous mixer
CTM	Cavity transfer mixer
EVAc	Ethylene-vinylacetate copolymer
FIM	Flow-induced morphology
GF	Glass fibers
HDPE	High-density polyethylene
HIPS	High-impact polystyrene
IR	Infra-red
MM	Motionless mixer
MW	Molecular weight
MWD	Molecular weight distribution
PA	Polyamide
PA-6	Polycaprolactam
PBT	Polybutyleneterephthalate
PE	Polyethylene

PEEK	Polyetheretherketone
PEI	Polyetherimide
PES	Polyethersulfone
PET	Polyethyleneterephthalate
POM	Polyoxymethylene (acetal)
PP	Polypropylene
PPS	Polyphenylenesulfide
PS	Polystyrene
PTFE	Polytetrafluoroethylene
PU	Polyurethane
PVC	Polyvinylchloride
QC	Quality control
QMC	Quick mold change
RIM	Reactive injection molding
UHMWPE	Ultra-high-molecular-weight polyethylene

Symbols

A	Equipment availability
B	Extrudate swell
b_i, m, m_1, m_2, M	Parameters
c_p	Specific heat
D	Screw diameter
d	Drop diameter
D^*	Striation thickness
De	Deborah number
D_E, D_S	Drop deformability in extensional and shear field
D_r	Rotational diffusion coefficient
e	Eccentricity
E, E^*	Supplied power; specific power consumption
g	Gravitational constant
G', G''	Storage and loss shear modulus
G_x	Cross-point modulus
$\dot{H}$	Total energy accumulation
$\dot{H}_{cond}, \dot{H}_{conv}$	Conducted, convected heat flux
$\dot{H}_\eta$	Heat generated in shear flow
$\dot{H}_H, \dot{H}_K$	Heat flux in and out of the system
k	Thermal conductivity coefficient
L	Screw length
M^*	Goodness of dispersion parameter
N	Rotational screw speed
n^*	Number of elements in MM
N_1, N_2	First and second normal stress difference
P	Pressure
p	Aspect ratio
Pe	Peclet number
Pr	Productivity
$\dot{Q}$	Volumetric flow rate
Re	Reynolds number

S	Striation number
T	Temperature
t	Time
t_p	Period of rotation
u	Exponent
V	Volume
w_i	Weight fraction
γ	Strain
$\dot{\gamma}$	Rate of shearing
$\dot{\varepsilon}$	Extensional strain rate
η'	Dynamic viscosity
η, η_0, η_r	Viscosity, zero-shear and relative viscosity
η_d, η_m, η_i	Viscosity of the dispersed phase, matrix, and the interphase
$\eta_E, \eta_{r,E}$	Extensional viscosity, relative extensional viscosity
$[\eta], \tilde{\eta}$	Intrinsic and reduced viscosity
κ	Capillarity number
λ	Viscosity ratio
ν	Interfacial tension coefficient
ρ	Density
σ_{ij}	Component of stress tensor
σ_y	Yield stress
τ	Relaxation time
ϕ, ϕ_m	Volume fraction, maximum packing volume fraction
χ_{23}	Polymer/polymer thermodynamic interaction coefficient
ω	Frequency

REFERENCES

1. Anonymous, *Chem. Market. Rep.*, **3**, 23, Nov. 23 (1987).
2. Utracki, L. A., *Polymer Alloys and Blends*, Hanser Publishers, Munich (1989).
3. Collyer, A. A., Clegg, D. W., Eds., *Rheological Measurements*, Elsevier Applied Sciences, London (1988).
4. Simha, R., *J. Appl. Phys.*, **23**, 1020 (1952).
5. Oldroyd, J. C., *Proc. Roy. Soc.*, **A218**, 122 (1953); **A232**, 567 (1955).
6. Fisa, B., Utracki, L. A., *Polym. Compos.*, **5**, 36 (1984).
7. Hinch, E. J., Leal, L. G., *J. Fluid Mech.*, **52**, 683 (1972).
8. Batchelor, G. K., *J. Fluid Mech.*, **44**, 419 (1970); **46**, 813 (1971).
9. Wildemuth, C. R., Williams, M. C., *Rheol. Acta*, **23**, 627 (1984).
10. Goldsmith, H. L., Mason, S. G., in *Rheology, Theory and Applications*, Vol. 4, Eirich, F. R., Ed., Academic Press, New York (1967).
11. Czarnecki, L., White, J. L., *J. Appl. Polym. Sci.*, **25**, 1217 (1980).
12. Utracki, L. A., in *Rheological Measurements*, Collyer, A. A., Clegg, D. W., Eds., Elsevier Applied Sciences, London (1988).
13. Han, C. D., *Multiphase Flow in Polymer Processing*, Academic Press, New York (1981).
14. Kanu, R. C., Shaw, M. T., *Polym. Eng. Sci.*, **22**, 507 (1982).
15. Goettler, L. A., *Rubber Chem. Technol.*, **56**, 619 (1983).
16. Singh, P., Kamal, M. R., *Polym. Compos.*, **10**, 344 (1989).

17. Anonymous, *Bayer Thermoplastics, Processing, Preparation of the Material*, Info. Brochure No. KU 48007, E129-888/845 324; Bayblend - Processing Guide for Injection Molding, 55-A43 (2.5)D.
18. Anonymous, *Moriyama Dispersion Mixers*, Moriyama Corp., Japan.
19. Crocker, Z., Canplast-'81, Montreal, 27 Oct. 1981.
20. Manas-Zloczower, I., Tadmor, Z., *Rubber Chem. Technol.*, **57**, 48 (1984).
21. Hold, P., *Adv. Polym. Technol.*, **4**, 281 (1984).
22. Valsamis, L. N., Canedo, E. L., Donoian, G. S., Farrel Corp. (1989).
23. Bulletin Nos. 234, 238, 239, Farrel Corp. (1988, 1989).
24. Hold, P., *Adv. Polym. Technol.*, **2**, 197 (1982).
25. Jakopin, S., Franz, P., *Plast. So. Africa*, 32-46, April (1989).
26. Anonymous, *Buss Kneader Lines*, Bulletin Nos. 0075, 2152, 2282, Buss AG, Basel (1989).
27. Anonymous, *Compounding and Granulating Plant for PVC*, Berstorff Techn. Inf. (1989).
28. Anonynus, *Planetary Roller Extruder*, Battenfeld Extrusiontechnik (1989).
29. Frederix, H., 5th Conf. Europ. Plast. Caoutch., Paris (1978).
30. Michaux, J., Intl. Seminar Energy Conserv. Res., Brussels, 23-25 Oct. (1979).
31. Anonymous, *Plast. Machin. Equip.*, 53, June (1981).
32. Renk, P., *U.S. Pat.* Nos. 4,128,342, Dec. 5 (1978); 4,253,771, March 3 (1981).
33. Anonymous, *Barmag 3DD Mixer*, Barmag A.G., Inf. Serv. No. 30 (1989).
34. Gale, G. M., *Plast. Comp.*, 70, May-June (1985).
35. Hindmarch, R. S., Gale, G. M., *Elastomerics*, 114(8), 20 (1982).
36. Alzner, B. G., Allen, W. F., Csongor, D., *SPE Techn. Pap.*, **32**, 883 (1986).
37. Housz, I., Polymer Processing Society, European Meeting, Kerkrade, 30 Oct. - 2 Nov. (1989).
38. LaVerne, L., *Plast Comp.*, 29, March-April (1984).
39. Ottino, J. M., *AIChE J.*, **29**, 159 (1983).
40. Suzaka, Y., *U.S. Pat.* No. 4,334,783, June 15 (1982).
41. Eise, K., Herrmann, H., Werner, H., Burkhardt, U., *Adv. Plast. Technol.*, **1**, 1 (1981).
42. Anonymous, *High Performance Twin—screw Compounding System*, Werner & Pfleiderer GmbH, Techn. Inf. Bull. Nos. 21E.8910A.1000, 05-098/2, 05-085/1, 05-100/2 (1989).
43. Anonymous, *Bitruders BT*, Reifenhäuser Techn. Inf. Bull. No. 0/3.2-8903 (1989).
44. Anonymous, *A No—Nonsense Guide to Extrusion Systems for Continuous Compounding*, Welding Eng., Technical Bulletin (1989).
45. Anonymous, *MPC/V, APV Chemical Machinery*, Baker Perkins, Techn. Inf. (1989).
46. Anonymous, *Twin—Screw Continuous Mixers*, ComacPlast, Techn. Inf. (1989).
47. Anonymous, *Bex 2—65/BEC 2000*, Battenfeld Extrusiontechnik, Techn. Inf. (1989).
48. Ultsch, S., Fritz, H.-G., Polymer Processing Society, Annual Meeting, Kyoto, Apr. 11-14 (1989).
49. Anonymous, *Production of Silane Crosslinkable PE—tubes*, Werner & Pfleiderer Techn. Inf. 05-075/1 (1989).
50. Maxwell, B., Scalora, A., *J. Mod. Plast.*, **37**, 107 (1959).
51. Thornton, B. A., Villasenor, R. G., Maxwell, B., *J. Appl. Polym. Sci.*, **25**, 653 (1980).
52. Westover, P. F., *SPE J.*, **1**, 473 (1962).
53. Tadmor, Z., *U.S. Pat.* Nos. 4,142,805 (1979); 4,194,841 (1980).
54. Tadmor, Z., Valsamis, L. N., Yang, J. C., Mehta, P. S., Duran, O., Hinchcliffe, J. C., *Polym. Eng. Rev.*, **3**, 29 (1983).

55.   Anonymous, *Diskpack Processor*, Farrel Mach. Group, Techn. Inf. Bull. No. 231 (1985).

56.   Anonymous, *Plast. Comp.*, 63, May-June (1985).

57.   Brewer, D., *Plast. Comp.*, 12, Jan.-Feb. (1987).

58.   Bigg, D. M., in *Science and Technology of Polymer Processing*, Suh, N. P., Sung, N.–H., Eds., MIT Press, Cambridge, MA (1979).

59.   Herrmann, H., *Kunstoffe*, 78, 876 (1988).

60.   Zeichner, G. R., Macosko, C. W., *SPE Techn. Pap.*, 28, 79 (1982).

61.   Fritz, H.-G., *Kunststoffe*, 75, 785 (1985).

62.   Fritz, H.-G., Stöhrer, B., *Intl. Polym. Proces.*, 1, 31 (1986).

63.   Hertlein, T., Fritz, H.-G., *Kunststoffe*, 78, 606 (1988).

64.   Zeichner, G. R., Patel, P. D., 2nd World Congress Chem. Eng. Montreal, Canada, 6, 333 (1981).

65.   Utracki, L. A., Schlund, B., *Polym. Eng. Sci.*, 27, 367, 1512 (1987).

66.   Bird, R. B., Armstrong, R. C., Hassager, O., *Dynamics of Polymeric Liquids*, 2nd ed., John Wiley & Sons, New York (1987).

67.   Rauwendaal, C., *Polymer Extrusion*, Hanser Publishers, Munich (1986).

68.   Cheremisinoff, N.P., *Polymer Mixing and Extrusion Technology*, M. Dekker, New York (1987).

69.   Elemans, P. H. M., *PhD thesis*, Technical University Eindhoven (1989).

70.   Meltzer, Y. L., *Foamed Plastics*, Noyes Data Corp., Park Ridge, NJ (1976).

71.   Levy, S., *Plastics Extrusion Technology Handbook*, Independent Press, New York (1981).

72.   Golba, J. C., Seeger, G. T., SPE RETEC, Huron, OH, March 10-11 (1986).

73.   Anonymous, *Bayer Thermoplastics Processing Extrusion*, Bayer Technical Ring Folder *Thermoplastics* No. E63-861/848736 (1989).

74.   Menges, G., *Makromol. Chem., Macromol. Symp.*, 23, 13 (1989).

75.   Michaeli, W., *Angew. Chem., Int. Ed.*, 28, 644 (1989); *World Plast. Rubb. Technol.*, 2, 187 (1990).

76.   Menning, G., Voltz, P., *Kunststoffe*, 70, 385 (1980).

77.   Schuele, H., Fritz, H.-G., *Kunststoffe*, 73, 603 (1983).

78.   Utracki, L. A., Fisa, B., *Polym. Compos.*, 3, 193 (1982).

79.   Utracki, L. A., Favis, B. D., Fisa, B., *Polym. Compos.*, 5, 227 (1984).

80.   Utracki, L. A., *Rubber Chem. Technol.*, 57, 507 (1984).

81.   Utracki, L. A., *Polym. Compos.*, 7, 274 (1986).

82.   Fisa, B., Utracki, L. A., *Polym. Compos.*, 5, 36 (1984).

83.   Freundlich, R. A., *Brit. Pat.* Appl. No. 2,055,680, March 11 (1981).

84.   Goettler, L. A., Lambright, A. J., *U.S. Pat.* Nos. 4,056,591, 4,057,610, Nov. (1977).

85.   Fisa, B., *Polym. Compos.*, 6, 232 (1985).

86.   Hagiware, K., Nakagawa, K., *U.S. Pat.* No. 4,304,539, Dec. 8 (1981).

87.   Hulbert, W. H., SPE RETEC, Huron, OH, March 10-11 (1986).

88.   Anonymous, *ZSK-40 Alloying, Reinforcing and Colouring*, Werner & Pfleiderer Techn. Inf. No. 05 085/1-1.5 (1989).

89.   Anonymous, *Modern. Plast. Intl.*, K89 Newsletters, Nov. 3, p. 4 (1989).

90.   Fillmann, W., *World Plast. Rubb. Technol.*, 2, 139 (1990).

91.   Anonymous, *Sandretto Series Eight Injection Molding Machines*, Techn. Inf. (1989).

92.   Anonymous, *Rapidomat Automatic Mold Changer*, Arburg Techn. Inf. (1989).

93.   Anonymous, *Enerpac QMC*, Techn. Inf. (1989).

94.   Anonymous, *PS20E5ASE*, Nissei Techn. Inf (1989).

95.   Anonymous, *Planification Technique du Moulage par Injection des Matieres Plastiques*, Metalmeccanica Plast., Milan (1979).

96.   Elliott, C., SPE RETEC, Montreal, Sept. 30 - Oct. 1 (1982).

97.   Anonymous, *Thermoplastic Cellular Molding*, Hettinga Techn. Inf. (1989).

98.   Macosko, C. W., *RIM, Fundamentals of Reaction Injection Molding*, Hanser Publishers, Munich (1989).

99.   Anonymous, *Plastiscope, Mod. Plast. Intl.*, **5**, Dec. (1989).

100. Utracki, L. A., *Polym. Plast. Technol. Eng.*, **22**, 277 (1984).

101. Sahto, M. A., M. Eng. thesis, Dept. Chem. Eng., McGill University (1983).

102. Kamal, M. R., Sahto, M. A., Utracki, L. A., *Polym. Eng. Sci.*, **22**, 1127 (1982); ibid., **23**, 637 (1983).

103. Dekkers, M. E. J., Hobbs, S. Y., Watkins, V. H., *J. Mater. Sci.*, **23**, 1225 (1988).

104. Borggreve, R. J. M., Gaymans, R. J., Luttner, A. R., *Makromol. Chem., Macromol. Symp.*, **16**, 95 (1988).

105. Heikens, D., 5th Annual Meet. Polym. Proc. Soc., Kyoto, April 11-19 (1989).

106. Hobbs, S. Y., *Polym. Eng. Sci.*, **26**, 74 (1986).

107. Anonymous, *Processing Injection Moulding*, Bayer Techn. Inf. No. E 129-886/845324 (1989).

108. Anonymous, *Processing Guide for Injection Molding*, Mobay Chem. Corp., Techn. Inf. No. 55-A431 (2.5)D (1986).

109. Anonymous, *Triax 2000*, Monsanto Plast., Data sheet No. 7072-A (1988).

110. Fisa, B., Dufour, J., Vu-Khanh, T., *Polym. Compos.*, **8**, 408 (1987).

111. Whelan, A., *Injection Moulding Materials*, Allied Science Publishers, London (1982).

112. Bevis, M. J., private communication (1982).

113. Allan, P. S., Bevis, M. J., 5th Annual Polym. Proces. Soc. Meeting, Kyoto, April 11-14 (1989).

114. Anonymous, *Mod. Plast.*, **12**, Jan. (1990).

115. Wübken G., in *Spritzgiess–Werkzeuge*, VDI, Dusseldorf (1980).

116. Mourgu, M., *Technologie du Moulage par Injection des Thermoplastiques*, Impr. Bosc Freres, Lyon (1980).

117. Hawley, R., *U.S. Pat.* Nos. 4,312,917, Jan. 26 (1982); 4,439,387, Mar. 27 (1984).

118. Quinn, K., O'Brien, G., *SPE Techn. Pap.*, **34**, 1617 (1988).

119. O'Brien, K. T., SPE RETEC, Rosemont, IL, Sept. 28-30 (1988).

120. Schmid, B., *Kunststoffe*, **79**, 624 (1989).

121. Xavier, S. F., Tyagi, D., Misra, A., *Polym. Compos.*, **3**, 88 (1982).

122. Fakirov, S., Fakirova, C., *Polym. Compos.*, **6**, 41 (1985).

123. Kenig, S., *Polym. Compos.*, **7**, 50 (1986).

124. Vincent, M., Agassant, J.-F., *Polym. Compos.*, **7**, 76 (1986).

125. Kamal, M. R., Song, L., Singh, P., *Polym. Compos.*, **7**, 323 (1986).

126. Darlington, M. W., Smith, A. C., *Polym. Compos.*, **8**, 16 (1987).

127. Kenig, S., Trattner, B., Anderman, H., *Polym. Compos.*, **9**, 20 (1988).

128. Zhou, C., Chen., S. J., *J. Polym. Eng.*, **8**, 39 (1988).

129. Utracki, L. A., SPE NATEC, Bal Harbour, FL, Oct. 25-27, 1982; *Polym. Eng. Sci.*, **23**, 602 (1983).

130. Anonymous, *SELAR Barier Resins*, DuPont Techn. Inf. E 73973 (1984).

131. Cifferri, A., Ward, I. M., Eds., *Ultra–high Modulus Polymers*, Applied Science Publishers, London (1979).

132. Zachariades, A. E., Porter, R. S., Eds., *The Strength and Stiffness of Polymers*, M. Dekker, New York (1983).

133.   Ward, I. M., *Mechanical Properties of Solid Polymers*, 2nd ed., John Wiley & Sons, New York (1983).

134.   Lee, I., Turner, S., Woodhams, R. T., *Polym. Compos.*, **3**, 212 (1982).

135.   Zachariades, A. E., Chung., B., *SPE Techn. Pap.*, **32**, 965 (1986).

136.   Utracki, L. A., Ed., *Panel Discussion on Thermoplastics vs. Thermosets in Advanced Composites, Polym. Compos.*, **8**, 437 (1987).

137.   Hojo, H., Yaguchi, H., Onodera, T., Kim, E. G., *Intl. Polym. Process.*, **3**, 54 (1988).

138.   Albers, A., Lotz, E., *Shear Roll Extruder; Continuous Processing Compounder for High Viscous Products*, Color Metal Dr.-Ing. Albers Mashinenbau GmbH, D-7843 Heitersheim (1989).

CHAPTER 2

THEORY AND PRACTICE OF POLYMER MIXING

by Peter Hold

32 Gulf View Court
Milford, CT 06460
U.S.A.

In Chapter 2, a broadbrush picture of the present understanding of mixing of materials, with a high viscosity fluid as a major component, is presented. As a result of research performed during the past few decades some of that knowledge can be formulated quantitatively by mathematical expressions. However, description of other phases of the mixing process still depends upon empirical evidence, awaiting more research for further clarification. Owing to the enormous number of particles involved in a mixing process, only a statistical definition of the state of mix can be derived. To facilitate the study of the mixing process and its relationship to the various industrial mixing problems, it is broken down into distinct mechanisms and the mixing problems are divided into a number of categories. Following the initial treatment of mixing as an isothermal process of power law fluids with identical rheologies is a discussion of the effects of nonuniform temperatures and rheologies, as well as of viscoelasticity. Finally, a look at potential future developments of industrial mixing equipment and methods is presented.

2.1   INTRODUCTION

In this chapter a concise overview of the present state of the art of mixing is presented with reference to the impact of that knowledge on the practice of polymer mixing.

To obtain polymeric materials with properties suitable for specific applications, in general, several components of the following generic types have to be added to the base polymer: other polymers (blending or alloying), solid and liquid additives (to modify the appearance, rheological, mechanical, electrical chemical, and thermodynamical properties), and fillers and fibers (to increase physical properties). The process of incorporating these components uniformly into the base polymer is referred to as mixing, an operation that is achieved by different mechanisms, depending upon the nature of the ingredient. In a broad sense modifications of the morphology and chemistry of the polymers performed in the process also come under that heading.

To facilitate the study of polymer processing it is useful to divide it into a number of elementary steps: solids transport, melting, mixing, devolatilizing, and pumping. Mixing is normally performed in conjunction with other processing steps; e.g., solids transport, melting, and pumping, which is characteristic of commercial continuous mixers.

The number of ultimate particles involved in a mixing process is enormously large and a definition of "uniform distribution," which is generally accepted as an indication that a mixture is perfectly random (which can only be given statistically), is the best we can do to characterize a mixture. Such a definition will be discussed later in this chapter.

Prior to the start of the mixing process the components are assumed to be completely separated and the particles of the individual components are not distinguishable from one another. The possible positional arrangement of the particles is therefore unity and the entropy of the state is zero. As the process of incorporation of the components proceeds the number of possible positional arrangements and with it the degree of disorder and the entropy increase until they reach their highest value as soon as all components are unified in one common mass. These arrangements would be equally probable if the mixing action were completely random.

An efficient mixing device, however, introduces actions leading to a systematic approach towards arrangements which satisfy the definition of uniformity. There exist a great number of these arrangements which form a subset of the overall number of possible arrangements. Whether the arrangements conforming with the definition of uniformity (randomness) equally help the achievement of specific properties is still an open question.

A visual comparison of a number of samples having a uniformly distributed color pigment with those representing the acceptance level specifications reveals that the appearance of the latter is superior to the former. A closer investigation shows that in the superior samples, which also conform to the definition of uniformity, the arrangement of the particles is spatially much more regular than in the other samples. An explanation of the superior quality of the color reference samples can be found in Section 2.3 at the end of the discussion of laminar distributive mixing.

From this brief introduction to mixing it can be concluded that the development of a theory of mixing and the simulation of practical mixing processes are rather complex subjects. The first attempt to develop a theory of mixing goes back to the early 1950s. Mixing machinery which is still industrially used, such as the Banbury Mixer and the Two-Roll Mill, has been developed based solely on empirical knowledge and intuition.

The ultimate objective of the development of a theory of mixing is a simulation of practical mixing processes that reliably predicts the results of the mixing process; the properties of the mix.

Figure 2.1 is a schematic presentation of a simulation of the mixing process. The information entering the simulation are: the action of the mixing device (mechanical and thermal), the rheological and the thermodynamic (occasionally also the chemical) properties of the constituents.

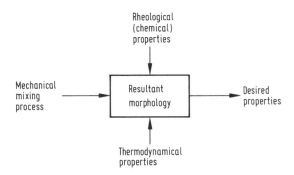

Figure 2.1 The mixing process.

A strategy to pursue that objective systematically has to include the following steps: 1) characterize the quality of a mixture, 2) define the actions that promote the progress of mixing, 3) establish quantitative relationships between these actions and the state (quality) of the mix, 4) develop methods describing the actions performed by a specific mixer geometry, and 5) combine the results of 3) and 4) to arrive at the mathematical model of a mixing device. This text will essentially follow the above sequence.

In Chapter 3 the principles discussed here will be used to explain the design principles of the twin-screw extruders. Due to reliability, versatility, high output (up to 60 tons/hr at present), controllability, and ease of automatization, these machines are steadily gaining popularity.

2.2 CHARACTERIZATION AND MEASUREMENT OF THE QUALITY OF A MIXTURE

As noted in the previous section the purpose of mixing is the development of materials that possess specific properties. The final answer to the question whether a mixing operation has been successful can therefore only be found by testing the properties of the mixture. In many cases such a procedure is complicated and time consuming and therefore not practical for quality control. It also does not provide insight into the structure of the mixture, which is essential for a study of the mixing process and for a decision how to improve it. It is therefore important to develop methods to quantitatively characterize mixtures. However, it must be realized that a relationship between the results of such a characterization and final properties of the mix has to be known. Unless the desired properties of the mixture are considered, a characterization of its structure will not tell us whether it is well or poorly mixed.

A complete characterization of the state of a mixture would require detailed information regarding the size, shape, orientation, and spatial position of every particle. That is, of course, beyond the realm of our capability. The best we can do is to characterize the degree of mixedness based upon the definition of a statistically random distribution of the minor component. In accordance with that definition the probability of finding an ultimate particle of the minor component is equal to the volume fraction of that component in the mixture. It is important to realize that a great number of spatial arrangements of the minor component satisfy the definition of randomness. In a specific case some might result in better and some in less desirable properties.

If a number of samples, all containing a total of n particles, are drawn from a mixture with a completely randomly distributed minor component with a volume fraction, ϕ, the probability of finding a concentration (k/n) of the minor component is given by the following binary frequency function:

$$b\,(k/n) = \frac{n!}{k!\,(n\,-\,k)!}\,\phi^k\,(1-\phi)^{n-k} \qquad (2.1)$$

The variation of that distribution:

$$\sigma^2 = \phi\,(1-\phi)/n \qquad (2.2)$$

The probability of finding a concentration of unity of the minor component in the sample taken from the completely segregated state is ϕ, its volume concentration. Samples with a concentration zero have a probability of $(1-\phi)$. The variance of the segregated state is therefore:

$$\sigma_0^2 = \phi\,(1-\phi) \qquad (2.3)$$

The average concentration found in N samples with concentration x_i is:

$$\bar{x} = \frac{1}{N} \sum_{i=1}^{N} x_i \qquad (2.4)$$

and the variance of these concentrations:

$$S^2 = \frac{1}{N-1} \sum_{i=1}^{N} (x_i - \bar{x})^2 \qquad (2.5)$$

To determine the closeness of the distribution of a mixture to that of the mixture with a perfectly randomly distributed minor component, mixing indices can be defined. One such index is:

$$M = (S/\sigma)^2 \qquad (2.6)$$

For the random mixture $M = 1$ and for the completely segregated state $M = n$.

To determine the *gross uniformity* of a mixture the samples have to be collected from a region defined by the *scale of examination*. The scale of examination depends upon the specific mixing operation. In a compounding/pelletizing operation, pellets will be taken during a certain period of time and investigated to determine the gross uniformity of the volume produced during that period of time. A comparison of the mixing indices found during different periods of time provides information related to gross uniformity of production runs at different times. In general, the scale of examination has to be determined so as to yield information of the gross uniformity of the mix crucial to its acceptance.

The size of the samples must be large relative to the ultimate particles, but small relative to the scale of examination. Otherwise, the test would indicate complete segregation or perfect randomness, respectively, as can be seen from Eq (2.2) for $n = 1$ and $n \to \infty$.

From the above discussion of randomness it is evident that even for the mixing index $M = 1$ (indicating perfect randomness) the *texture* of the mix might have nonuniform appearance because of particles, stripes, and streaks of different concentration. Random sampling at different points cannot characterize the texture. To characterize texture the relative position of the points has to be considered. Danckwerts proposed two statistically defined quantities as a measure of texture: *scale of segregation*, the size of segregated regions, and *intensity of segregation*, the relative difference in concentration between these regions [1,2]. Obviously, a higher intensity of segregation could be tolerated if the scale were small and vice versa.

According to Danckwerts the *scale of segregation*, Z, is the integral of the coefficient of correlation $R(r)$ between concentration x'_i, x''_i, at two points separated by a distance r:

$$Z = \int_0^\beta R(r)\,dr \qquad (2.7)$$

The integral is to be taken over values $r = 0$ $[R(0) = 1]$ to a value at which there is no correlation between the two concentrations.

$$R = \left[\sum_{i=1}^{N} (x_i^1 - \bar{x})(x_i^n - \bar{x}) \right] / NS^2 \tag{2.8}$$

and S^2 is to be calculated from the concentration at all points.

$$S^2 = \left[\sum_{i=1}^{2N} (x_i - \bar{x})^2 \right] / (2N - 1) \tag{2.9}$$

The *intensity of segregation* is defined as follows:

$$I = S^2 / \sigma_0^2 = S^2 / \bar{x}(1 - \bar{x}) \tag{2.10}$$

It is the ratio of the measured variance divided by the variance of the completely segregated system. The intensity of segregation reflects gross uniformity reduced to the scale of examination.

Also in the case of investigation of the texture, the scale of examination depends upon the criteria of acceptance of the mixture. As an example, consider a colored injection-molded frame for a TV set. The lower limit of the range of the scale of examination is given by the perception of the human eye at a certain distance from the frame and the upper limit is the entire frame. For a complete description of the texture the scale of examination should be varied over the entire range. A plot of the scale of segregation as a function of the scale of investigation provides a good characterization of texture.

In applications where the order of the minor component at the level of the ultimate particles is significant the scale of examination has to be extended down to the *local structure*. Examples are rubber containing carbon black, a mixture of PE and carbon black or other additives to increase UV resistance, and blends of polymers. The local structure can be classified by macroscopic, pattern recognition, and a variety of other methods, related to different properties of the mixture.

In summary, a mixture can be characterized by its *gross uniformity, texture,* and *local structure*. The discussion of the methods of characterization provides not only the background for an understanding of the methods of characterization applied in industry and research but it also aids in understanding the make-up of a mix which is a prerequisite for a study of the mixing process.

2.3   ANALYSIS OF THE MIXING PROCESS

Depending upon the physical properties of the components the *distributive* (extensive) or the *dispersive* (intensive) mixing (in combination with distributive mixing) has to be applied for achieving a uniform mixture.

In the absence of cohesive forces, as in the mixing of compatible melts, the regions of different temperatures or those containing nondispersive fillers (which do not significantly affect the rheological properties), the mixing is determined by distributive mixing associated with the strain history of the components and the orientation of the minor elements.

In the presence of components whose ultimate particles are subject to cohesive forces, stress associated with the rate of strain has to be exerted to break up the aggregates. This type of mixing is referred to as dispersive mixing. Examples are granular solids below a certain particle size, liquids and gases with surface tension, and viscoelastic domains relevant to blending and alloying of immiscible polymer melts. In the latter case the rate of stress build-up also appears to be critical.

All the ultimate particles of the mixture have to be put in relative motion to bring about the transition from the completely segregated to the uniformly mixed state. Three basic types of motion are involved in this process. In most mixing problems pertaining to liquids and gases, *eddy diffusion* (turbulence) markedly increases the mixing action. However, with high-viscosity polymer melts it is not possible to create turbulence. The *molecular diffusion* is another important factor in mixing of gases and low molecular fluids. However, with polymers, diffusion takes place at such a low rate that it is only of limited importance, as will be explained later. *Convection, plug convection,* and *laminar convection* are the predominant motions in the mixing process.

2.3.1 Molecular Diffusion Mixing

If the ultimate particles of the minor component are of molecular size, the interface between them at the initial stage is small and surface tension does not play an important role. Once the mixing process has progressed and the interfacial areas have increased sufficiently the interdiffusion can aid the mixing process down to the molecular level provided that the components are miscible. For the interdiffusion to take place the fundamental criterion is that the Gibbs free energy decreases as a result of the process:

$$\Delta G = \Delta H - T\Delta S < 0 \qquad (2.11)$$

ΔS, the change in entropy is rather small for polymer systems. For polymers with large molecules the increase in disorder during mixing is very small and $\Delta S \to 0$. Thus most polymer blends are immiscible. Partially miscible mixtures can be obtained for small values of ΔG. In such cases, from a thermodynamical viewpoint, the components are not miscible but the driving force towards phase separation is not very large and is overcome by the kinetic impulse forces. The minor component is broken down during the mixing action and if the mix is solidified within a reasonable period of time the separation will not take place. Blends of semi-compatible polymers (e.g., polyolefins) exhibit properties which combine the properties of both components. In the case of a mixture of a liquid with small molecules (lubricant, plasticizer, or other liquid additives) chances of miscibility are correspondingly better.

2.3.2 Bulk Convective Mixing

Plug or *bulk convective mixing* can be *ordered* or *random*. Bulk convective mixing involves a plug-type flow. Chunks of material are separated from the bulk of the material and are re-introduced changing the arrangement with or without deformation. The process involves an increase of the interfacial area between the components since otherwise there would be no progress in the degree of mix.

Random bulk convective mixing takes place in most commercial mixers as a stage in the mixing action, e.g., within the zone between the rotors of a Banbury mixer. The mixing action is very complex. In order to make progress in modeling the entire mixing action one has to gain further insight into this type of flow. Recent research has been focused on the analysis of chaotic mixing, which is a step in that direction [3]. Another example of random plug convective mixing is dry blending. In order to predict the progress of random bulk mixing, at best, we can use the following empirical expression:

$$M = 1 - \exp\{-kt\} \qquad (2.12)$$

where t is the mixing time and k is a constant which has to be experimentally determined.

Static or motionless mixers represent examples of *ordered plug convective mixing*. Figure 2.2 is a schematic presentation of their action. Between stages, the material is divided into k parts which are deformed in the following state and recombined. The striation thickness, r, is a measure of the degree of mix:

$$r = 2A/V \qquad (2.13)$$

decreases from one state to the next. In Eq (2.13) V is the total volume and A is the total interfacial area. The decrease in striation thickness is given by:

$$r = r_0 / k^n \qquad (2.14)$$

Where n is the number of stages and r is the initial striation thickness. If at the onset of the process, the components are not evenly divided (as shown in the picture) the efficiency of the mixer is reduced.

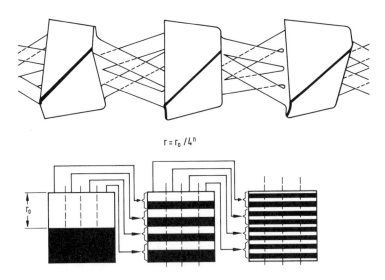

$$r = r_0 / 4^n$$

Figure 2.2 *Ordered bulk mixing.*

In summary, the plug convection mixing is a distributive process capable of improving gross uniformity, texture, and, theoretically in the case of the static mixer, also the local structure. The importance of plug convective mixing can be better understood after the discussion of laminar mixing.

2.3.3 Laminar Convective Mixing

The most important polymer mixing action in industrial mixers is *laminar convective mixing*, associated with both the distributive and the dispersive mixing.

To gain further insight into *laminar distributive mixing*, a laminar flow pattern with the velocity components v_x, v_y, and v_z is considered as shown in Fig. 2.3. A small sphere of the fluid containing a smaller sphere of the minor component with identical rheological properties will undergo a translation, a rotation, and a deformation into an ellipsoid whose axes are the principal axes of stretching. As a result of that deformation, the interface between the two components is increased with a concurrent reduction in striation thickness. The degree of mixing is therefore increased. Figure 2.4a,b are simulations of a mixing process in simple shear flow v_x, $v_y = v_z = 0$ demonstrating the effect of the orientation of the minor component. In the mixing process shown in Fig. 2.5 the direction of the shearing action is changed 90° halfway during the action. A much greater increase in interfacial area has been achieved for the same strain γ. The reason will become evident from the following analytical investigation of laminar mixing. Figure 2.6 demonstrates mixing in uniaxial flow v_x, $v_y = -v_x$, $v_z = 0$ (also called pure shear flow because the velocity pattern can also be considered as the resultant of two simple shear flow at 90°). In this case the direction of the shearing actions remains constant at 45° to the major axis of deformation. No rotation of the major axis takes place. The progress in mixing is even greater than in the previous two cases. The flow pattern will be analyzed later.

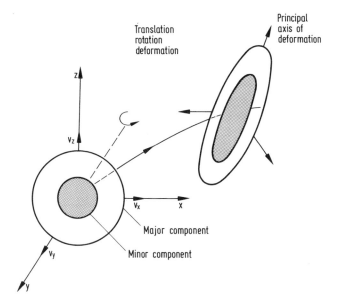

Figure 2.3 *Distributive mixing in laminar flow.*

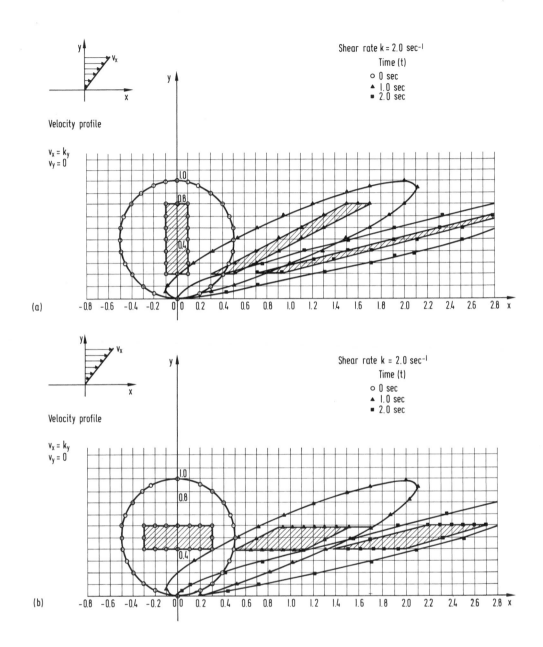

Figure 2.4 *Simulation of laminar mixing in simple shear flow: a) orientation of the minor component perpendicular to the direction of shear; b) orientation of the minor component in the direction of shear.*

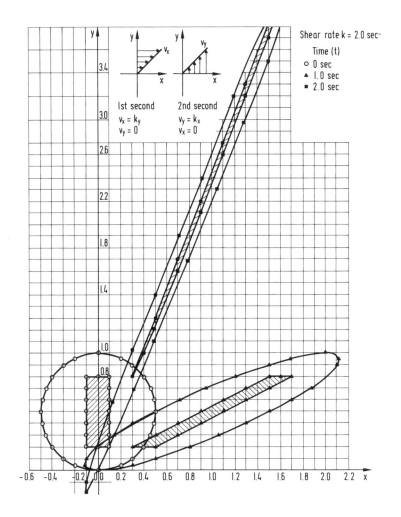

Figure 2.5 *Simulation of laminar mixing in simple shear flow. Change of the direction of shear.*

The laminar action in most commercial mixers can be approximated by a two-dimensional flow between parallel plates or concentric cylinders. Spencer and Wiley were the first ones to develop a mathematical model of the interfacial growth in a simple shear flow [4]. Consider an arbitrarily oriented surface A_0 in simple shear flow as shown in Fig. 2.7, $v_x = \dot\gamma y$, $v_y = v_z = 0$, $\gamma = \dot\gamma \Delta t = v_x \Delta t / y$. The orientation of the normal to the surface is given by α_x, α_y, α_z. It can be shown [1,2] that the surface A_0 after a time Δt (strain γ) has assumed the value:

$$A = A_0 (1 - 2 \cos \alpha_x \cos \alpha_y - \gamma^2 \cos^2 \alpha_x)^{1/2} \tag{2.15}$$

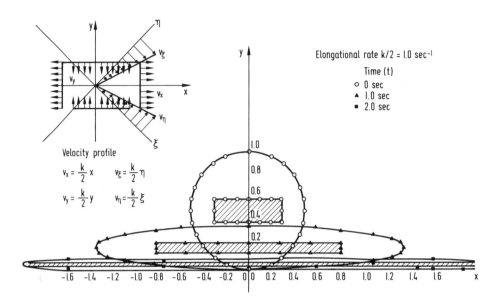

Figure 2.6 *Simulation of laminar mixing in extensional (pure shear) flow.*

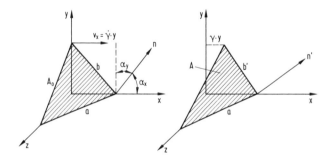

Figure 2.7 *Analysis of the increase of interfacial area in simple shear flow.*

The change of the interfacial area therefore depends upon the original orientation of the surface and the shear strain γ. For large deformations:

$$A = A_0 \, \gamma \cos \alpha_x \tag{2.16}$$

In the general case, for the random orientation of the surfaces, the progress of mixing is reduced by one half [2]:

$$A = (1/2) \, A_0 \, \gamma \cos \alpha_x \tag{2.17}$$

Therefore the degree of mixing is considered to be proportional to the shear deformation. If the expression given by Eq (2.15) is differentiated by γ, the surface A_0 is considered to be perpendicular to the x-y, plane $\alpha_z = 0$, and $\gamma = 0$ is assumed, then the instantaneous rate of increase of the surface A can be expressed as [5]:

$$(dA/d\gamma) = -A_0 \cos \alpha_x \cos \alpha_y \tag{2.18}$$

That value reaches a maximum for $\alpha_y = 45°$, $\alpha_x = 135°$:

$$(dA/d\gamma)_{max} = A_0/2 \tag{2.19}$$

and upon integration:

$$A = A_0 \exp\{\gamma/2\} \tag{2.20}$$

To achieve that rate of increase of the degree of mixing the direction of the interface would have to be turned back after each shear unit to the original $45°$ direction. The uniaxial extensional flow (pure shear flow) in Fig. 2.6 is representative of that condition, which, however, cannot be maintained continuously in a mixing apparatus. This result explains the substantial increase in the rate of mixing that was observed in the simulation Fig. 2.5 and emphasizes in general the importance of frequent changes in the direction of the shearing action and the great mixing efficiency of extensional flow. If the original surfaces in extensional flow are randomly oriented, the increase of interfacial area is reduced to one half of the maximum value:

$$A = (A_0/2) \exp\{\gamma/2\} \tag{2.21}$$

The flow between two concentric cylinders (circular Couette flow) can also be considered as a basis for simulation of the action of a mixer: Fig. 2.8a. For a ratio β of the outside to the inside radius, assuming that the inside cylinder R_i rotates evenly while initially the minor component is uniformly distributed in the radial direction, after N revolutions the striation thickness is not uniform. The strain $\gamma(N)$, after N revolutions as a function of the radial position, y, is given by:

$$\gamma(y,N) = 4\pi N R_i \beta^2 / y (\beta^2 - 1) \tag{2.22}$$

Since the increase in the interfacial area is proportional to γ, the striation thickness decreases with increasing radial position, y.

To complete the description of the laminar distributive mixing it should be pointed out that streamlines in laminar flow never cross (Fig. 2.8b). An initially nonuniform distribution of the minor component across the streamlines can never be washed out by laminar flow. Random plug flow can not only improve such a condition but it also changes the orientation of the interfacial areas relative to the direction of shearing. Random plug flow is therefore an important feature of an efficient mixer. Since it is easier to generate that motion in a twin rotor mixer compared to a single rotor machine, in general, the former is a better mixer.

As a rule of thumb, for adequate mixing 18000 ± 6000 shear strain units are required. If laminar mixing is continued beyond the perfect gross uniformity (a distribution measured on the scale of investigation defined before, which conforms with the variance for that condition), the regularity of the arrangements of the minor particles could increase, in which case we would find an even narrower variance of the samples. However, the sampling test for gross uniformity is not very sensitive on that scale since the gross uniformity remains random. On the other hand, the texture, characterized by the scale and intensity of segregation, will definitely show an improvement as will the local structure. The reference color sample mentioned in Section 2.1, which satisfied the definition of randomness, was mixed considerably longer than the samples whose gross uniformity also satisfied that definition.

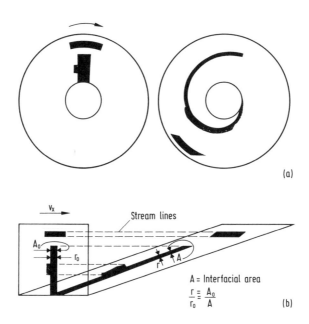

Figure 2.8 *The effect of the initial distribution of the minor component: a) circular Couette flow; b) simple shear flow.*

Laminar dispersive mixing is associated with the rate of deformation and in some cases also with the rate of stress build-up. This type of mixing ought to be used when the ultimate particles form agglomerates caused either by the presence of cohesive forces or surface tension. The understanding of that process is still very limited. In the theoretical study of dispersive mixing the process is divided into three sequential steps: rupture of agglomerates, followed by separation of the fresh fragments away from each other, and finally distribution of the ultimate particles throughout the melt [6]. A great number of simplifying assumptions were made in the development of that mixing model. Tadmor and co-workers [7] reported a promising experimental approach for gaining further insight into the dispersive mixing process. They built an apparatus that features the action of a dispersive high shear mixing zone in a commercial mixer. A mixture of rubber and carbon black was used in the experiment. The mixture was passed through the high-shear zone, collected at the exit and inspected. After three or four passes the agglomerates became visible under the microscope and their number decreased with the number of further passes at a diminishing rate.

Our ultimate objective is the establishment of a qualitative relationship between the flow pattern in the mixing zone and the progress of dispersive mixing, which would aid in the design of dispersive mixing zones and permit prediction of the mixing efficiency of existing mixing apparatus, in a manner similar to the information obtained for distributive laminar mixing.

The motion of the agglomerates imparted by the flow is complex, the break-up mechanism is little understood and the flow pattern in the mixing zone is complicated by the fact that the pressure flow due to the pressure build-up in the converging zone ahead of the narrow-gap, high-shear zone, is superimposed on the shear flow in that zone, increasing the shear rate at the walls but canceling it in the center. Figure 2.9 depicts the velocity distribution in a dispersive mixing zone.

Concluding the discussion on various mixing mechanisms, Fig. 2.10 summarizes the effect the different flow motions have on the mixing process.

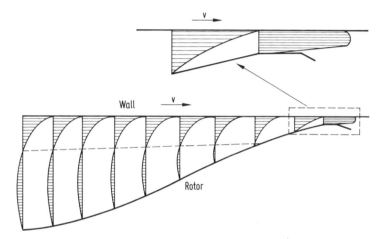

Figure 2.9 *Velocity distribution in the high shear zone of an internal mixer for a power law flow.*

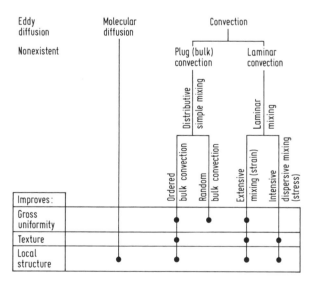

Figure 2.10 *Summary of the various mixing actions.*

2.4   NONISOTHERMAL OPERATION AND POWER CONSIDERATIONS

So far it was assumed that all components in the mixer have the same rheological properties and are at the same temperature (although in discussing dispersive mixing it was stated that stresses associated with the rate of strain cause break-up of agglomerates). In the presence of stresses caused by motion, an irreversible conversion of mechanical energy into heat takes place resulting in the temperature gradients. The energy equation formulates that statement [8]:

$$\rho C_v \frac{DT}{Dt} = -(\nabla . \mathbf{q}) + (\underline{\sigma} : \nabla \mathbf{v}) \tag{2.23}$$

The above expression applies to incompressible fluids and neglects the work done on the fluid by the pressure forces. The term on the left represents the rate of temperature change of a fluid element. The first term on the right expresses the rate of heat gain or loss by the element, and the second term represents the rate of viscous dissipation within the element.

2.4.1   Steady-State Mixing

For steady-state conditions the term $\partial T/\partial t$ in the total differential of T is zero. If a two-dimensional flow pattern is considered the energy equation becomes:

$$\rho C_v \left[v_x \frac{\partial T}{\partial x} + v_y \frac{\partial T}{\partial y} \right] = - \left[\frac{\partial q_x}{\partial x} + \frac{\partial q_y}{\partial y} + \frac{\partial q_z}{\partial z} \right] -$$

$$\tag{2.24}$$

$$- \left[\sigma_{xx} \frac{\partial v_x}{\partial x} - \sigma_{yy} \frac{\partial v_y}{\partial y} \right] - \sigma_{xy} \left[\frac{\partial v_x}{\partial y} + \frac{\partial y}{\partial x} \right]$$

Finally, considering a simple shear flow of a power law fluid with constant coefficients of heat conduction, K, and neglecting the heat dissipation by the normal stress, the energy equation takes the following form:

$$\rho C_v \left[v_x \frac{\partial T}{\partial x} \right] = K \left[\frac{\partial^2 T}{\partial x^2} + \frac{\partial^2 T}{\partial y^2} + \frac{\partial^2 T}{\partial z^2} \right] + m \left[\frac{\partial v_x}{\partial y} \right]^n \tag{2.25}$$

with n being the power law exponent and m the consistency index approximated by:

$$m = m_0 \exp \{-a (T - T_0)\} \tag{2.26}$$

The equation of motion and that of continuity are characterized by the velocity, pressure, and temperature distribution and can also be expressed theoretically. However, only for very simple conditions is it possible to find a closed mathematical solution for the temperature distribution Eq (2.25). To arrive at solutions for more complex conditions some advanced computer techniques have to be employed. In this case it is advantageous to consider the fluid to be pseudo-Newtonian. The Newtonian constitutive equation for incompressible fluid can be introduced into Eq (2.25). The last term in the relation becomes $\eta(\partial v_x/\partial v_y)^2$ and the shear rate and temperature dependence of the viscosity (stored as a result of regression analysis of the experimental data) are introduced into the computation in an iterative way. According to Eq (2.25) an equilibrium will be reached at a temperature level where in every volume element the heat transport by conduction and convection equals the viscous dissipation. The temperatures in the mixer must not exceed a level determined by the heat sensitivity of the material, which indicates the need for effective external heat removal.

2.4.2 Energy Dissipation

For a power law fluid the viscous dissipation in terms of total strain, γ, and time, t, is given by:

$$\dot{E} = \sigma\gamma/t = m(\gamma/t)^{n+1} \tag{2.27}$$

where σ is the shear stress. It follows that the rate of energy dissipated for a given total strain, required to achieve a certain degree of dissipative mixing, increases with decreasing time or with the speed of mixing. Reducing the residence time in the mixer at constant total shear (constant quality of the mix) will result in higher temperatures. The minimum time for distributive mixing is therefore determined by the efficiency of the cooling system. On the other hand, excessive cooling, especially for materials whose viscosity is strongly temperature-dependent, has a diminishing affect because the value of m increases, thereby enhancing the heat generation.

If, however, a certain level of stress is required, as in the case of dispersive mixing, the rate of energy dissipation becomes:

$$\dot{E} = \sigma^{(n+1)/n}/m^{1/n} \tag{2.28}$$

In that case it is best to operate at the highest possible temperature to maintain a high value of m.

2.5 MACROSCOPIC ENERGY BALANCE

The macroscopic energy balance provides important information for evaluation of the mixer efficiency. Two elements ought to be considered: specific power input (to attain a certain quality of mixture) and efficiency of the cooling system.

A continuous mixer operates in a steady-state condition. To perform a macroscopic energy balance the following data have to be collected during operation: the mass output rate $\dot{Q}$, pressure and temperature at the inlet and the outlet, P_1, T_1, P_2, T_2, the rate of heat removal by the coolant $\dot{W}$, and the rate of mechanical energy input $\dot{E}$ (after deduction of the losses in the drive system). With the rate of enthalpy difference $\Delta \dot{H}$ becomes:

$$\Delta\dot{H} = C_v\dot{Q}\Delta T + \dot{Q}(\Delta P/\rho) = \Delta\dot{u} + \dot{Q}(\Delta P/\rho) \tag{2.29}$$

where u is the total energy of the system. The macroscopic energy balance can be stated as follows:

$$\dot{E} = \dot{W} + \Delta\dot{H} \tag{2.30}$$

If the measurements of the above listed quantities have been exact Eq (2.30) will be satisfied.

The batch mixer operates in a nonsteady condition. The rates of the quantities measured have to be integrated over the time duration of the mixing cycle and Q is the total mass in the mixer. Since pressures at the beginning and end of the mixing are the same the macroscopic energy balance becomes:

$$E = W + C_v Q\Delta T = W + \Delta u \tag{2.31}$$

2.6   INFLUENCE OF RHEOLOGY

Laminar mixing frequently involves components that have significantly different properties, including viscosity. Another reason for differences in viscosity is a nonuniform temperature distribution, the causes of which can be understood from the discussion in Section 2.4. Finally the rheology of the mixture also changes with the progress of the state of mixing. Locally, nonuniform mixing can therefore lead to differences in viscosity. In the batch-type mixer with the progress of mixing and in the continuous mixer with the change in position along the mixer the differences in viscosity ought to be expected.

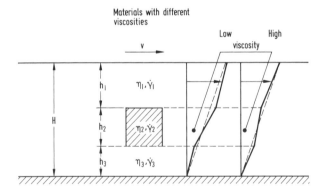

Figure 2.11 Velocity distribution in simple shear flow in the presence of materials with different viscosity.

To study the shear flow between two parallel plates of immiscible components with different viscosities in which the minor component is sandwiched between two layers of major components (as shown in Fig. 2.11), we assume that the shear stress is constant throughout the system. Using the notations in Fig. 2.11 and with $S = h/H$ we arrive at expressions for the shear rates $\dot{\gamma}_1$ and $\dot{\gamma}_2$.

$$\dot{\gamma}_1 = (V/H) / [(1 - S) + S\eta_1/\eta_2]$$

$$(2.32)$$

$$\dot{\gamma}_2 = (V/H) / [(1 - S)\eta_2/\eta_1 + S]$$

the resulting velocity distribution for a minor component with viscosities lower or higher than the viscosity of the major component are shown in Fig. 2.11. In the first case, the shear rate in the major component is reduced in comparison to the shear rate, V/H, for the uniform viscosity while the shear rate in the minor component increases (vice versa in the second case). As mentioned before, the stress levels are not affected but in the case of the low viscosity of the minor component the average velocity of the major component increases. The minor component acts as a lubricant. The analysis of the bicomponent flow presented here is simplified and should be treated with caution. The bicomponent flow is of great commercial interest in conjunction with mixing, mold filling, and coextrusion. Analyses of this type of flow showed that deformation of the interface takes place during the initial phase, which can be explained by the above-mentioned velocity distribution. In other cases, it was found that one of the components broke down into a noncontinuous phase. Other researchers [9], using the "marker and cell" calculation method to simulate the flow of a pair of immiscible liquids of different viscosity in a rectangular channel, found good agreement of the simulation with their experimental results. They found that increasing the viscosity ratio of the two components causes an increase in the interface between components, i.e., mixing takes place.

As far as the change of rheology with mixing is concerned, an empirical exponential expression of the specific energy input during mixing in co-rotating twin-screw extruders [10] was proposed. The experimental parameters for that expression were established in a small-batch mixer. The changing rheology of the mixture during the mixing necessitates the specific energy input determination and its use in an iterative fashion in the mathematical model.

A more detailed discussion on rheology of two-phase systems can be found in following chapters; foams in Chapter 5, blends in Chapters 6 and 7, and fiber-filled systems in Chapters 11 and 12.

2.7   LINEAR VISCOELASTICITY

A crude presentation of the molecular mechanism responsible for the viscoelastic behavior of linear amorphous polymers is provided by placing a Maxwell and a Voigt model in series (Fig. 2.12). The compliance, $J(t)$, of the model in a creep experiment at constant stress is given by [11]:

$$J(t) = \frac{\gamma(t)}{\sigma_0} = \frac{1}{G_1} + \frac{t}{\lambda_1 G_1} + \frac{1}{G_2}(1 - e^{t/\lambda_2})$$

$$(2.33)$$

with the relaxation time:

$$\lambda = \eta/G$$

$$(2.34)$$

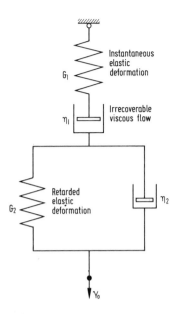

Figure 2.12 *A crude model of viscoelastic materials.*

The relaxation modulus $G(t)$ for a stress relaxation experiment at the constant strain is given by:

$$G(t) = \frac{\sigma(t)}{\gamma_0} = G_1 e^{-t/\lambda_1} + G_2 \gamma_0 \qquad (2.35)$$

If a constant stress is applied to the model, the strain increases with time, t, according to Eq (2.33). If the stress is suddenly removed at time, t, a portion of the strain is immediately recovered, followed by an exponential decay down to the level of strain which cannot be recovered.

In materials with linear viscoelastic behavior the relaxation modulus is only a function of time, not the strain history. More complex models with a spectrum of relaxation times, λ, describe the behavior of linear viscoelastic polymers more closely than the above example.

If we consider the mixing process as a successive application of incremental stresses at times $t_0, t_1, \ldots t_n$ then at a time, t (which is a measure of the degree of mixing), by application of Boltzmann's superposition principle, the strain is:

$$\gamma(t) = J(t - t_0)\,[\sigma_0] + J(t - t_1)\,[\sigma_1 - \sigma_0] + \ldots$$

$$(2.36)$$

$$\ldots + J(t - t_i)\,[\sigma_{i-1} - \sigma_i] + \ldots + J(t - t_n)\,[\sigma_n - \sigma_{n-1}]$$

where $J(t - t_i)$ is the compliance between the time t and t_i in the past, and $\sigma_{i-1} - \sigma_i$ is the incremental stress applied at the time, t_i. For continuous loading:

$$\gamma(t) = \int_0^t J(t - \Theta)\, \dot{\sigma}(\Theta)\, d\Theta \qquad (2.37)$$

$\dot{\sigma}(\Theta)$ describes the stress history.

In summary, the following conclusions can be drawn: 1) the degree of mixing of an element of viscoelastic material after time, t, depends upon the stress history of that element of the mixture from the beginning of the mixing process to t; 2) a mathematical model of the mixing process must contain time as an additional variable; 3) when the mixing process is finished and the stresses are removed the relaxation continues with a detrimental effect on the state of mix; 4) the effect of visoelasticity on the mixing process depends upon the ratio material relaxation time and the time scale of the process; and 5) in response to rate of strain, the viscoelastic materials exhibit the normal stress difference which influences the velocity distribution and therefore can also affect the mixing action.

2.8 RESIDENCE TIME DISTRIBUTION AND GENERALIZATION OF THE DISTRIBUTION FUNCTIONS

The external residence time distribution function, f(t)dt, is defined as the fraction of the flow rate exiting from a continuous processing equipment with a residence time in the apparatus between t and t + dt. The cumulative external residence time function, F(t), is the integral of f(t) between 0 and t:

$$F(t) = \int_0^t f(t)\, dt \qquad (2.38)$$

F(t) represents the fraction of the total exiting flow rate with a residence time between 0 and 1. From the definition of f(t) it follows that:

$$F(\infty) = 1 \qquad (2.39)$$

The mean residence time is:

$$\bar{t} = \int_0^\infty tf(t)\, dt = V/\dot{Q} \qquad (2.40)$$

where V is the volume of the apparatus and $\dot{Q}$ is the volumetric flow rate.

To measure F(t) a step change in tracer concentration at the inlet and measurement of the concentration as a function of time has to be performed. Differentiation of F(t) renders f(t) which is easier to interpret. An impulse tracer injection results directly in the f(t) function.

Depending upon the type of mixing machine and process to be performed, a narrower or wider residence time distribution will be beneficial in achieving optimum results. In the majority of applications continuous mixers are fed with preblended material, or by feeding devices metering the various components continuously at the correct proportions. In that case, provided no mixing in the axial direction takes place, the material moves in a plug fashion through the machine with the correct fractions of components in the cross sections. Only mixing in the direction perpendicular to the material movement is required. Longitudinal mixing (backmixing), which, in general broadens, the residence time distribution, would be detrimental. But a narrow residence time might be desirable for other reasons as well. If some constituents of the primarily polymer mixture are temperature sensitive, prolonged exposure to high temperatures may result in degradation. Other cases which require knowledge and control of the residence time distribution are: processes with temperature-activated additives (foaming agents, crosslinking agents), and continuous polymerizations.

Like the residence time distribution function, other distribution functions can be defined rendering information about the performance of the mixer. A general function, $f(x)dx$, can be defined as a fraction of the exiting flow rate with a certain property between x and $x + dx$. The variable can be the total strain as an indicator of the uniformity and the degree of mixing, the temperature, or a product of time and temperature as a measure of heat history.

2.9   FUTURE DEVELOPMENT TRENDS IN MIXING

All we have to do to glean ideas of what is going to happen to the development of mixing is to look back and recall what happened to the development of extrusion technology after the first workable computer simulation program created by Tadmor and Klein became available to industry [12]. A systematic engineering approach to the design and operation of the extruder gradually replaced the empirical approach to solution of these problems. That and other similar programs since developed have been improved and today we are in position to engineer the design of extruders tailored to specific materials with a predictable performance.

Exactly the same development is already under way for mixing. However, it has to wait for progress in computing methods and computer performance. In the case of an extruder we are primarily interested in data such as pressure, temperature, flow rate, and fraction of the flow rate of melt, which are common to a cross section of the material. As far as the mixer is concerned it is evident from the discussion in the previous parts that much more detailed information is required to characterize the performance.

In the meantime the finite element method (FEM), which had been used only for the analysis of solids, has been adapted to handle fluid flow problems. Faster computers with increased memory, capable of handling these programs, have become available at affordable cost. The FEM which is easily adaptable to complex mixer geometries is now extensively applied to analyze their operation. But we still have to find solutions to some basic mixer problems, namely: the dispersive mixing, mixing of materials of differing rheology, mixing of viscoelastic materials, and modeling of the random plug mixing, before we can assemble all that know-how into a complete model of the mixing process. Such a model for a continuous mixer will probably consist of a number of successive computing steps. Beginning with a mesh which only has to fit the mixer geometry, a first velocity distribution will be established with simplified conditions. The next steps will be performed with a new mesh following discrete stream tubes. In successive computational steps more complex rheologies for the components will be introduced. The results of the analysis will then be summarized by a number of distribution functions of the properties characterizing the performance of the mixer, as discussed in Section 2.8.

The advanced understanding of various phases of the mixing process and our capacity to predict their performance has already, in a number of cases, led to improvements. As in the case of the extruder, the principal design of the various elements of existing mixers will, in general, remain the same, but their detailed design will be optimized with respect to the specified material and operating objectives. The design of the continuous mixer will be segmented. Each segment will be designed to optimally perform a specific task. To gain independent control of the design parameters and operating conditions, the mixer will, in some instances, have to be dived into two units with individually controlled drives, possibly with a third unit responsible for the generation of pressure only.

A complete mathematical model of the batch-type mixer is still one step further removed from realization than that of the continuous mixer. The operation of a batch mixer is an unsteady state and it is much more difficult to keep track of individual volume elements during the mixing operation than in the continuous mixer. Furthermore, arrangements of the components at the beginning of the mixing cycle are poorly defined as distinguished from their fairly uniform composition at the inlet of the continuous mixer.

A complete review of the existing industrial mixers is beyond the scope of this text. Interested readers are referred to [12,13].

NOTATION

A	Interfacial area between components of the mixture
A_0	Initial interfacial area
a	Temperature dependence coefficient of the consistency index in the power law model
$b(k/n)$	Probability of finding a concentration (k/n) in a sample drawn from a mixture with a randomly distributed minor component
C_v	Specific heat at constant volume
E	Energy, total mechanical energy input during a batch-type mixer cycle
$\dot{E}$	Energy rate (power), rate of viscous dissipation, mechanical power input to the continuous mixer
$f(t)dt$	External residence time distribution function
$F(t)$	Cumulative external residence distribution
$f(x)dx$	Fraction of exiting flow with a certain property between x and dx
ΔG	Difference in Gibbs free energy
G_1, G_2	"Modulus" in the model for viscoelastic response
$G(t)$	Relaxation modulus for the stress relaxation experiment
ΔH	Difference in enthalpy
$\Delta \dot{H}$	Difference in enthalpy flux between input and output in the continuous mixer
H	Channel width in parallel plate flow
h	Width of minor component in parallel plate flow of fluids with different viscosities
I	Intensity of segregation
$J(t)$	Compliance in the creep experiment
$J(t-t_i)$	Compliance between time t and t_i
K	Coefficient of heat conduction
k	Number of minor particles found in a perfectly randomly distributed sample
k	Empirical constant in the mixing index for random plug mixing
M	Mixing index
m	Consistency index in the power law model
m_0	Consistency index at temperature T_0
n	Parameter of the power law model

n	Number of ultimate particles in the sample
n	Number of stages in the static mixer
N	Number of samples drawn from the mixture
N	Number of revolutions of the circular Couette mixer
P	Pressure
ΔP	Pressure difference
$\mathbf{q}$	Heat flux vector
Q	Total mass in the batch mixer
$\dot{Q}$	Mass output rate of the continuous mixer
R_i	Coefficient of correlation between points with concentrations x''_j and x''_i
r	Distance between points with the concentrations x''_i and x''_j
r	Striation thickness
r_0	Initial striation thickness
R_i	Radius of the inner cylinder in Couette flow
S^2	Variance of the concentration of the drawn samples
ΔS	Difference in entropy
S	Fraction of the gap occupied by the minor component in parallel plate flow of fluids with different viscosities
T	Temperature
ΔT	Temperature difference
u	Total energy of the system
V	Volume
V	Velocity of upper plate in parallel plate flow of fluids with different viscosities
$\mathbf{v}$	Velocity vector
v_x, v_y, v_z	Components of the velocity vector
W	Total heat removed by the coolant during a cycle of the batch mixer
$\dot{W}$	Rate of heat removed by the coolant from the continuous mixer
x, y, z	Cartesian coordinates
x_i	Concentration found in the ith sample drawn from the mixture
$\bar{x}$	Average concentration of the samples drawn from the mixture
y	Radial position of a point in circular Couette flow
Z	Scale of segregation
$\alpha_x, \alpha_y, \alpha_z$	Angles of the directional cosines
β	Ratio of the outer to the inner cylinder in circular Couette flow
ϕ	Fraction of the minor component in the mixture
γ	Strain
$\dot{\gamma}$	Rate of strain
γ_0	Initial strain
λ	Relaxation time
η	Viscosity
η_1, η_2	Viscosities in the model for viscoelastic response
ρ	Density
σ^2	Variance of the concentration of the samples drawn from a mixture with a perfectly randomly distributed minor component
σ_0^2	Variance of the completely segregated state of the components of the mixture
$\underline{\sigma}$	Stress tensor
$\sigma_{xx}, \sigma_{xy}, \ldots$	Components of the stress tensor

$\dot{\sigma}$ Stress history for continuous loading
$[\sigma_i - \sigma_{i-1}]$ Incremental stress

REFERENCES

1. McKelvey, J. M., *Polymer Processing*, John Wiley & Sons, New York (1962).
2. Tadmor, Z., Gogos, C., *Principles of Polymer Processing*, John Wiley & Sons, New York (1979).
3. Khakhar, D. V., Rising, H., Ottino, J. M., *J. Fluid Mech.*, **172**, 419 (1986).
4. Spencer, R. S., Wiley, R. N., *J. Colloid Sci.*, **6**, 133 (1951).
5. Erwin, L., *Polym. Eng. Sci.*, **18**, 738 (1978).
6. Manas-Zloczower, J., Feke, D. L., *Int. Polym. Proc.*, **2**, 182 (1988).
7. Cohan, R. K., David, B., Nir, A., Tadmor, Z., *Int. Polym. Proc.*, **1**, 13 (1987).
8. Bird, R. B., Armstrong, R. C., Hassager, O., *Dynamics of Polymeric Liquids*, Vol. 1, 2nd Edition, John Wiley & Sons, New York (1987).
9. Bigg, D. M., Middleman, S., *Ind. Eng. Chem. Fundam.*, **13**, 184 (1974).
10. Kalyon, D. M., Gufsis, A., Gogos, C., Tsenoglou, C., *SPE Techn. Pap.*, **34**, 64 (1988).
11. Williams, D. J., *Polymer Science and Engineering*, Prentice Hall, New York (1971).
12. Matthews, G., *Polymer Mixing Technology*, Applied Science Publishers, London (1982).
13. Tadmor, Z., Klein, I., *Engineering Principles of Plasticating Extruders*, Van Nostrand Reinhold, New York (1970).

CHAPTER 3

COMPOUNDING IN TWIN-SCREW EXTRUDERS

By David B. Todd

APV Chemical Machinery Inc.
901 Durham Avenue
S. Plainfield, NJ 07080
U.S.A.

Two-phase polymer systems are usually prepared by a melt compounding process. The properties of such systems may depend as much on the processing history as on the constituent base polymer and minor phase components. Single-screw extruders have long been used for compounding - the provision of homogeneous mixtures as a result of the dispersive and distributive actions occurring within the extruder between screw and barrel. As a need for more sophisticated morphologies has developed, twin-screw extruders are finding increased usage because of the additional advantage of screw-to-screw interaction. This chapter describes the features of commercially available twin-screw extruders, and how their unique features affect polymer processing.

3.1   INTRODUCTION

Twin-screw extruders are either co-rotating or counter-rotating, and either tangential or intermeshing. The intermeshing varieties may vary in the degree of intermesh to the point of being fully self-wiping. The main feature which distinguishes twin-screw from single-screw extruders is the interaction between the screws. Generally, the twin-screw extruders are starve-fed; a feature that provides independence between feed rate and screw speed, and thus an additional level of control over the compounding operation. The intermeshing twin-screw extruders have splined or keyed shafts and a great variety of slip-on screw and kneading paddle sections that permit a seemingly endless variation of feeding, mixing, venting, and pressure generation arrays (Fig. 3.1). The historical development of twin-screw extruders was recently reviewed by White et al. [1].

The monograph by Martelli [2] described much of the early applications of twin-screw extruders, with emphasis on counter-rotating intermeshing screws for profile extrusion rather than for general compounding. Janssen [3] explored thoroughly the flow phenomena involved in twin-screw extruders, particularly the leakage flows in intermeshing counter-rotating screws.

General extrusion texts, like those of Rauwendaal [4] and Tadmor and Gogos [5], contain portions which are relevant to twin-screw extruders as well. The general principles of compounding were presented in Chapter 2.

3.2   INTERMESHING CO-ROTATING TWIN-SCREW EXTRUDERS

Several equipment manufacturers (e.g., APV Chemical Machinery, Berstorff, Ikegai, Japan Steel Works, Leistritz, Toshiba, Werner & Pfleiderer) offer co-rotating intermeshing twin-screw extruders. There are many similarities between the different suppliers. The major differences are in the variety of barrel sections, screw, and kneading paddle offerings.

A specification of fully intermeshing configuration imposes a constraint on the shape of the screw channel. The channel depth, h, is fixed by the centerline distance (C) between shafts to barrel diameter (D_b) ratio:

$$h/D = 1 - C/D_b \qquad (3.1)$$

where D is the screw diameter.

Thus, the compression ratio achieved in single-screw extruders by changing flight depth, in fully intermeshing twin-screw extruders can be obtained only by a change in helix angle.

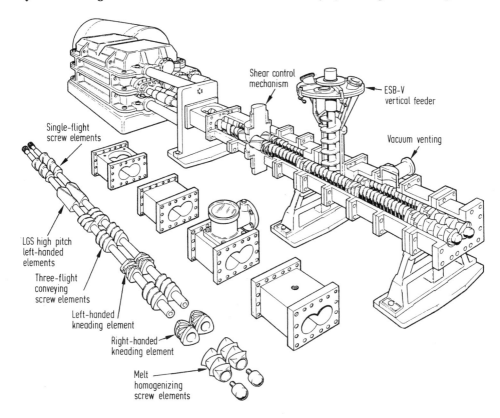

Figure 3.1 *Modular intermeshing co-rotating twin-screw extruder (courtesy Werner & Pfleiderer Corp.).*

Choice of the C/D ratio represents a compromise between a high value for greater torque capability (larger root diameter) and a low value for greater feeding and venting capacities (larger flight depth). The minimum C/D ratio for self-wiping also depends on the number of screw starts. For a triple start (tri-lobe) screw: $(C/D)_{min} = \sqrt{3}/2$; whereas for a dual start (bi-lobe) screw: $(C/D)_{min} = \sqrt{2}/2$. Generally, twin screws are not provided at the minimum C/D ratio, as then the flight tip is a knife edge.

Cross sections of self-wiping co-rotating screws are shown in Fig. 3.2. The manufacturers of these extruders also supply kneading components of the same cross section as the screws, either as individual paddles or in ganged arrays. As can be seen in Fig. 3.3 the arrays can vary in axial width and angular offset. They can be arranged in a forwarding array (shown in Fig. 3.3) or in neutral or reverse arrays. In the forwarding array, they have a drag-flow forwarding capability similar to regular screws. However, because of the angular gaps between paddle tips, the material being processed has an easy leakage path backward.

The variation of channel and flight-tip geometry as a function of C/D ratio has been fully detailed by Booy [6]. Useful plots of relative process cross sections and lateral and "rubbing" surfaces are shown in Figs. 3.4 and 3.5 [7].

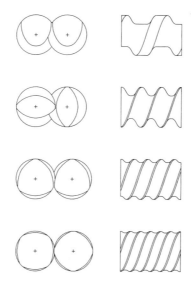

Figure 3.2 Screw profiles.

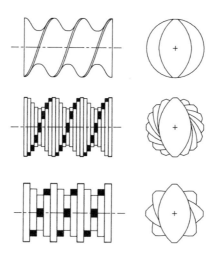

Figure 3.3 *Kneading paddles, staggered to simulate a screw profile.*

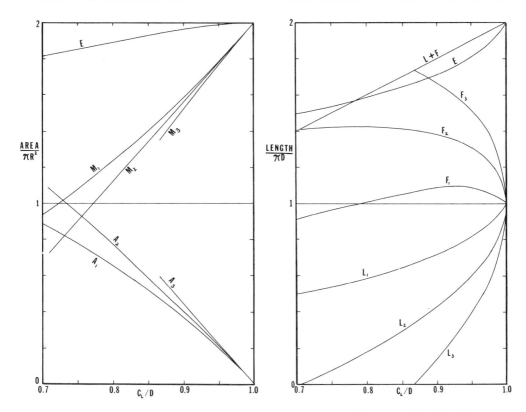

Figure 3.4 *Relative cross-sections for single-, double-, and triple-start screws: E, empty barrel; M, occupied by metal; A, available for process [7].*

Figure 3.5 *Relative peripheral surfaces for single-, double-, and triple-start screws: E, empty barrel periphery; F, flank length not contacting the barrel; L, land length contacting the barrel [7].*

3.2.1   Feeding

One of the advantages of these type extruders is that the larger feed-port dimensions and the self-wiping characteristics permit starve feeding (see Fig. 3.6). These features usually eliminate the need for either a preblending stage or for crammer feeders. However, if each constituent is to be independently fed to the extruder feed throat, each stream must be accurately metered. Curry et al. [8] have shown how the kneading blocks are useful in dampening the consistency variance that otherwise exists with non-uniform feeders.

The mechanism for solids melting in the intermeshing twin-screw extruders is distinctly different from that in single-screw machines. The solids are conveyed forward by interaction of the two screws independently of solids friction against the barrel.

Kneading paddles are the most effective elements for both melting and dispersion. The dispersion face (Fig. 3.7) of the paddle promotes breakup of agglomerates. The dispersion angle, ϕ, is determined by the C/D ratio:

$$\mathrm{Cos}\ \phi = C/D \qquad (3.2)$$

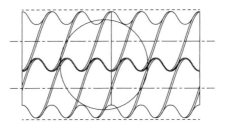

Figure 3.6 *Large feed port in a twin-screw extruder.*

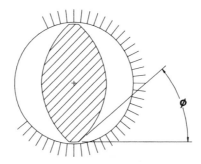

Figure 3.7 *Dispersive mixing face of a kneading paddle.*

Szydlowski et al. [9] presented calculated pressure profiles that can be generated owing to this lubrication wedge effect.

In the kneading paddle sections, a compression/expansion action occurs that is unique to the intermeshing co-rotating twin-screw machines. Figure 3.8 illustrates how a portion of the material contained as one crescent is expanded to a larger cross-section, and then recompressed to its original crescent shape. This kneading, massaging action is most beneficial in causing solids to melt by rubbing them against each other. The action also enhances mixing of fully melted volumes by repeated reorientation of the laminar layers.

As the polymer begins to melt from the kneading action, each solid particle is soon surrounded by melt. As the contents of any one cross section are being squished around, the unmelted solids are being warmed and the outer edges being either abraded or sloughed off. Melting thus occurs in a short axial distance without dependence of solids friction against the barrel, and with less likelihood of overheating some portions while others remain unmelted.

3.2.2   Mixing

In addition to the dispersive and distributive mixing arising from the compression/expansion effect, an additional mixing derives from the scissor-slicing action between the paddle on one shaft and the next upstream and downstream paddles on the opposite shaft, as illustrated in Fig. 3.9.

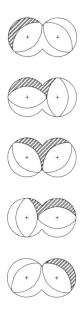

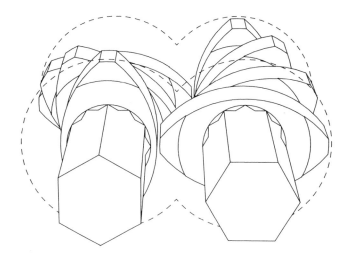

Figure 3.8 *Compression–expansion effect in an intermeshing co–rotating twin–screw extruder.*

Figure 3.9 *Slicing interaction between kneading paddles.*

The extent of mixing action in the kneading section depends on how full that section is. An assured complete filling can be obtained by utilizing reverse screw or paddle elements, as shown in Fig. 3.10. An alternate method is to use blister rings which require some pressure development for polymer to pass over, thus causing complete filling for some portion of the upstream screw or paddles (Fig. 3.11). Because of the intermeshing, the blister rings require some axial offset. The amount of resistance is determined by the axial length and radial clearance of the blister rings.

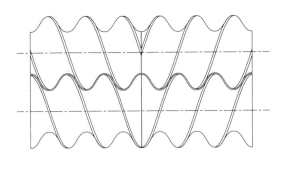

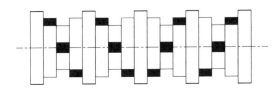

Figure 3.10 *Reversing screw or paddle arrays to provide holdback.*

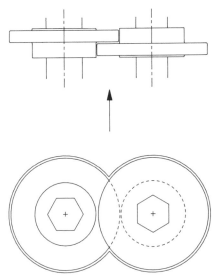

Figure 3.11 *Overlapping full-bore blister rings.*

Both reverse elements and blister rings provide fixed resistances. To provide variability, several devices have been offered. The barrel valve [10] depicted in Fig. 3.12 can be fitted at any (or several) axial location(s) along the barrel. It is placed in a cavity in the barrel, the cavity being a bypass over a pair of blister rings. The valve contains a vane, which, by simple rotation, varies the amount of opening over the blister ring restriction. With such a device, a product temperature variation of about 30°C has been reported [11]. A similar effect can be achieved by valve elements that move radially [12].

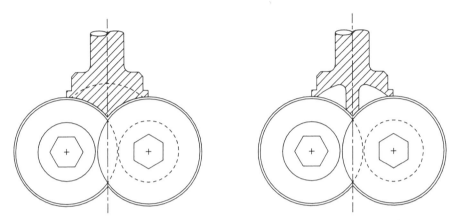

Figure 3.12 *Barrel valve for use in conjunction with full-bore blister rings (courtesy APV Chemical Machinery Inc.) [10].*

Other devices involve matching tapered sections on screws and barrel, which, by axial displacement of the screws with respect to the barrel, provide a variable restriction (Fig. 3.13) [13]. Placing a valve at the extruder's exit provides additional flexibility.

Providing variability extends the range of products that can be optimally processed without necessitating a rotor configuration change. With on-line measurement of some critical product property, such as product viscosity, closed-loop control can be imposed [12]. A general approach to optimization of compounding has been outlined by Colbert [14]; incremental changes are to be made in throughput and rotor speed to obtain improved quality at the maximum production rate.

In addition to the paddle type kneading elements illustrated in Fig. 3.9, other specific elements may be provided to enhance homogenization, perhaps at the sacrifice of some self-wiping. For example, gearlike homogenizers (Fig. 3.14) have been used to increase distributive mixing without as much energy consumption as by kneading paddles [15].

3.2.3   Venting

The modular barrel approach makes it easy to provide vent ports along the barrel. The self-wiping action of the screws is effective in generating fresh surface. Because of the continuous figure-eight passage around the screws, the effect of vacuum venting extends axially over a longer barrel length than just at the vent opening. Vent ports may in turn be provided with either single- or twin-screw stuffers to protect vacuum systems from the consequences of overflow foaming or polymer entrainment following an upset condition.

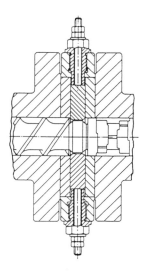

Figure 3.13 *Throttle device (courtesy Berstorff) [13].*

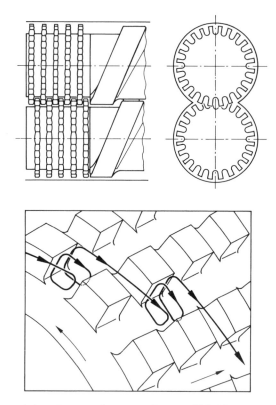

Figure 3.14 *Gear type mixing elements (courtesy Berstorff) [13].*

3.2.4 Pressure Generation

Intermeshing screws, especially single start, impede pressure flow and permit pressure generation over a short barrel length. For modest pressure requirements (1 to 10 MPa) in-line extrusion is generally feasible. Eise et al. [16] described direct sheet extrusion of $CaCO_3$ or talc-filled polypropylene with excellent properties and at a cost advantage.

In some instances, a reduced bore discharge can be utilized. By using two parallel single-screw extensions, Todd [17] has shown that this diameter reduction can lower to one-third the power required for pressure generation, and the consequent rise of the polymer temperature. An additional advantage of reduced bore extrusion is the halving of the thrust load on the twin-screw gear box.

Since intermeshing co-rotating extruders frequently are operated at high rotor speed (200-500 rpm), it is sometimes desirable to separate the mixing function from the pressure generation function. Thus, a gear pump may be used in tandem with the twin-screw extruder for pressure generation above 10 MPa.

3.2.5 Power and Heat Transfer

Most of the power required for compounding is needed to supply the enthalpy change in melting the polymer and bringing all ingredients to a temperature suitable for extrusion. If there is a difficult dispersion task in breaking up obstinate agglomerates or a second insoluble polymer, the additional energy to mix can be significant. Once molten, the power consumption, G, depends on barrel length, L, screw diameter, D, rotor speed, N, and effective viscosity, η:

$$G = f_1 D^2 L N^2 \eta \tag{3.3}$$

Since the viscosity is shear dependent ($\eta \propto N^{n-1}$, with n being the "power-law" exponent):

$$G = f_2 D^2 L N^{n+1} \tag{3.4}$$

Secor [18] has confirmed the (n + 1) dependency of power on rotor speed for a partially filled 305 mm twin-screw extruder.

Heat transfer in extruders is notoriously poor, especially for cooling. Cooling polymer at the barrel wall increases its viscosity, and any gain in heat removal may be more than offset by the increase in power consumption.

Davis [19] has addressed the overall problems concerned with heat transfer in extruders. Todd [20] presented an overall correlation for heat transfer in twin-screw extruders:

$$HD/k = 0.94 \, (D^2 N \, \rho/\eta)^{0.28} \, (c_p \eta/k)^{0.33} \, (\eta/\eta_w)^{0.14} \tag{3.5}$$

where H is the film transfer coefficient, k, thermal conductivity, c_p, heat capacity, ρ, density, and η_w, the effective viscosity at the wall. Surprisingly, the author found no significant effect of the degree of barrel filling nor of the rotor clearance.

3.3   FLOW MECHANISM

Herrmann et al. [21] described the general flow principles, melting mechanism, and pressure buildup capacity for multiscrew extruders. Bigio and Erwin [22] studied mixing of similar viscous fluids in both screw and kneading block sections at very low rotor speeds.

Szydlowski and White [9,23-25] presented a series of papers modeling the flow in intermeshing co-rotating twin-screw extruders. The pressure peak generated by the dispersion face of a kneading paddle promotes polymer reorientation and backmixing. Both the metering (drag) flow and pressure flow have been characterized [23,24]. The gaps in the offset kneading paddles exhibit a great sensitivity to pressure.

Kalyon et al. [26,27] described experimental techniques and simulation methods which are helpful in analyzing and characterizing the mixing occurring in twin-screw extruders. Their experimental studies confirmed the theoretical approach in calculating velocity, temperature, and stress profiles, and in simulating the pathlines of the interface generation in such systems as highly filled poly(butadiene-co-acrylonitrile)/epoxy formulations.

3.3.1   Throughput

Loomans et al. [28] described the flow distribution in intermeshing co-rotating twin-screw extruders in terms of the traditional drag (Q_d) and pressure (Q_p) flow geometric factors:

$$Q = Q_d - Q_p \tag{3.6}$$

$$Q = AN - B\Delta P / \eta L \tag{3.7}$$

The A and B terms can be defined in terms of the screw helix angle and the apparent forwarding and reversing helix angles of tips of the kneading paddles.

The pressure flow through static arrays of intermeshing screws or paddles was studied by Todd [29]. The drag flow term was independently determined and confirmed to be proportional to N:

$$Q_d = (a\,Z\,N\,\cos\,\theta) / 2 \tag{3.8}$$

where Z is the lead length, θ is the helix angle, and a is the open cross section of the figure-eight barrel available for polymer flow.

3.3.2   Residence-Time Distribution

Backmixing in starved intermeshing co-rotating extruders was correlated by Todd [30] in terms of a Peclet number describing the extent of axial dispersion. In general, flow through screws alone approaches plug flow. With kneading paddles, more axial mixing occurs with either an increase of offset gap between adjacent paddles, or with a degree of fill. Kao and Allison [31] injected a pulse of yellow die or carbon black into a styrene based copolymer in a 30 mm twin-screw extruder, with product color analysis by UV spectrophotometry.

Altomare and Anelich [32] studied the effect of pitch, number of kneading paddles, stagger angle, and forwarding or reverse arrays on the residence-time distribution in a twin-screw cooking extruder. The Peclet numbers thus determined were in the range of 5 to 50, but the spread in the distribution was not represented by a single parameter.

Jakopin [33] defined a self-cleaning time as the width (b) of the overall residence-time distribution curve less the width (w) between the two inflection points (Fig. 3.15) of a pulse injection test. In most instances, it may be difficult to ascertain the time when the last bit of tracer has left the extruder. If polymer truly adheres to either barrel or screw, that film of material will interchange with the main stream of polymer flow at a much slower rate than the bulk conveying and intermixing between screw and paddle components.

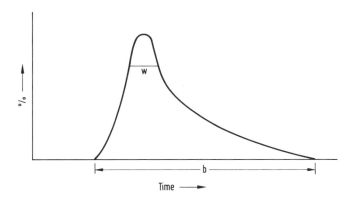

Figure 3.15 *Residence-time distribution: w, width between inflection points; b, total width.*

Agur [34] measured residence-time distributions in two twin-screw extruders (30 and 53 mm diameter), along with power input and product temperature. Based on these measurements, the author proposed a scale-up method. Where heat transfer through the barrel is minimal, the capacity can be scaled up in proportion to cube of diameter and first power of rotor speed. However, most extruder manufacturers recommend lowering the top allowable rotor speed as diameter is increased.

3.4   APPLICATIONS

3.4.1   General Compounding

Descriptions of compounding processes have appeared frequently, but generally without complete descriptions of either the polymer and product properties or the rotor configuration and operating conditions.

Incorporation of glass fiber is frequently used as an example of the need to optimize distributive and dispersive mixing functions. A staged operation that allows the polymer to be fully fluxed before subsequent addition of the glass fiber minimizes glass breakage [33,35]. After glass fiber addition, the mixing must be distributive enough to debundle and wet out the individual fibers while not being so dispersive as to grind the fibers back to fine sand.

Eise and co-workers [36,37] have presented specific examples of polypropylene filled with 25% glass fiber (from either rovings or chopped fibers). Mauch [38] showed variation in distribution of the glass-fiber length as a result of incorporation in screws or kneading paddle sections. Wall [39] determined that the mixer configuration was more critical than rotor speed for obtaining the correct balance between fiber length preservation and fiber-matrix bonding. Unfortunately, good fiber length in pellets from compounding may be destroyed in the remelting that occurs with injection molding.

The importance of staged addition has been illustrated by Eise and Mielcarek [40] where the polyethylene additive, such as of slip agent or antiblock additive, affects the apparent viscosity.

One of the available intermeshing co-rotating extruders has a split clam-shell barrel (Fig. 3.16). Karian [41] took advantage of this feature to deadstop and open the barrel quickly to measure the course of compounding of polyvinylchloride formulations.

Figure 3.16 *Clam-shell twin-screw barrel (courtesy APV Chemical Machinery Inc.).*

Plochocki and co-workers [42,43] have reported studies of the effect of mixing history on the morphology of polymer blends formed from immiscible polymers (polystyrene and polyethylene). Addition of a compatibilizer, such as a block styrene-butadiene copolymer, stabilized a smaller sized dispersed phase. Kapfer [44] reviewed those features of the intermeshing co-rotating twin-screws that are most relevant to the preparation of polymer blends. A new feature described was the provision of serrations in the screw flight periphery to allow some gentle backmixing.

Incorporation of various fillers into high-performance polymers such as polyphenylene sulfide, polyimide, and liquid crystal polymers has also been described by Kapfer [45]. In the latter case, temperatures of 400 to 450 °C were required.

Energetic material, such as solid rocket propellant, represents a special case of an exceedingly highly filled mixture. Kowalczyk et al. [46] described particular features of intermeshing co-rotating twin-screw extruders developed for this specific task.

3.4.2   Devolatilizing

Early descriptions of devolatilization in twin-screw extruders [47] described performance along the same lines as with single-screw extruders. The approach to equilibrium was shown to vary inversely with the throughput and directly with the square root of rotor speed.

Collins et al. [48,49] conducted nitrogen-stripping tests for the removal of Freon 113 from a bubble-free polybutene solution in a 50.8 mm diameter co-rotating twin-screw test extruder. The data were calculated in terms of a mass transfer coefficient [49] and the length of a transfer unit [48], the latter being proposed as an appropriate measure of devolatilizing effectiveness. With the foaming, that is normally present during devolatilization, the length of the transfer unit would be expected to be considerably shorter.

Werner and Curry [50] described the removal of monomers from polymethylmethacrylate and polystyrene in 30 and 90 mm diameter extruders. Staging of volatile removal was necessary for efficient use of the extruder, as was providing an adequate degree of starved flights to allow for unimpeded vapor discharge.

Anders [51] described an extruder variant with an expanded barrel diameter at the discharge end to facilitate starved operation and more effective devolatilizing, as shown in Fig. 3.17. Mack [52,53] elaborated on advantages of rear degassing, special mechanical seals, and screw shaft cooling for some applications. Comparisons were made between single- and twin-screw devolatilizers. One of the advantages of the latter is a greater ability for post venting addition of compounding ingredients.

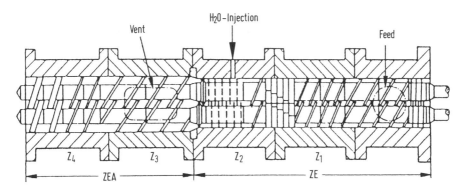

Figure 3.17 *Two-stage barrel with expanded downstream section for venting (courtesy Berstorff).*

Notorgiacomo and Biesenberger [54] conducted devolatilization tests comparing single- and twin-screw devolatilizers. If the two extruders have equal surface generation capabilities, such as using a triple-flighted single-screw extruder to duplicate the three parallel passageways in a co-rotating twin-screw extruder, the extent of devolatilization is equivalent.

3.4.3   Reactive Extrusion

The intimate mixing possible with intermeshing twin-screw extruders, combined with effective staging, allows many types of reactions involving viscous polymers to be carried out. Betz [55] patented co- and ter-polymerizations in a 60 mm diameter twin-screw extruder. Polymerization of trioxane all the way to a powdered solid polyoxymethylene discharge was described by Todd [56] where the self-wiping feature was essential to prevent complete blockage of the screws.

Mack and Chapman [57] described the advantages of the gear type homogenizers (Fig. 3.18) in achieving continuous polymer reactions. Curry et al. [58] illustrated the peroxide breaking of polypropylene. Product viscosity was continuously monitored and used to control the addition of peroxide.

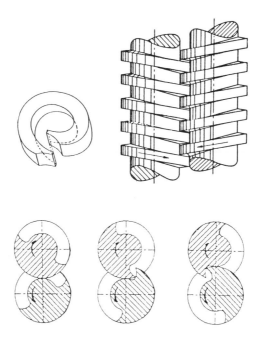

Figure 3.18 *Intermeshing counter-rotating twin-screw extruder.*

3.5   INTERMESHING COUNTER-ROTATING TWIN-SCREW EXTRUDERS

Although intermeshing counter-rotating twin-screw extruders have long been used for pipe and profile extrusion (particularly for polyvinylchloride) [2], their use in other compounding applications has been increasing. Although material is conveyed in closed C-shaped flight chambers down the barrel, mixing does arise from the circulation pattern generated in the C-shape chamber, and by the leak flows over the flights and between the screws (Fig. 3.18). Janssen [3] has described these leakage flows in detail. The flow in the calendar gap between the two screws is particularly effective for dispersive mixing.

Since conveying is positive and not dependent upon drag flow, the intermeshing counter-rotating twin-screw extruders are more efficient in pressure development. As such, there is less product temperature rise, which is critical for temperature-sensitive polymers. The high pressures can be generated at low screw speeds. Absence of an easy leakage path minimizes pressure flow, making these extruders ideal for profile extrusion.

Intermeshing counter-rotating twin-screw extruders used just for compounding tend to be longer and operate at higher screw speeds. Thiele [59,60] described the features that may offer particular advantages in compounding. The feed flight channel width can be selected not only to provide easy positive feeding, but also to control the degree of fill in subsequent downstream channels (with lower helix angles).

The material subjected to calender gap dispersion is remixed in the C-chamber. A given level of dispersion capability is achieved without high heat generation. Additional shearing can be provided with overlapping blister rings similar to those in Fig. 3.11. When these are serrated, they are similar to the gear homogenizers used in co-rotating screws (Fig. 3.14), and can provide distributive mixing with less heat input. Speur and co-workers [61] have modeled the flow patterns in the calendar gap by finite element analysis. A vortex may form at the entry, but its presence is determined uniquely by the relative leakage flow.

Venting in counter-rotating extruders has an advantage in containing the melt within the screws (less likelihood of vent port fouling) but (because of the closed C-chambers) is effective only below the vent port itself. Sakai and co-workers [62,63] have described venting of mixed octane-hexane solvent from linear low-density polyethylene from an initial 10% down to 500 ppm, and removing chlorinated solvents from a rubber slurry. Comparisons were also made with other twin-screw extruders.

Poltersdorf et al. [64] have modelled the foam- and diffusion-controlled degassing which occurs in intermeshing counter-rotating screw extruders. Shah and co-workers [65] reported removal of tetramethylsulfone from a polyamide in a 34 mm diameter extruder. The authors also determined the residence-time distribution with MnO tracer. The apparent Peclet number was greater than 10, even with a 5 min average residence time.

Menges et al. [66] proposed model laws to describe the performance of 80.8 mm and 58.5 mm diameter extruders. The model laws are correlative but not predictive, as their limits of applicability have not yet been confirmed. Feistkorn and Rengshausen [67] have considered the effects of changing the operating conditions on the model laws, based on their experience with polyvinylchloride in 34 and 85 mm diameter extruders.

Dey et al. [68] discussed the criterion for assessing mixing performance wherein the extent of mixing is defined by the striation thickness. Dey and Biesenberger reported results for the reactive extrusion of methyl methacrylate in a 34 mm extruder run at low rotor speed. The product quality was somewhat difficult to control, perhaps because of inadequate heat removal [69].

The formation of polymer blends was investigated by Lipomi [70] in a 34 mm diameter extruder. Tensile, flexural, impact, and hardness properties were obtained for various blend ratios of acrylonitrite-butadiene-styrene terpolymer with polycarbonate, polymethylmethacrylate, polyurethane, and polysulfone.

3.6   TANGENTIAL COUNTER-ROTATING TWIN-SCREW EXTRUDERS

Twin-screw extruders that are not intermeshing, such as shown in Fig. 3.19, exhibit features that have both similarities and differences when compared to both single-screw and the other

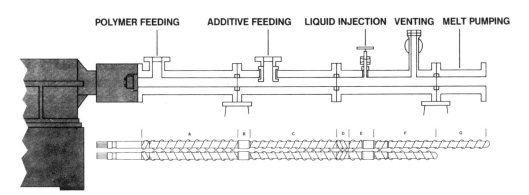

Figure 3.19 *Tangential counter-rotating twin-screw extruder (courtesy Welding Engineers).*

intermeshing twin-screw extruders. Even though this type extruder has been on the market for a long time, little technical information other than in the patent literature was published until the 1980s.

3.6.1   General Design

Here one screw is generally longer than the other. The feeding, melting, mixing, and venting functions are all performed in the twin-screw sections. Pressure development for discharge is confined to the longer screw. This feature allows placement in the gear box of a large thrust bearing on the long shaft without interference from the short shaft.

Since the screws do not intermesh, axial tolerance is not critical. These extruders can be constructed with longer L/D ratios (over 100) than is possible with other twin-screw machines. Timing of the two screws with respect to each other is also not critical. Thus, the angular twist imposed by the additional torque on the longer screw can be tolerated. Versatility can be provided by complete replacement of screw-shaft sections rather than by individual elements on a keyed shaft, again because of not having to maintain axial intermesh and angular timing.

The figure-eight barrel has truncated apexes for both mechanical and process reasons. The structural problem associated with the knife-edge web is avoided. The openness between the two barrel halves promotes beneficial interchange between the screws, albeit at the expense of more difficult melt sealing between the process zones. The two screws are usually staggered to maximize the interchange (Fig. 3.20). Nichols and Yao [71] described the leakage flow through the apex region with forward and reverse screws, and with blister rings. The difference in drag flow for two parallel single extruders and the tangential counter-rotating twin screw was attributed to longitudinal backmixing, as shown in Fig. 3.21. Nichols [72] confirmed the flow vs. pressure characteristics of screws with different flight depths using two high-viscosity silicone oils. The pressure flow response was shown to depend on the extent of apex opening.

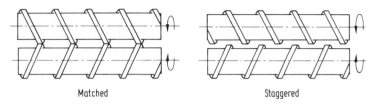

Matched   Staggered

Figure 3.20 *Matched and staggered screw arrays, tangential counter-rotating twin-screw extruder (courtesy Welding Engineers).*

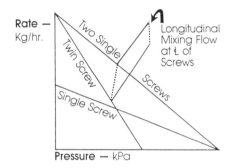

Figure 3.21 *Tangential counter-rotating twin-screw characteristics.*

3.6.2 Melting and Mixing

With matched screw flights, melting is comparable to that in a single-screw extruder. With the more common staggered array, Nichols and Kheradi [73] showed that the solid bed was soon broken up by interchange between the screws, and the melting occurred by dissipative mixing and heat transfer between melt and solids.

Min et al. [74] reported better pumping resulting from matched flights (observed in an extruder with windows). With staggered flights, the solid bed was rapidly broken up as melting took place. Howland and Erwin [75] conducted striation-thickness experiments with black and white curable silicone rubbers under full barrel conditions. They showed that reorientation continually occurred in the apex nip region with staggered arrays.

Walk [76] studied residence-time distribution of polymethylmethacrylate in a 50.8 mm twin-screw extruder, using x-ray fluorescence to determine the amount of the tracer (cadmium selenite) in the product. Walk found that a model, based on a combination of plug flow and perfect mixing, fitted his data. The fraction of plug flow varied with the screw configuration. Tucker and Nichols [77] reported dependence of the residence time on the flow rate and screw speed of both staggered and matched flight arrays.

Bigio and co-workers [78,79] have developed a two-dimensional model for evaluating mixing in the nip region of tangential counter-rotating twin-screw extruders. The rate of change of the interface was directly related to the rate of change of the velocity field and reorientation of the interface in fully filled channels.

Jerman [80] developed a two-dimensional model for non-Newtonian non-isothermal flow over blister rings. The most critical factor was the radial clearance, but blister-ring length and screw speed also affected the results. The temperature and melt viscosity altered the response of the melt seal as well.

The finite element modeling of this type extruder was performed by Nguyen and Lindt [81]. Their model better takes into account the effects of curvature of the flow field and of flow in the nip region. Nichols and Lindt [82] examined scale-up considerations. Although some similarity exists with single-screw scale-up rules, the unique intermixing between screws does not permit exact prediction of the performance. The empirical factor of capacity with the 2.5 power of diameter appears justified by practical experience. Scaling relations suggested by other approaches were also tabulated. Nichols and Steller [83] have summarized some experiences with incorporating fiber glass into polymers.

Removal of mixed cyclohexane/benzene solvent from 24 and 35% solutions of styrene-butadiene copolymer was reported by Nichols and Lubiejewski [84]. Back venting was used along with three stages of vacuum.

Foster and Lindt [85-87] investigated the removal of ethylbenzene from polystyrene and methylmethacrylate from polymethylmethacrylate in a 72 L/D, 20 mm extruder. The presence of foam limited the net vapor flow. The authors have developed a model for foaming devolatilization which takes into account the three regimes of bubble growth in both partly and fully filled screw channels, and the drag flow induced surface renewal.

The anionic polymerization of caprolactam to polyamide-6 was investigated by Nichols et al. [88] in a 20 mm extruder at screw speeds up to 520 rpm. The molecular weight was independent of screw speed, and no ultra-high-molecular-weight gels were formed. Other reactive modifications in tangential counter-rotating twin-screw extruders were described by Nichols and Kheradi [89], such as visbreaking of polypropylene with peroxide, and grafting of acrylic acid on polypropylene to improve adhesion to glass-fiber reinforcement.

3.7   CONCLUSIONS AND OUTLOOK

As the requirements for more advanced two-phase polymer systems have arisen, the twin-screw extruder manufacturers have provided increasing sophistication in the hardware and controls. Intermeshing twin-screw extruders, both co-rotating and counter-rotating, are supplied with slip-on screw sections of various pitch, and kneading elements of various widths and offset angles. Various screw sections are also available with tangential twin-screw extruders. Reverse sections of either screw or kneading elements can be used to provide more intense mixing or greater residence time.

The flexibility of rotor configurations allows for staging of the many functions frequently demanded, wherein the appropriate balance between forwarding, dispersing, and distributing actions can be attained. The availability of differing barrel sections permits location of vents or additional feed ports along the barrel length.

The interaction between the two screws of twin-screw extruders enhances reorientation and distributive mixing. Twin-screw extruders are generally starve-fed. As such, the ability to vary screw speed independently from feed rate provides flexibility in the mixing energy imparted to the polymer.

NOTATION

A	Drag flow dimensional constant
a	Process cross-sectional area of extruder
B	Pressure flow dimensional constant
b	Overall width of residence-time distribution
C	Centerline distance between shafts
c_p	Heat capacity
D	Screw diameter
D_b	Barrel diameter
f	Constant in power equation
G	Power consumption
H	Heat transfer coefficient
h	Channel depth
k	Thermal conductivity
L	Barrel length
N	Screw speed
n	Power-law viscosity exponent
P	Pressure
Q	Volumetric flow rate
w	Width between inflection points of residence-time distribution
Z	Screw lead
ϕ	Dispersion angle between screw and barrel
θ	Screw helix angle
ρ	Polymer density
η	Polymer viscosity

Subscripts

d	Drag
p	Pressure
w	Wall

REFERENCES

1. White, J. L., Szydlowski, W., Min, K., Kim, M. H., *Adv. Polym. Tech.*, **7**, 295 (1987).
2. Martelli, F., *Twin−screw Extruders: A Basic Understanding*, Van Nostrand Reinhold, New York (1983).
3. Janssen, L. P. B. M., *Twin−screw Extrusion*, Elsevier, New York (1978).
4. Rauwendaal, C., *Polymer Extrusion*, Hanser Verlag, Munich (1986).
5. Tadmor, Z., Gogos, C. G., *Principles of Polymer Processing*, John Wiley & Sons, New York (1979).
6. Booy, M. L., *Polym. Eng. Sci.*, **18**, 973 (1978).
7. Todd, D. B., Leach, E. A., *2nd Intl. Conf. on Reactive Processing of Polymers*, Pittsburgh (Nov. 2, 1982).
8. Curry, J., Kiani, A., Dreiblatt, A., *AIChE Meeting*, Washington, D.C. (Dec. 1, 1988).
9. Szydlowski, W., Brzoskowski, R., White, J. L., *Intl. Polym. Proc.*, **1**, 207 (1987).
10. Todd, D. B., *U.S. Pat.* 4,136,968 to Baker Perkins (1979).

11. Todd, D. B., *SPE Techn. Pap.*, **26**, 220 (1980).
12. Herrmann, H., Nettelnbreker, H. J., Dreiblatt, A., *SPE Techn. Pap.*, **33**, 157 (1987).
13. Anders, D., Hunziker, P. F., *Plast. Comp.*, **7**, (6), 54 (1984).
14. Colbert, J., *Plast. Comp.*, **12**, (1), 25 (1989).
15. Uhland, E., Dienst, M., *SPE Techn. Pap.*, **26**, 36 (1980).
16. Eise, K., Herrmann, H., Kapfer, K., *SPE Techn. Pap.*, **30**, 10 (1984).
17. Todd, D. B., *Plast. Comp.*, **11**, 33 (Jan./Feb. 1988).
18. Secor, R. M., *Polym. Eng. Sci.*, **26**, 969 (1986).
19. Davis, W. M., *CEP*, **84**, (11), 35 (1988).
20. Todd, D. B., *SPE Techn. Pap.*, **34**, 54 (1988).
21. Herrmann, H., Burkhardt, U., Jakopin, S., *SPE Techn. Pap.*, **23**, 481 (1977).
22. Bigio, D., Erwin, L., *SPE Techn. Pap.*, **31**, 45 (1985).
23. Szydlowski, W., White, J. L., *Adv. Polym. Techn.*, **7**, 177 (1987).
24. White, J. L., Szydlowski, W., *Adv. Polym. Techn.*, **7**, 419 (1987).
25. Szydlowski, W., White, J. L., *Intl. Polym. Proc.*, **2**, 142 (1988).
26. Kalyon, D., Gotsis, A., Gogos, C., Tsenoglou, C., *SPE Techn. Pap.*, **34**, 64 (1988).
27. Kaylon, D. M., Gotsis, A. D., Yilmazer, U., Gogos, C. G., Sangani, H., Aral, B., Tsenoglou, C., *Adv. Polym. Techn.*, **8**, 337 (1988).
28. Loomans, B. A., Kowalczyk, J. E., Jones, J. W., *ADPA Compability and Processing Symp.*, (Apr. 18, 1988).
29. Todd, D. B., *SPE Techn. Pap.*, **35**, 168 (1989).
30. Todd, D. B., *Polym. Eng. Sci.*, **15**, 437 (1975).
31. Kao, S. V., Allison, G. R., *Polym. Eng. Sci.*, **24**, 645 (1984).
32. Altomare, R. E., Anelich, M., *AIChE Meeting*, Washington, D.C. (Dec. 1, 1988).
33. Jakopin, S., *Compounding of Fillers*, A.C.S. Adv. Chem. Ser., No. 134, 114 (1974).
34. Agur, E. E., *Adv. Polym. Techn.*, **6**, 225 (1986).
35. Todd, D. B., Baumann, D. K., *Polym. Eng. Sci.*, **18**, 321 (1978).
36. Sirabian, S., Eise, K., Curry, J., *Plast. Eng.*, **40**, (4), 27 (1984).
37. Eise, K., Curry, J., Farrell, J., *Polym. Eng. Sci.*, **25**, 497 (1985).
38. Mauch, K., *Kunststoffe*, **71**, 266 (1981).
39. Wall, D., *SPE Techn. Pap.*, **33**, 778 (1987).
40. Eise, K., Mielcarek, D. F., *CEP*, **78**, 62 (1982).
41. Karian, H. G., *J. Vinyl Techn.*, **7**, (4), 154 (1985).
42. Plochocki, A. P., Dagli, S. S., Curry, J. E., Starita, J., *SPE Techn. Pap.*, **32**, 874 (1986).
43. Karian, H.G., Plochocki, A.P., *SPE Techn. Pap.*, **33**, 1334 (1987).
44. Kapfer, K., *SPE Techn. Pap.*, **34**, 96 (1988).
45. Kapfer, K., *Kunststoffe*, **77**, 377 (1987).
46. Kowalczyk, J. E., Loomans, B. A., Jones, J. W., A.D.P.A. *Compatibility and Processing Symp.* (Apr. 18, 1988).
47. Todd, D. B., *SPE Techn. Pap.*, **20**, 472 (1974).
48. Collins, G. P., Denson, C. D., Astarita, G., *Polym. Eng. Sci.*, **23**, 323 (1983).
49. Collins, G. P., Denson, C. D., Astarita, G., *AIChE J.*, **31**, 1288 (1985).
50. Werner, H., Curry, J., *SPE Techn. Pap.*, **27**, 623 (1981).
51. Anders, D., *SPE Techn. Pap.*, **30**, 15 (1984).
52. Mack, M. H., *SPE Techn. Pap.*, **32**, 855 (1986).
53. Mack, M. H., *Plast. Eng.*, **42**(6), 47 (1986).
54. Notorgiacomo, V., Biesenberger, J. A., *SPE Techn. Pap.*, **34**, 71 (1988).
55. Betz, R. K., *Europ. Pat. App.* 0143894 (1985).

56. Todd, D. B., *SPE Techn. Pap.*, **33**, 128 (1987).
57. Mack, M. H., Chapman, T. F., *SPE Techn. Pap.*, **33**, 136 (1987).
58. Curry, J., Jackson, S., Stoehrer, B., Van Der Veen, A., *AIChE Annual Meeting*, New York, (Nov. 18, 1987).
59. Thiele, W., *Plast. Comp.*, **4**, (4), 23 (1981).
60. Thiele, W., *SPE Techn. Pap.*, **29**, 127 (1983).
61. Speur, J. A., Mavridis, H., Vlachopoulos, J., Janssen, L. P. B. M., *Adv. Polym. Techn.*, **7**, 39 (1987).
62. Sakai, T., Hashimoto, N., *SPE Techn. Pap.*, **32**, 860 (1986).
63. Sakai, T., Hashimoto, N., Kobayashi, N., *SPE Techn. Pap.*, **33**, 146 (1987).
64. Poltersdorf, B., Schmidt, S., Han, C. S., *Kunststoffe*, **76**, 36 (1986).
65. Shah, S., Wang, S. F., Schott, N., Grossman, S., *SPE Techn. Pap.*, **33**, 122 (1987).
66. Menges, G., Feistkorn, W., Knierbein, B., *Kunststoffe*, **74**, 167 (1984).
67. Feistkorn, W., Rengshausen, H., *Kunststoffe*, **75**, 920 (1985).
68. Dey, S. K., Kiani, A., Plochocki, A. P., Curry, J. E., *Kunststoffe*, **76**, 27 (1986).
69. Dey, S. K., Biesenberger, J. A., *SPE Techn. Pap.*, **29**, 628 (1983).
70. Lipomi, L. M., *SPE Techn. Pap.*, **29**, 628 (1983).
71. Nichols, R. J., Yao, J., *SPE Techn. Pap.*, **28**, 416 (1982).
72. Nichols, R. J., *SPE Techn. Pap.*, **29**, 130 (1983).
73. Nichols, R. J., Kheradi, F., *SPE Techn. Pap.*, **29**, 134 (1983).
74. Min, K., Kim, M. H., White, J. L., *Intl. Polym. Proc.*, **3**, 165 (1988).
75. Howland, C., Erwin, L., *SPE Techn. Pap.*, **29**, 113 (1983).
76. Walk, C. J., *SPE Techn. Pap.*, **28**, 423 (1982).
77. Tucker, C. S., Nichols, R. J., *SPE Techn. Pap.*, **33**, 117 (1987).
78. Bigio, D., Penn, D., *SPE Techn. Pap.*, **34**, 59 (1988).
79. Bigio, D., Zerafati, S., *SPE Techn. Pap.*, **34**, 85 (1988).
80. Jerman, R. E., *SPE Techn. Pap.*, **32**, 995 (1986).
81. Nguyen, K., Lindt, J. T., *SPE Techn. Pap.*, **34**, 93 (1988).
82. Nichols, R. J., Lindt, J. T., *SPE Techn. Pap.*, **34**, 80 (1988).
83. Nichols, R. J., Steller, M. A., *Plast. Comp.*, **9**, 14 (June/July 1986).
84. Nichols, R. J., Lubiejewski, P. E., *SPE Techn. Pap.*, **31**, 12 (1985).
85. Foster, R. W., Lindt, J. T., *AIChE Annual Meeting*, New York (Nov. 18, 1987).
86. Foster, R. W., Lindt, J. T., *SPE Techn. Pap.*, **35**, 150 (1989).
87. Lindt, J. T., Foster, R.W., *SPE Techn. Pap.*, **35**, 146 (1989).
88. Nichols, R. J., Golba, J. C., Johnson, B. C., *Polym. Proc. Soc., 2nd Annual Meeting*, Montreal (1986).
89. Nichols, R. J., Kheradi, F., *3rd Chemical Congress*, Toronto (June 5, 1988).

CHAPTER 4

STRUCTURE-PROPERTY RELATIONSHIPS IN PLASTICS STRUCTURAL FOAMS

by Peter R. Hornsby

Department of Materials Technology
Brunel University
Uxbridge, Middlesex UB8 3PH
UNITED KINGDOM

This chapter discusses the formation, structural characterization, and mechanical properties of integral skin or structural foam plastics. Particular emphasis is placed on thermoplastics compositions containing chemical and physical expansion systems, processed by modified injection molding and extrusion. Structural integrity is considered in terms of density variations within the foam, the extent of surface skin formation, the nature of surface irregularities, and the morphology of both the cellular and polymeric phases (in semi-crystalline variants). The structural role of short glass fiber reinforcement is also highlighted. The interrelationship between structural features and mechanical performance is described by reference to the measurement and prediction of flexural and impact properties. Specific attention is given to developments in injection-molding technology, which enable greater design flexibility in the preparation of structural foam artefacts.

4.1   INTRODUCTION

Structural plastics foams, also known as integral skin plastic foams, are characterized by an outer skin region with density close to that for unfoamed plastic, surrounding a relatively low density foamed plastics core.

The skin and core regions are usually formed from the same polymer, (Fig. 4.1). However, using variations in processing technology, skin and core may be made from different polymers to produce so-called two-component structural foams. An important distinguishing technological feature of structural foams is that they are generally produced by a single manufacturing step involving melt or by reactive processing, which is frequently based on modified forms of injection molding, but sometimes extrusion, reaction injection molding, or rotational molding techniques. This is in contrast to laminated foam structures, which require the joining together of prefabricated components, such as solid external sheets with a low-density foamed polymeric core.

As will be described in more detail later, the manufacturing route and method of expansion used to produce structural foam artefacts, can have a profound influence on the structural design of the material and ultimately its physical properties. It is pertinent, therefore, to overview appropriate aspects of structural foam processing technology which define the structural character of finished components. In this context, particular emphasis will be given to modified forms of thermoplastics injection molding technology employing chemical and physical expansion systems.

Structural foams can be formed from a variety of thermoplastics, including polyolefins, polyvinylchloride, modified polyphenylene ether, and ABS. Often the properties of the base polymer are further changed by inclusion of additives such as glass-fiber reinforcement, mineral fillers, pigments, flame retardants, or hollow inorganic microspheres, also used in the production of syntactic foams. Incorporating such components into structural foam can significantly complicate the understanding and prediction of end performance of an already complex material, particularly in connection with the fiber-reinforced variants.

Thermosetting resins formed from monomers or low-molecular-weight oligomers, may also be converted into structural foams. In these systems, combined polymerization and foaming takes place, with stabilization of the foam structure occurring at the same time. Hence stringent process control is required to achieve well-defined and reproducible foam structures.

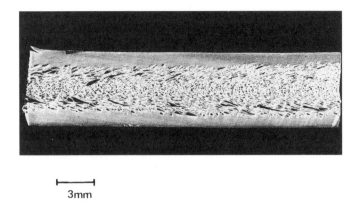

3mm

Figure 4.1 *Transmitted light micrograph of single-component integral skin thermoplastics structural foam.*

To this end, reaction injection molding (RIM) technology has been widely applied to manufacture structural foam parts, most commonly from polyurethane formulations. Components with a dense integral skin and a lower density cellular core can be made in rigid or flexible (elastomeric) form, although in the latter case, the skin may be less well-defined. A detailed discussion of polyurethane formulation and RIM technology is beyond the scope of this chapter and the interested reader is directed elsewhere [1,2].

Structural foams may be made from other thermosetting resins, including epoxy, polyester, and phenolic formulations, using either liquid resin systems cured at ambient or low temperatures, or from recently developed hot cure methods with polyester or phenolic molding compositions, employing modified injection and compression molding techniques.

The integral skin foamed core structure in plastics structural foams generally confers a higher stiffness to weight ratio than found in their solid counterparts, enabling them to be used in load-bearing situations, often as substitutes for wood and metal. Prediction of physical properties, such as flexural rigidity is therefore of major interest to the designer and user of plastics structural foams, if their end performance is to be optimized. In this context, an understanding of the interrelationship between internal structure and properties is paramount and is the subject of extensive discussion in this communication.

4.2   FORMATION OF PLASTICS STRUCTURAL FOAMS

As mentioned earlier, many methods exist for the manufacture of structural foams from both thermoplastic and thermosetting plastics materials. Although a detailed account of the engineering design and process technology of these techniques is outside the scope of this chapter, nevertheless, since the structural characteristics of the product can be very dependent on processing, some discussion of process technology is considered necessary. Greater details of the technology developed for structural foam preparation have been reported elsewhere [3-5] and only a more general summary is presented below.

4.2.1   Expansion Systems

The starting point for any commentary on manufacture of plastics structural foams concerns the selection of the expansion system. In this respect, a distinction can be made between physical and chemical methods.

The most widely used physical blowing agents are compressed inert gases (in particular nitrogen and carbon dioxide for foaming thermoplastics), and volatile liquids (such as chlorinated fluorocarbons, CFCs, used to produce integral skin polyurethane structural foams). It is claimed that the use of CFCs, as an alternative to nitrogen in thermoplastics, can result in substantial density reductions in the cellular core and can result in improved surface appearance. Owing to environmental concerns, the CFCs are being phased out. They still find some use in polyurethane foams, although even for those many manufacturers have already switched to other foaming agents. Nitrogen, by itself, has been widely used in the fabrication of thermoplastics structural foams by injection molding with the so-called Union Carbide process.

Chemical blowing agents thermally decompose over a narrow temperature range to yield large volumes of inert gas or a mixture of gases. The use of organic chemical blowing agents, in particular, is in widespread use for the preparation of structural foamed plastics by various extrusion and injection molding techniques. Table 4.1 summarizes the properties of selected blowing agents and the polymers recommended for use with each [6,7]. However, it should be noted, that their range of application can be considerably extended through the incorporation of additives, such as zinc stearate or urea, which activate the blowing agent and lower the decomposition temperature.

One undesirable characteristic of chemical blowing agents is the residue remaining in the foam, which may influence the physical properties of the polymer, or through interaction with iron or ferrous alloys used in tool and die construction, can lead to plate-out of hard insoluble deposits. This has been observed with cyanuric acid, a major component of the azodicarbonamide decomposition residue [8].

Gaseous products from decomposition of azodicarbonamide blowing agents comprise mainly nitrogen (62%) and carbon monoxide (35%), together with small quantities of carbon dioxide and ammonia (3%). The liberated ammonia can cause corrosion of beryllium-copper tooling and problems of degradation with certain polymers (such as polycarbonates, polyamides, and polysulphones). This led to development of ammonia-free blowing agents based on hydrazide compounds [9].

Further information regarding expansion systems can be found in Chapter 5.

4.2.2. Injection Molding

A large number of techniques are available for the preparation of integral skin foamed thermoplastics injection moldings. Generally these processes differ according to the cavity pressures experienced by the material, the quality of finish produced on the molding surface, and the number of materials introduced into the cavity to produce solid skin/foamed core structures.

Low-pressure methods generally give rise to cavity pressures of up to about 5 MPa. Polymer melt containing dissolved inert gas, is rapidly transferred from the injection cylinder, where it is maintained under pressure to prevent premature expansion, to partially fill the mold cavity. The depressurized melt from this short shot expands to completely fill the mold to produce a molding with a solid integral skin, frequently with a rough surface texture (Fig. 4.2).

Table 4.1 *Properties of Chemical Blowing Agents Used in the Fabrication of Thermoplastics Structural Foams*

Chemical blowing agent[a]	Operating temperature range (°C)	Gas yield (ml/g)	Polymer type
P'p' - oxybis (benzene sulphonhydrazide)	140–160	125	Low-density polyethylene, EVA copolymers
Azodicarbonamide (Azobisformamide)	170–230	200–230	Polypropylene, polystyrene, high density polyethylene, ABS
Ammonia-free hydrazides	220–250	125	Polyamides, polycarbonates, polysulphones, ABS
P-toluene sulphonyl semi-carbazide	216–146	146	ABS, high-impact polystyrene, polypropylene, high-density polyethylene, polyamides, modified-polyphenylene ether, polycarbonates
Trihydrazinotriazine	265–290	190	Polyamides, polycarbonates, polysulphones

[a] *Concentration of blowing agent typically varies from 0.1 to 2 wt%, depending on desired density, component geometry, and processing conditions.*

Gas needed to expand the melt is generated either from decomposition of a chemical blowing agent in the injection cylinder, or by direct metering of nitrogen into the polymer melt along the barrel.

Efforts directed towards eliminating the surface irregularities introduced by the low-pressure method have resulted in a number of processes in which the cavity pressures are considerably higher (up to 15 MPa) than in the short-shot method, but well below pressures used in solid injection molding, where the material is packed in the cavity at pressures up to 50 MPa, as it cools and contracts.

Different principles can be applied to achieve a dense outer skin with improved surface quality. In one approach, the mold is completely filled with gas saturated melt using high pressures to ensure good contact of the polymer with the wall. After surface cooling and formation of an integral skin, the core is expanded by partial opening of the cavity using a

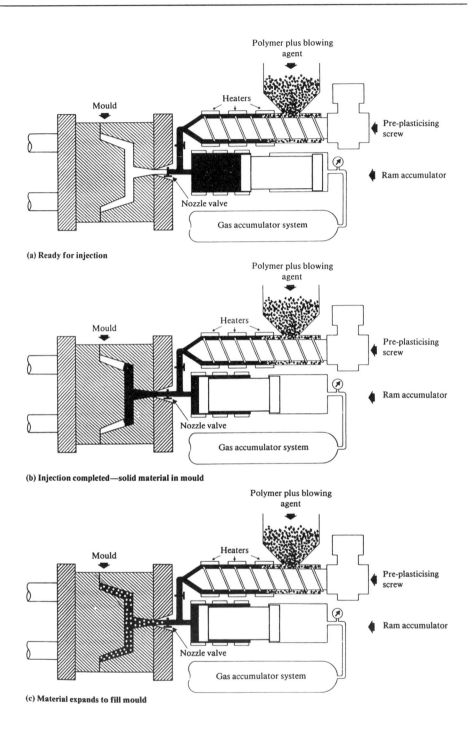

Figure 4.2 *Stages in low-pressure short-shot structural foam molding process.*

vertical flash tool design or by moving certain core sections of the mold outwards, to allow internal foaming.

An alternative approach utilizes a gas counter-pressure in the cavity to prevent melt expansion during filling. The cavity pressure and resulting molding density is then reduced either by relying on melt contraction during cooling or by deliberate withdrawal of molten polymer from the core, allowing expansion to occur.

In another modification, directed towards improving surface finish of low-pressure moldings, the mold surface temperature is maintained above the polymer melting point during filling. After surface imperfections have been smoothed out, the mold and material are then cooled.

The aforementioned techniques are normally used to manufacture single component moldings, where polymer in both skin and foamed core have the same chemical composition. However, in two-component structural foam molding different materials can be used for skin and core, extending the possible scope for property design in the finished part.

Two principal variants exist. In one method the mold cavity is completely filled by sequentially injecting two different materials through the same sprue. After complete filling and pressurization of the cavity, one of the melts, which forms the molding core and contains dissolved gas, is then allowed to expand by partial mold opening.

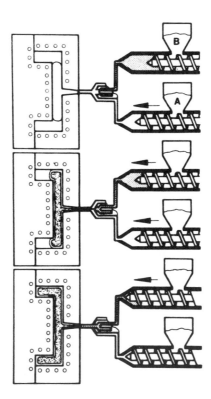

Figure 4.3 *Two-component structural foam injection molding.*

Two-component skin-foamed core structures may also be produced by low-pressure injection molding, whereby the two melts which form skin and core are introduced simultaneously through a special shut-off nozzle comprising two or three separate flow channels. A short shot of foamable material is injected and contained within the outer skin, then expands to fill the mold cavity, developing a cellular core (Fig. 4.3). A variety of materials can be combined for use as skin and foamed interior. However, attention must be given to ensure that the shrinkage characteristics of the two melts are similar in order to avoid problems of delamination between skin and core or the creation of over stressed moldings [10].

It should be appreciated from this brief overview that distinct differences in surface quality, achievable skin thickness and molding density profile are possible, depending on the processing route [3-5].

4.2.3   Continuous Extrusion

Extruded structural foam profile may also be made by several different techniques, the simplest being so-called free-expansion of foamable polymer melt as it emerges from the extrusion die into a cooled sizing unit. Very limited skin formation is possible by this method and a mat surface generally results, although very fine uniform cell structures can be achieved.

Much thicker skins with smooth surfaces can be combined with low-density cellular cores using the Celuka process [11]. In this method an expandable thermoplastics formulation is extruded over a profiled mandrel in the die, to form a hollow section. This passes immediately into a cooled sizing unit, which has similar dimensions to the extrusion die. A smooth solid integral skin is produced as the outer region of the extrudate cools, whilst the molten central region is allowed to expand inwards to produce a foamed core. Although more complex than the free-expansion method, this technique has found widespread application for the manufacture of structural foam profile with well-defined skin-core structures, particularly from unplasticized polyvinylchloride. A detailed description of the process and product properties is given in Chapter 5.

In an analogous way to two-component injection molding, feedblock co-extrusion technology has been successfully applied to produce continuous profile (sheet and pipe), with integral skin cellular core sandwich structures. A foamable melt stream is combined between unmodified melt skin layers, expansion of the core taking place as the melt composite exits from the die [12,13].

4.3   MORPHOLOGICAL CHARACTERISTICS OF THERMOPLASTICS STRUCTURAL FOAMS

In common with unfoamed thermoplastics, the physical properties of structural foams are highly dependent on the form of structure produced in the processed artefact, which in turn is influenced by the particular manufacturing route and processing conditions. Hence there are a large number of variables, many being interdependent, which can have an effect on the ultimate mechanical performance of this class of material.

Whereas with solid amorphous thermoplastics, properties may be influenced by the occurrence of molecular orientation or internal stresses introduced in the material during processing [14], in semi-crystalline thermoplastics these factors are also relevant, together with additional complexities arising from the extent of crystallization introduced and the considerable variations in crystal type that may occur, for example, throughout the thickness of an injection-molded part [15,16]. Inclusion of additives such as fillers, pigments, and fibrous reinforcements may create further changes in the crystalline morphology and properties of the material [17]. Anisotropic properties resulting from preferred alignment of fibers can also be very significant in thermoplastics composites made by melt processing [18].

These factors, which are known to be so significant in solid thermoplastics components, potentially have a similar role in determining the properties of thermoplastics structural foams. However, superimposed on these microstructural parameters are other factors, which exert gross changes in structure, arising from the foaming process. These include the ratio of solid skin to foamed core material, the presence of surface defects introduced by collapse of expanding polymer melt, and the form of cellular morphology produced in the expanded core region. The latter parameter, together with the thickness and weight of the foamed part, influences the overall structural foam density, which in turn has a dominant effect on physical properties. However, since local density changes occur across the section of structural foam product, the consequences of the density profile on physical properties must also be realized. Table 4.2 summarizes the structural parameters which are most likely to influence the mechanical properties of polymeric structural foams.

Table 4.2 *Structural Parameters Relevant to the Physical Properties of Plastics Structural Foams*

Density (density profile)
Ratio of skin to cellular core thickness
Cell size, shape, and distribution
Skin integrity (presence of microvoids/blowing agent residue)
Surface quality (sites for crack inition)
Crystal morphology in semicrystalline variants (skin and foames core regions)
Fiber alignment (short-fiber-filled grades)

The role of these factors will be considered, emphasizing flexural and impact behavior, since these properties are generally of greatest relevance to the practical application of structural foam materials and have consequently been most widely studied.

4.3.1 Cell Structure and Foam Density

Examination of the cross section of injection-molded structural foams reveals a non-uniform foam structure, with larger nearly spherical cells evident in the center, becoming smaller and elongated (or flattened) in the direction of flow, as the surface skin is approached (Fig. 4.4). This form of cell morphology reflects the changing shear and thermal history experienced by the foaming polymer during mold filling and cooling, cell size tending to decrease with degree of undercooling of molten polymer and cell elongation occurring in regions of higher melt flow strain to the walls of the molding.

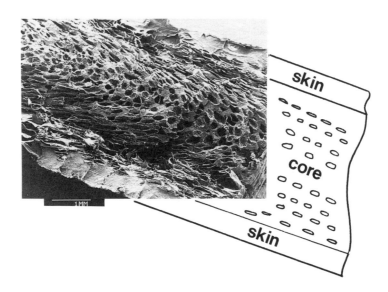

Figure 4.4 *Skin-core morphology of an injection-molded polypropylene structural foam.*

The variation in cell geometry from skin to core results in changing density across the thickness of a structural foam part. The use of X-ray radiography combined with microdensitometry can provide an effective non-destructive means for quantifying the form of this density profile, in both thermoplastics and thermoset structural foams [19,20].

Figure 4.5 shows close agreement between density distributions across a 20 mm thick polypropylene structural foam molding, determined by X-ray radiography and also by direct measurement of density, calculated from mass and volume measurements on sections taken across the foamed molding [20].

The density profile across structural foam bars can be modelled using a polynomial function of the form [21]:

$$\bar{\rho}/\rho_S = R + (1-R)\left[\frac{(C+1)\,Z^C}{(1-\varepsilon)^C} - \frac{CZ^{C+1}}{(1-\varepsilon)^{C+1}}\right]$$
$$\text{for}\ \ 0 < Z < (1-\varepsilon)$$

$$(4.1)$$

$$\bar{\rho}/\rho_S = 1, \qquad\qquad\qquad \text{for}\ \ (1-\varepsilon) \le Z < 1$$

where: $\bar{\rho}$ is the local density, ρ_S is the solid polymer density, R is the ratio of core to solid density, Z is the ratio of distance away from the center-line to half the bar thickness, C is the a shape factor which may vary from 3 to infinity depending on form of the density profile, and ε is the ratio of skin thickness to half bar thickness.

Using an appropriate value for the shape factor, Eq (4.1) can be applied to give close agreement with experimentally measured density profiles [22]. Figure 4.6, for example, shows density profiles determined by X-ray radiography and calculated from Eq (4.1), for injection-molded structural foam bars of different thickness and density.

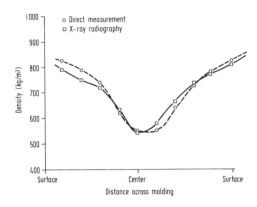

Figure 4.5 *Density profiles across a 20 mm thick polypropylene structural foam molding [20].*

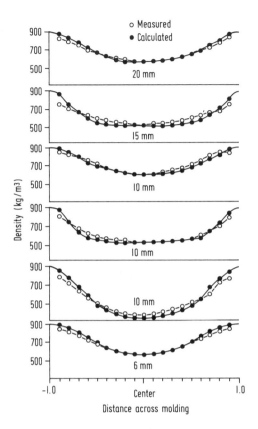

Figure 4.6 *Measured and calculated [from Eq (4.1)] density profiles across injection-molded polypropylene structural foams, with thicknesses ranging from 6 to 20 mm [20].*

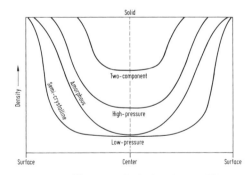

Figure 4.7 *Characteristic density profiles across structural foams produced by two-component, high-pressure and low-pressure injection molding techniques [23].*

It is evident from these and other results, that the shape of density profile is influenced not only by the molding thickness and extent of overall density reduction achieved, but also by the process. Figure 4.7 distinguishes between typical density distributions across the thickness of parts made by three different forms of structural foam molding techniques described earlier [23].

Noticeable variations in apparent molding density, density profile, and cell structure are frequently observed in different regions of a single molding. Often this is manifested by a reduction in cellular core structure (and hence an increase in apparent density) towards the molding periphery where the polymer flow path is greatest [24].

Cell structure throughout a single molding may also vary at a particular plane through the section thickness, depending on the flow length of the expanding melt and the processing conditions adopted, in the case of short-shot injection-molding techniques. In this context, the rate of injection is particularly relevant, very high injection speeds tending to yield greater uniformity in cell structure.

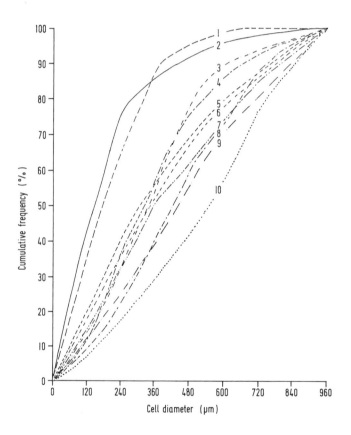

Figure 4.8 *The influence of processing parameters on cell size distribution in 6 mm thick polypropylene structural foam injection moldings made by the short-shot technique and a chemical blowing agent (measurements were determined on a plane 2.5 mm below the sample surface) [24].*

It has been shown for polypropylene expanded using a chemical blowing agent, that high injection rates, high melt pressures (in the injection cylinder), and low melt temperatures, give a reduction in average cell size (Fig. 4.8). Finer more uniform cell structures also result from using a powdered rather than a granular polymer feedstock, and by the addition of short glass-fiber reinforcement. This is attributed to improved blowing agent distribution and the stabilizing influence of the glass fibers on cell growth. In this study surface-modified glass fibers were employed. They were found not to act as cell nucleation sites.

a

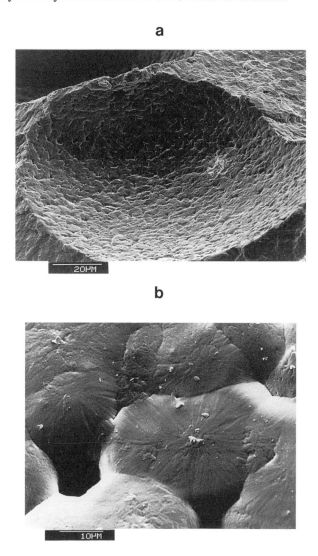

b

Figure 4.9 *Cell-wall morphology in the core of polypropylene structural foam injection molding showing: a) general view of funnel-shaped holes and b) flattened spherulites [25].*

As mentioned earlier, as is the case with unfoamed semi-crystalline polymers, the possible influence of micromorphology on the physical properties should also be considered. Microscopic examination of polypropylene structural foam made using a short-shot injection-molding procedure and an azodicarbonamide chemical expansion system, has revealed a distinct form of spherulitic morphology evident in the cell walls, which was attributed to the unusual conditions of melt expansion and polymer crystallization occurring during filling and subsequent cooling within the mold cavity [25,26].

A network of troughs and funnel-shaped holes have been observed covering the cell walls and circumscribing what were identified as flattened spherulites (Fig. 4.9). Spherulites close to the solid skin-foamed core boundary were found to be smaller than those forming the walls of the cells in the center of the molding, as is generally the case with solid polypropylene injection moldings [16]. The spherulites became more distinct and rounded toward the center of the molding where almost complete hemispheres could be found projecting into the cells (Fig. 4.10). Many of these had flat or slightly concave tops with sharply defined edges (Fig. 4.11). When microtomed sections cut from the core of the molding were viewed in transmission through a polarizing microscope (Fig. 4.12), spherulite nucleation was evident in the bulk of the polymer and not at the cell-wall surface [26].

A proposed explanation for these effects is depicted schematically in Fig. 4.13. After injection of the melt, some spherulitic growth occurs as expansion takes place in the mold cavity (Fig. 4.13a). Spherulites growing close to an expanding gas cell are constrained from growing and develop a flattened or concave surface at the melt-cell interface. As the spherulites continue to grow, eventually they impinge on one another, thus restricting their movement (Fig. 4.13b). Further expansion of the cell exposes flat-topped structures, protruding into the cell, and reveals underlying spherical spherulites, which were growing beneath the melt-cell interface (Fig. 4.13c and Fig. 4.10). This occurs until the mold cavity is full and/or the melt has cooled sufficiently to prevent further cell expansion. The remaining amorphous polymer continues to crystallize causing shrinkage and formation of troughs and funnel-shaped holes, observed at the spherulite boundaries (Fig. 4.13d and Fig. 4.9).

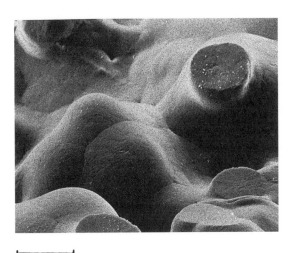

├────────────┤
30μm

Figure 4.10 *Rounded and flat-topped cell wall spherulites in polypropylene structural foam [26].*

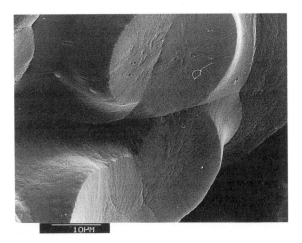

Figure 4.11 *Concave surface of polypropylene spherulites found in the cell wall of a polypropylene structural foam [26].*

The occurrence of this secondary shrinkage between the previously formed spherulites is estimated to increase cell diameter by at least 7% in the interspherulitic regions toward the center of the molding. The process depends on factors such as the degree of density reduction achieved in the structural foam and the amount of bulk material surrounding the interface region [26].

In solid injection molding, relatively high pressure is maintained in the cavity as the polymer cools and crystallizes, enabling more material to be packed into the mold to compensate for contraction of the melt. However, although this action minimizes shrinkage, it also introduces residual stress into the molding. Since these conditions are not imposed on low-pressure structural foam moldings and the polymer in the core of the material is free to contract as it crystallizes, the incidence of residual or molded-in stress is likely to be minimal.

4.3.2 Skin Characteristics

The integral skin in plastics structural foams has an overriding effect on its ultimate physical properties, particularly flexural and impact performance. In a single component structural foam, the skin is created from the same material composition as the cellular core and therefore may contain small amounts of residue from decomposed chemical blowing agent. Additionally, the skin-core boundary may be ill-defined (cf. Fig. 4.1), complicating precise determination of skin thickness. The skin itself may contain microvoids or gas bubbles.

These factors are pertinent to predicting physical properties, such as flexural stiffness, since, as discussed later, it is often necessary to ascribe values for density and modulus for material present in the skin region. It has been shown that certain blowing agents or their residues can noticeably reduce the mechanical properties of the host polymer [27], thus the skin cannot necessarily be assumed to have properties equivalent to those of a solid resin without blowing agent present. The significance of microvoids in the skin layer is more difficult to determine, although it is generally assumed that increased embrittlement of material will result.

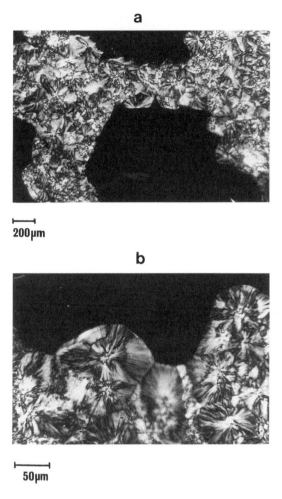

Figure 4.12 Polarized light micrographs of spherulite morphology around the cell wall of a polypropylene structural foam molding: a) general view; b) magnified view of hemispherical cell-wall spherulites [26].

Knowledge of the relative amounts of skin and core material through the structural foam cross section is especially relevant to the prediction of mechanical properties. This may not only vary locally within a particular test sample, necessitating multiple measurements of a single specimen to yield a reliable average value, but it may also change within a large structural foam molding. Figure 4.14, for example, shows X-ray radiographs of polypropylene specimens cut from the same injection molded plaque, and illustrates the variation in skin thickness. Evaluation of skin properties is further complicated in two-component structural foams, since the skin may comprise elements from two distinct materials each having different properties.

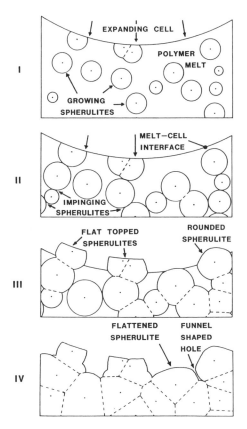

Figure 4.13 *Stages in the deformation of cell-wall spherulites seen in polypropylene structural foam [26].*

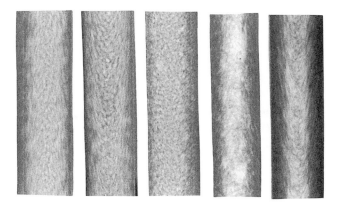

Figure 4.14 *Variation in skin-core structure within a single 10 mm thick polypropylene structural foam plaque made by a short-shot injection molding technique [20].*

Skin formation in single component thermoplastics structural foams made by short-shot low-pressure injection molding is considered to occur, in part, by collapse and densification of expanding melt within the mold cavity as it meets the cold mold surface (aided by internal gas pressure within the melt). Additionally, melt densification may also take place in the cooler skin region by redissolution of expanded gas. However, it should be noted that in several of the commercial processing routes outlined earlier, an important requirement for producing thick, densified outer skin layers is to inhibit or limit the extent of melt expansion prior to solidification of the molding or extrudate periphery.

In this context, the control of skin formation is of great significance in optimizing the physical properties of the finished component (including the quality of surface finish). It depends on the operational details of the particular process adopted. However, as shown in Fig. 4.15, for polypropylene structural foam made by a short-shot low-pressure molding, there is a clear relationship between overall density and the ratio of skin to foamed core material. The scatter in the results is due to non-uniformities at the skin–core boundaries which existed in these specimens.

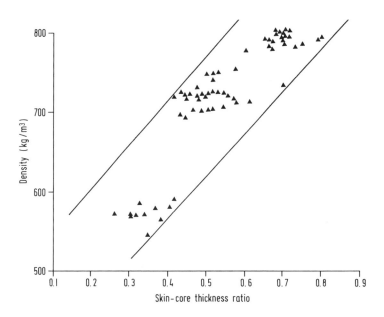

Figure 4.15 *The influence of molding density on the ratio of skin-to-core thickness in injection-molded polypropylene structural foam [20].*

4.3.3 Surface Texture

Due to the high cavity pressures employed during mold packing and the close contact that results between mold wall and polymer melt in conventional injection molding, it is readily possible to achieve a high-quality glossy surface finish. However, as has already been explained, introduction of an expanding polymer melt into a cavity (or from the exit of an extrusion die) can result in an irregular mat surface often with a swirl-like appearance, common for moldings made by short-shot low-pressure injection molding, or in free-expanded extrusion.

Poor surface appearance in early structural foam products was seen as a major limitation to their successful market acceptance, thus necessitating costly secondary surface finishing and painting procedures. This drawback provided the main stimulus for technological development of alternative fabrication methods (outlined earlier), free from attendant problems of poor surface appearance and necessary post-forming surface treatment. The aforementioned is particularly relevant to the present discussion due to the possible influence of surface defects on mechanical properties, notably toughness.

Figure 4.16 shows the surface features present on a polypropylene structural foam made by the short-shot low-pressure molding, produced as the expanding melt collapses and folds at the cavity surface. It is evident that microscopic fissures and surface imperfections of this type could lead to a reduction in observed fracture initiation energy under impact loads (this point will be discussed more fully in Section 4.4.3).

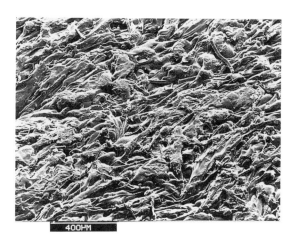

Figure 4.16 *Sruface irregularities on a polypropylene structural foam injection molding made by a short-shot process.*

Quantification of surface texture on structural foams can be estimated as a center-line average roughness value, using a surface roughness meter which employs a traversing stylus located on a pick-up arm [22]. Although this technique has obvious limitations in relation to accuracy (particularly an inability to measure surface fissures with radius less than the stylus tip; typically 2.5 μm), it can provide a rapid means for determining relative level of surface finish between structural foam products and for highlighting the influence of the processing parameters.

Figure 4.17 shows a decrease of polypropylene surface quality with reducing foam density, Fig. 4.18 indicates the relationship between surface roughness and injection speed at increasing distance from the central point of injection, on a polypropylene structural foam

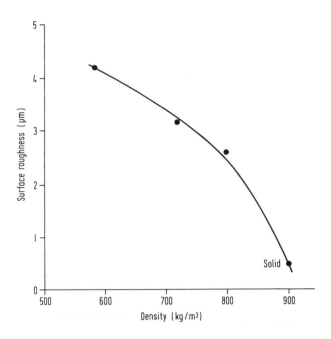

Figure 4.17 *The relationship between surface roughness and molding density for structural foam made by a short-shot technique [20].*

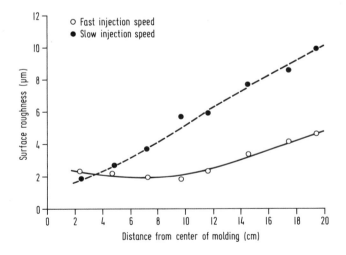

Figure 4.18 *Variation in surface roughness across the surface of glass-fiber-reinforced polypropylene structural foam moldings, prepared by a short-shot procedure using slow (2s) and fast (0.6s) injection times [24].*

plaque made by the short-shot process. It is clear that, for this material, large variations in surface texture can exist within a single molding and that surface quality is enhanced as the injection rate is increased.

4.4.   INTERRELATIONSHIP BETWEEN STRUCTURE AND PROPERTIES

4.4.1   Flexural Properties

Perhaps the most frequently cited benefits of structural foam in relation to solid plastic materials, is the increased stiffness-to-weight ratio. In this respect, the skin integrity is critical, in order to support the imposed bending stresses.

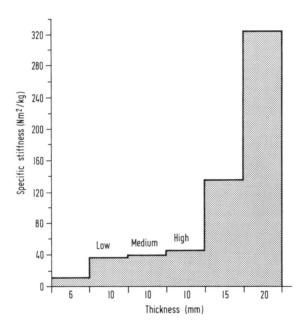

Figure 4.19 *The relationship between specific flexural stiffness and thickness of polypropylene structural foam bars. The 10 mm thick specimens were tested at three different molding densities at 23° C [20].*

For the development of a discernible cellular foam core, structural foam moldings generally have a thickness of at least 4 mm. However, for a given weight of material, increasing the section thickness can result in a substantial increase in bending stiffness, as shown in Fig. 4.19 for polypropylene bars with thicknesses ranging from 6 to 20 mm. This is in agreement with the law of bending for a straight beam loaded at its midpoint and supported at its ends, where the deflection, δ is inversely proportional to the beam thickness h, according to the relation:

$$\delta = PL^3/4E_s\, bh^3 \qquad\qquad (4.2)$$

where P is the load at the center of the beam, L is the beam span, b is the beam width, and E_s is the elastic modulus. In practical terms, the benefits of increased stiffness and sink-free moldings introduced by increasing molding thickness, are offset by the greatly prolonged cooling times necessary during molding. In addition to thickness, the flexural properties of structural foam depend on the internal structure of the material, in particular the density and the quality of skin.

The effects of apparent molding density on flexural modulus of polypropylene structural foam is illustrated in Fig. 4.20 at specific locations across the thickness. The results were obtained by careful sectioning and mechanical testing of specimens cut from a 20 mm-thick polypropylene structural foam plaque [22]. The importance of skin development on specific modulus is clearly demonstrated in Fig. 4.21, for a wide range of values of polypropylene skin-to-core-thickness ratio.

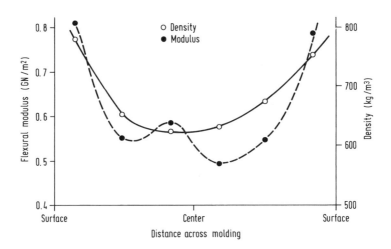

Figure 4.20 *Changes in flexural modulus and density across a 20 mm thick polypropylene structural foam molding at 23° C [22].*

Whereas the overall molding density and local density within a structural foam have a great influence on mechanical stiffness, effects from the size of cells produced appear less significant. Figure 4.22 presents results from the analysis of 6 mm thick polypropylene structural foam moldings made by a short-shot injection-molding technique, in which the foam core cell size distribution was determined by an image-analysis procedure at a distance of 1.5 mm beneath the outer skin surface [22]. The samples were prepared using molding conditions that yielded different cell structures, and specimens were tested with and without their outer skins. There appears to be no correlation between measured values for the median cell diameters and specific modulus, although a marked reduction in specific modulus is seen for the samples without skins.

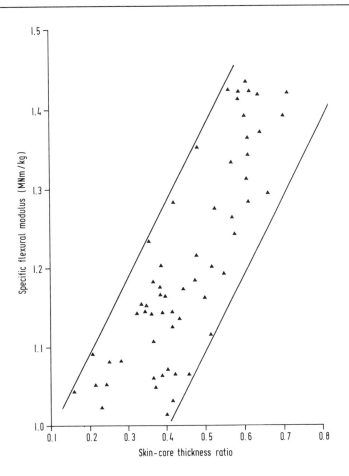

Figure 4.21 *The influence of skin-to-core thickness ratio on the specific flexural modulus of polypropylene structural foam at 23° C [20].*

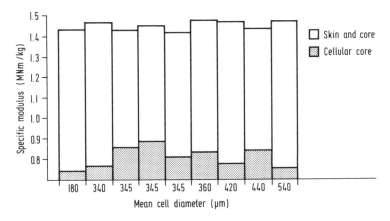

Figure 4.22 *The effect of cell size on the specific flexural modulus of 6 mm thick polypropylene structural foams. The samples were tested at 23° C with and without their outer skins [20].*

The presence of reinforcing short glass fibers in a structural foam may also significantly modify resistance to flexural deformation according to the fiber location and direction of alignment.

In solid short-fiber reinforced injection moldings, the preferred orientation of fibers can create highly anisotropic properties, which can vary according to the flow conditions experienced during mold filling [18]. Similar effects are seen with short–fiber reinforced polypropylene structural foams where a high degree of flexural anisotropy can result, showing great sensitivity to injection speed. This is evident from analysis of circular discs cut from molded plaques and tested at various angles of bending relative to the flow direction (Fig. 4.23). These differences have been shown to result from changes in fiber orientation, principally in the outer skins of the moldings (fibers in the core structure exhibiting a mostly random arrangement).

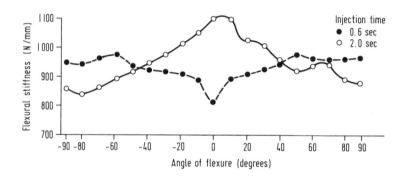

Figure 4.23 *The influence of injection time on the extent of flexural stiffness anisotropy in 8 mm thick short glass-fiber-reinforced polypropylene structural foam moldings (at 23°C) (fiber loading 30% by weight) [24].*

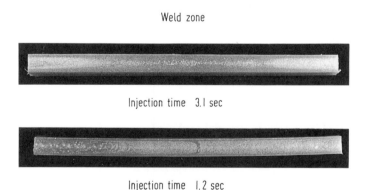

Figure 4.24 *The influence of injection time on weld line structure in 6 mm thick short fiber reinforced polypropylene structural foam injection moldings at 23°C (fiber loading 30% by weight) [28].*

Similarly, in fiber-reinforced structural foams containing weld regions, the mechanical properties in this zone are also strongly influenced by processing conditions, notably the injection rate. Table 4.3 shows average physical property data obtained for fiber-reinforced polypropylene structural foam analyzed across the weld lines and at a location adjacent to it. At a slow injection time (1.2 sec) the values for flexural modulus, apparent density, and the ratio of skin-to-core thickness are all lower across the weld zone than at positions away from it, whereas with a faster injection time the opposite situation is observed. Macroscopic reflected light micrographs taken across the weld region (Fig. 4.24), reveal examples of the large differences in skin-core structure quantified in Table 4.3.

Table 4.3 *Physical Properties of Weld Lines in Short−Fiber−Reinforced Polypropylene Structural Foam [28]*

Property	$t_I = 1.2$[a]		$t_I = 3.1$[a]	
	Bulk[b]	Weld[c]	Bulk[b]	Weld[c]
Flexural modulus (GPa)	2.65	2.18	2.56	2.57
Impact failure energy (Nm)	1.12	0.57	0.80	0.84
Apparent density (kg/m^3)	980	961	952	1009
Skin to core thickness ratio	0.55	0.27	0.56	2.79

[a] t_I, *injection time (s).*
[b] *Samples taken from nearby locations away from weld line.*
[c] *Samples taken across weld line.*

Moldings made at the slow injection speed exhibited a region of high skin thickness and limited foamed core structure. The weld was thus bridged by a large band of solid (skin) material, in contrast to material made under more rapid injection conditions. Fiber alignment was found to be predominantly parallel to the weld line (i.e., normal to the direction of flow) and was not greatly influenced by the injection speed [28]. This is analogous to the situation observed in solid short-fiber reinforced thermoplastics injection moldings (e.g., see Chapters 11 and 12), where fibers generally lie tangentially to the flow front during mold filling and subsequently parallel to the weld line, substantially reducing strength in this area [18].

4.4.2   Prediction of Flexural Stiffness

The behavior of structural foam plastics under flexural loading is of great interest to the designer and end user of these materials. Considerable attention has been given to development of accurate means for calculating the flexural rigidity of integral skin foamed beams from a knowledge of structural parameters, particularly the densities of solid and foamed polymer and ratios of skin and foamed core thickness. From the extensive literature on this subject [3,21,29-33], a summary containing principles of the most prominent approaches is given below.

Table 4.4 *Models for Predicting the Flexural Stiffness of Structural Foam, $(EI)_{sf} =$*

$$E_s I_e = E_s \left[\frac{b_c\, h_c{}^3}{12} + \frac{b h_s{}^3}{6} + \frac{b_c\, h_s(h_c + h_s)^2}{2} \right] \tag{4.4}$$

$$E_s I_e = E_s \left[\frac{b_c\, h_c{}^3}{12} + \frac{b h_s{}^3}{6} + \frac{b_c\, h_s(h_c + h_s)^2}{2} - \frac{9}{144}\frac{(b - b_c)h_c{}^3)}{} \right] \tag{4.5}$$

$$E_s I_e = E_s \left[\frac{b_c h_c{}^3}{12} + \frac{b h_s{}^3}{6} + \frac{b_c h_s(h_c+h_s)^2}{2} + \frac{(b-b_c)h_s{}^3}{18} + h_s(b-b_c)\left[\frac{h_c}{2} + \frac{2h_s}{3} \right]^2 \right] \tag{4.6}$$

$$E_s I_e = E_s \left[\frac{b_c(h_c + 2h_s)^3}{12} \right] \tag{4.7}$$

$$E_s I_e = E_s I_s \left\{ 1 - \left[1 - \frac{E_c}{E_s} \right]\left[1 - 2\left[\frac{h_s}{h_s + h_c} \right]^3 \right] \right\} \tag{4.8}$$

$$E_s I_s \left[3(1-e)^3 \left[\frac{R^2}{3} + 2R(1-R)\left[\frac{(C+1)}{(C+3)} - \frac{C}{(C+4)} \right] + (1-R)^2 \left[\frac{(C+1)^2}{(2C+3)} - \frac{2C(C+1)}{(2C+4)} + \frac{C^2}{(2C+5)} \right] \right] \right.$$
$$\left. + (1-(1-e)^3) \right] \tag{4.9}$$

$$E_s I_s \left[(1-e)^3 R^2 + (1 - (1-e)^3) \right] \tag{4.10}$$

$$E_s I_s\, (\rho_{sf}/\rho_s)^{1.8} \tag{4.11}$$

Due to the variable density existing across the thickness of structural foams and the relatively high scatter of density in commercial moldings, models to describe the mechanical behavior of uniformly foamed low-density plastics are seldom applicable [34]. By considering flexural rigidity of structural foams, $(EI)_{sf}$, as the product of flexural modulus (E) and moment of inertia (I), then:

$$(EI)_{sf} = E_s\, I_e = E_e\, I_s \tag{4.3}$$

where E_s and I_s are the modulus and moment of inertia for solid material and E_e and I_e are the modulus equivalent and equivalent moment of inertia respectively, modified to take account of the fact that the material has been foamed.

A commonly adopted approach is to consider the foamed beam in terms of a solid beam having the same thickness but with reduced cross-section normal to the axis of symmetry, and to alter the moment of inertia term, I_e, accordingly [30-32]. Figure 4.25 shows various forms of beam cross section represented by equations given in Table 4.4. These include the I-beam approximation, Eq (4.4), (Fig. 4.25a) often used to model two-component or sandwich moldings, and applied here to single-component parts and cross sections, in which there is either a linear gradient from the skin to the center of the core to account for the observed decrease in foam modulus towards the center of the sample, Eq (4.5), (Fig. 4.25b) or a decrease in modulus of the skin, Eq (4.6). In the latter case, the core width increases to take account of material removed from the skin (Fig. 4.25c). The last equivalent beam represents a situation where skin–core effects are not considered and a linear density modulus relationship is assumed across the sample (Fig. 4.25d). This over-simplification is represented by Eq (4.7).

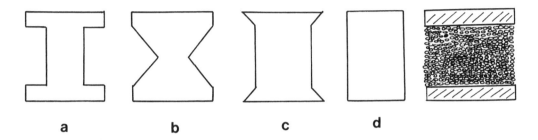

Figure 4.25 *Idealized forms of beam cross sections used to model the flexural stiffness of structural foam.*

A further approach considers that the structural foam has a uniform density foam core sandwiched between solid skins Eq (4.8) [33]. In order to calculate I_e, values of both skin and foam-core thickness as well as of moduli are needed. Although several models exist to calculate values of core modulus, experimental results derived from structural foams have been shown to be most closely described by the so-called density squared relationship [34]:

$$E_c/E_s = (\rho_f/\rho_s)^2 \qquad (4.12)$$

where E_c and E_s are the elastic moduli of foamed core and skin, with ρ_f and ρ_s being the corresponding densities. Eq (4.12) can be derived from first principles, by considering the process of cell-edge bending during linear elastic deflection of a cubic model structure for an open-all foam [35].

In principle, a more accurate approach for describing flexural rigidity in structural foams has to take into account the known variation of density occurring across the foam cross section, illustrated in Fig. 4.6. This may be modelled by Eq (4.9), which comprises an extension of the expression mentioned earlier for calculation of density profile across a structural foam beam [21]. In the case of uniform core density and a finite skin thickness, the equation reduces to Eq (4.10).

The validity of these expressions in relation to experimental measurements of flexural rigidity is presented in Table 4.5. The data refer to polypropylene structural foam moldings with thicknesses ranging from 6 to 20 mm. The calculated values of stiffness also include results obtained from the empirical Eq (4.11), in which the modulus of the structural foam is related to that for solid material by the density ratio to the power 1.8 [22].

Table 4.5 *The Relationship Between Experimental and Predicted Flexural Stiffness of Polypropylene Structural Foams of Different Thickness [20]*

	Flexural Stiffness (N/mm)						
	Specimen thickness (mm)						
Equation	6	10^a	10^b	10^c	15	20	δ^d
(4.4)	7.9	22.9	30.02	35.85	104.1	267.4	0.13
(4.5)	8.33	35.1	36.95	38.41	125.43	307.2	0.27
(4.6)	6.31	19.35	27.03	34.29	79.46	234.1	0.07
(4.7)	3.84	15.39	23.67	34.29	79.46	202.3	0.29
(4.8)	7.9	23.0	30.0	35.85	104.0	267.4	0.13
(4.9)	7.85	21.9	29.25	35.54	101.9	256.0	0.11
(4.10)	8.5	34.0	37.9	38.4	130.5	304.6	0.26
(4.11)	6.77	18.5	26.7	31.2	86.5	215.7	0.04
Measured	7.42	19.7	26.5	32.6	85.3	216.4	——

[a] *low density,* [b] *medium density,* [c] *high density,* [d] *average percentage deviation =*

$$Abs\left[\sum_i^n (x_i - \bar{x})/\bar{x}\right]/(n-1).$$

Comparison of calculated and measured values indicates large discrepancies. Eqs (4.5) and (4.10) overstate the stiffness, while Eq (4.7) substantially underestimates it for most of the beam thicknesses. Closest agreement was noted for Eq (4.6) (an equivalent section method which assumes decreasing modulus across the skin), Eq (4.9), (which models the form of density profile between skin and core), and, perhaps unexpectedly, the simple empirical relationship, described by Eq (4.11).

4.4.3 Impact Properties

Detailed analysis of the impact performance of structural foam plastics requires consideration of a variety of parameters, related to both the structure of the material and the form of test method used for evaluation. The complexities involved have precluded a comprehensive understanding of the interrelationship between structure and material toughness, although several useful guidelines have emerged.

As is the case of flexural stiffness, impact properties show a marked dependence on structural foam density and on skin thickness. Figure 4.26 shows the relationship between falling weight impact strength, determined using an instrumented facility, and apparent density of 10 mm thick polypropylene structural foam bars. Both the energy required to initiate cracks and the total energy to failure, increase with foam density.

Figure 4.27 demonstrates the anticipated enhancement of the specific falling-weight impact energy as a function of polypropylene skin-to-core thickness ratio. Similar trends

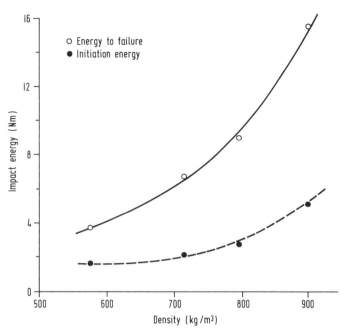

Figure 4.26 *The relationship between falling-weight impact strength and density for 10 mm thick polypropylene structural foam bars at 23° C [20].*

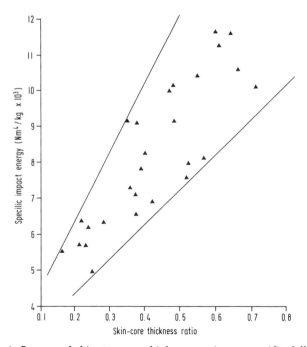

Figure 4.27 *The influence of skin-to-core thickness ratio on specific falling-weight impact strength of polypropylene structural foam. The values of total impact enrgy to failure were obtained at 23° C [20].*

between impact strength and skin thickness, have been reported for high-density polyethylene [29], where the Gardner impact strength of the structural foam (IS_{sf}) at a given plaque thickness and molding density, was represented by the expression:

$$IS_{sf} = IS_c \exp\{b\,h_s\} \qquad (4.13)$$

Here IS_c is the impact strength of the foam core, h_s is the skin thickness, and b is the slope of the curve between IS_{sf} and h_s.

Further modifications have been proposed with inclusion of a term for the reduced foam-core density (R) and the structural foam molding thickness (h_{sf}), viz:

$$IS_{sf}/IS_s = (Rh_{sf}/h)^2 \exp\{b\,h_s\} \qquad (4.14)$$

where IS_s and h are the respective values for impact strength and molding thickness, in the unfoamed state [29]. However, this equation has not been validated experimentally.

An empirical relationship has been proposed to describe the falling-weight impact energy to failure of polypropylene structural foam (IS_{sf}) from a knowledge of apparent density (ρ_{sf}) and total specimen thickness (h_{sf}) of the structural foam, together with values for the impact energy (IS_s), density (ρ_s), and thickness (h), of the corresponding solid polypropylene [20]:

$$IS_{sf}/IS_s = (\rho_{sf}/\rho_s)^3 (h_{sf}/h) \qquad (4.15)$$

Good correlation has been found between experimental and predicted impact energies using this expression, with structural foam thicknesses ranging between 6 and 20 mm (Fig. 4.28).

There are some reports describing the mechanism of failure of polymeric structural foams under impact loads. Often polymers which show a ductile failure in the bulk form, embrittle when foamed. Polypropylene falls into this category when tested at 23°C, [20,36] together with other semicrystalline materials including polybutylene terephthalate and polyoxymethylene [29].

Of the types of failure observed, brittle failure is manifested either by a star-shaped cracking pattern on the under-side of the specimen in a falling-weight impact test, and/or by complete removal of material punched out from the sample at a 45° fracture angle [36].

A form of ductile failure has been seen in ABS structural foam where an impact ductile failure of the top skin occurs and cells immediately below the surface are flattened, deforming and failing under compression, leaving an impactor depression in the top surface of the sample. Crack propagation through the underlying cellular material is then thought to occur with some localized yielding within the cell walls. Some bottom-skin deformation may result without generation of an extensive crack pattern [29].

In some structural foams a combination of brittle and ductile failure has been seen. Under some conditions of testing, polypropylene has shown evidence of stress-whitening or localized yielding at the point of failure on the lower specimen surface, combined with an otherwise brittle form of failure [20].

The cause of embrittlement in structural foams made from ductile polymers may be due to several factors. The influence of blowing agent residue on mechanical properties of the base polymer was discussed earlier. In some cases this could lead to a reduction in elongation to break of the material.

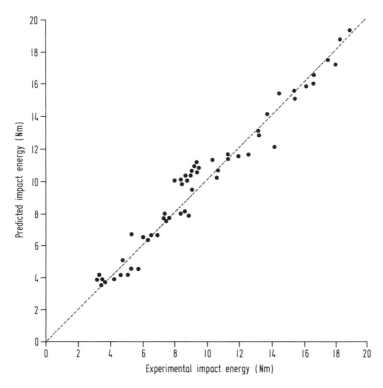

Figure 4.28 *Relationship between predicted [by Eq (4.15)] and experimental values for falling-weight impact energy at 23°C of polypropylene structural foam [20].*

The effect of surface finish on impact performance is generally thought to be significant, particularly in the case of structural foam moldings made by a low-pressure molding technique. It has been suggested that the characteristic swirl surface, such as shown in Fig. 4.16 for polypropylene structural foam, may contain fissures with the necessary geometry to produce a significant reduction in fracture initiation energy.

Estimates relating Charpy impact energy to the critical fracture surface energy, G_c, have shown that fissure depths of 0.1 to 0.2 mm would be sufficient to cause large reductions in impact energy in unnotched specimens [37]. Measurements on the surface of structural foam moldings have revealed defects with this order of depth. Also the presence of microvoids within or adjacent to the skin region will contribute to a lowering in impact strength, although the extent of this effect has not been quantified.

In semicrystalline polymers, changes in polymer micromorphology may account for some loss in impact resistance. It has already been shown that the foaming process can create unusual morphological features in and around the cell walls of polypropylene structural foam. In this material, the presence of propagating cracks or funnel-shaped holes at the boundaries of impinging spherulites may weaken the cellular core structure, providing an easier passage for propagating cracks.

In high-density polyethylene structural foams, polarized light microscopy has revealed the presence of transcrystalline morphology immediately around the foam bubbles surrounded by regions of spherulitic morphology further behind this region. This observation has prompted the suggestion that zones of transcrystallinity within the structural foam will lower its resistance to impact loads, particularly between adjacent cells [29].

Shear-induced transcrystallinity has also been observed in polypropylene structural foams made by a low-pressure short-shot molding technique. Figure 4.29 shows crystalline morphology in the outer regions of the skin zone formed at very high injection rates with the presence of transcrystalline layers. Moving away from the skin zone, normal spherulitic

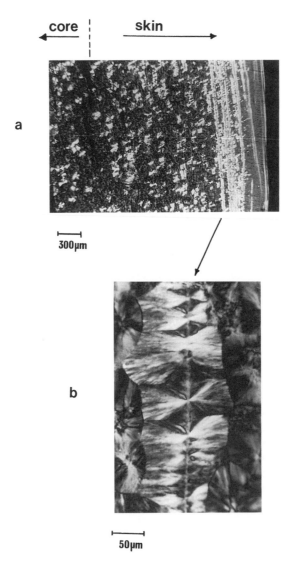

Figure 4.29 *Spherulitic morphology in the skin of a polypropylene structural foam made by a low-pressure injection molding technique: a) general view; b) magnified view of flow-induced transcrystalline layer.*

structures were developed with the expected increase in size towards the center of the molding. In the polypropylene surrounding the cells in the foam core there was no evidence of transcrystalline structure (see Fig. 4.12), but only well-developed spherulites, often with the unusual rounded or flat-topped forms, as discussed earlier [26].

Crack-induced fracture studies of the cellular core region of polycarbonate structural foams has demonstrated that tensile fracture of single edge-notched specimens underestimated fracture toughness and gave large variations in the value of the strain energy release rate, G_S. Values of G_S determined from the notched, double-cantilever beam specimens were significantly larger for foamed polycarbonate with pore size between 0.2 and 0.3 mm, than for corresponding bulk material, probably due to localized yielding of the polymer between adjacent bubbles [38]. It was inferred that the increase in crack extension force observed in foamed material containing 5% by weight of short glass fibers, resulted from an increased level of crazing along the glass fibers and branching of the crack tip.

Reductions in cell size in integral skin thermoplastics foams may also lead to an increase in toughness, as seen in microcellular foams with cell sizes of the order of 10 μm or less. These materials are produced by saturating a polymer with a gas, then nucleating cells by utilizing thermodynamic instabilities, which result when the polymer is heated and the pressure is reduced [39]. Although not "structural" in the context of the present chapter, microcellular foams have been shown to possess superior impact properties when compared to their solid counterparts.

4.5   FUTURE TRENDS

As explained earlier, the structural features and physical properties of integral skin-foamed core plastics depend strongly on the fabrication route. Future developments in this field are likely to result from novel or modified forms of processing technology (particularly concerned with injection molding), in order to impart specific material characteristics or design features, which are not achievable by the more established procedures for structural foam fabrication.

Recent developments are discussed below, which illustrate these trends and show particular promise for the future preparation of structural foam materials.

4.5.1   Thermosetting Plastic Structural Foams

Foaming of thermoset materials, including phenol-formaldehyde, urea-formaldehyde, epoxy or polyester resin compositions using liquid-based systems, and cold-cure reactions is already well known [40]. Furthermore, although it is often possible to achieve very low foam densities by these methods, well-defined skin-core morphologies are difficult to obtain, greatly limiting the load-bearing capability and potential applications of the resulting products.

Only recently techniques have been developed for fabricating thermoset structural foam artefacts using modified forms of injection or compression molding, combined with hot-cured thermoset resins, based primarily on polyester and phenolic formulations [41,42].

Design of the foamable thermoset composition is critical, requiring incorporation of a chemical blowing agent which decomposes thermally during molding to evolve gas necessary for resin expansion. Since stabilization of the foam structure is achieved by chemically curing the resin, a fine balance exists between the onset and rate of foaming and the commencement and rate of cure.

This is achieved, either by adjusting the decomposition temperature of the blowing agent and/or by tailoring the curing characteristics of the resin through judicious selection of the catalyst system. As discussed earlier in relation to thermoplastics foam molding, a wide variety of chemical blowing agents are available which decompose over a narrow and well-defined temperature range. These may also be chemically activated using compounds, such as urea or triethanolamine, to provide additional flexibility in temperature of gas evolution.

Figure 4.30 shows a DSC thermogram for expandable polyester molding compound containing chemical blowing agent, where exotherms at positions a and b represent the temperatures of maximum rates of gas evolution and resin cure respectively.

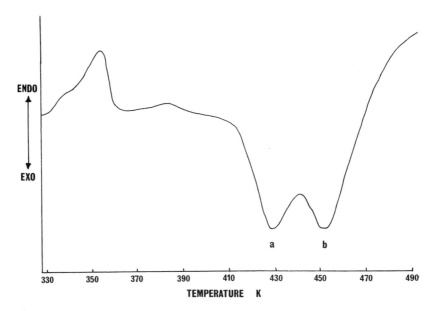

Figure 4.30 DSC thermogram of foamable polyester resin formulation at heating rate of 20 K/min: a) blowing agent decomposition; b) resin cure [42].

Molding and structure formation is achieved by introducing the thermoset compound into a hot mold, then heating the mixture under compression, such that parts of the mixture in contact with the mold cavity walls solidify to form a skin. The mold is then enlarged to a predetermined size to allow mixture beneath the skin to foam and cure. This procedure is readily achieved using vertical flash tooling and a programmed injection-compression-decompression clamping sequence on a thermoset screw injection-molding machine (c.f. Fig. 4.31). The variation of apparent molding density and skin thickness may be adjusted, either through the compression time after mold filling or by the extent of decompression stroke during foaming (Fig. 4.32). As with thermoplastics structural foams there is an inverse relationship between skin thickness and molding density.

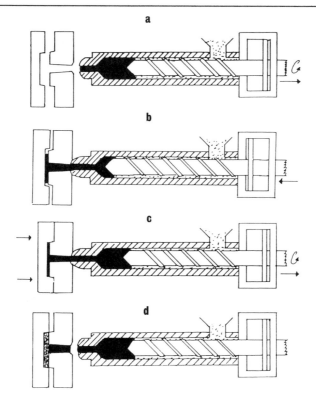

Figure 4.31 *Stages in the injection molding of thermosetting plastics structural foams [42].*

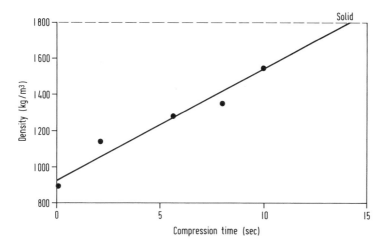

Figure 4.32 *The relationship between compression time and molding density, for injection–molded thermosetting plastics structural foams [42].*

Figure 4.33 shows the distinct skin-core profile of a dough molding compound (DMC) polyester structural foam produced by this technique, with a density of approximately 85% that of the corresponding material in the bulk. By contrast, a much thinner skin surrounding a more fully developed foam-core structure is seen when the material is foamed to give a 50% density reduction (Fig. 4.34). A two-phase structure has been observed in the core of these materials (Fig. 4.35), with microcells of the order of 1 μm in diameter present in much larger cell walls.

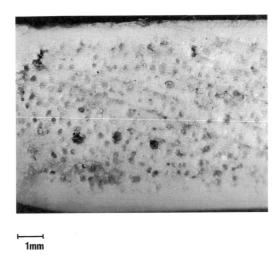

1mm

Figure 4.33 *Skin-core structure in a high-density DMC structural foam molding [42].*

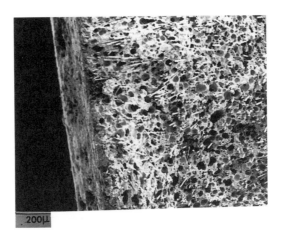

200μ

Figure 4.34 *Skin-core structure in a low-density DMC structural foam molding [42].*

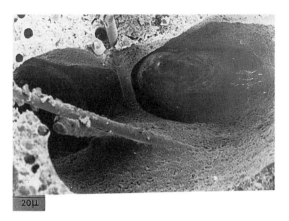

Figure 4.35 *Two-phase cellular microstructure in the core of a DMC structural foam [42].*

As with thermoplastics structural foams, a principal attraction of thermoset variants is the high stiffness-to-weight ratio and the ability to produce controlled skin-core structures useful in load-bearing situations. Table 4.6 lists selected mechanical properties of phenolic and polyester structural foams containing short-glass-fiber reinforcement.

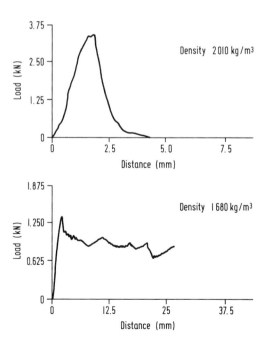

Figure 4.36 *Falling–weight impact energy load–distance traces for solid and structural foam phenolic moldings at 23° C [42].*

Although resistance to crack initiation is reduced under impact conditions, the fiber–reinforced thermoset foams of optimum density can exhibit a substantial increase in total energy absorption to failure relative to solid material. These enhanced energy-absorbing characteristics can be clearly seen in load-deflection traces for foamed and solid fiber-reinforced phenolic injection moldings tested on a falling-weight impact testing machine (Fig. 4.36). The differences are even more pronounced when results are compared on an equal weight basis (Table 4.6).

Table 4.6 *Selected Mechanical Properties at 23°C of Fiber–Reinforced Phenolic and Polyester Resin Structural Foams [42]*

I. Flexural Strength and Modulus

	Average density (kg/m^3)	Flexural strength (MPa)	Flexural modulus (GPa)	Specific flexural strength (kNm/kg)	Specific flexural modulus (MNm/kg)
Unfoamed DMC	1910	92.1	10.9	48	5.7
Foamed DMC	1460	74.9	5.8	51	4.0
Unfoamed phenolic	1820	133.7	12.7	74	7.0
Foamed phenolic	1550	126.9	8.8	82	5.7

II. Falling Weight Impact Strength

	Average density (kg/m^3)	Peak force (Nm)	Failure energy (Nm)	Specific failure energy (MNm^4/kg)
Unfoamed DMC	1900	3.1	11.0	5.8
Foamed DMC	1460	3.3	11.1	6.6
Unfoamed phenolic	2190	4.2	5.7	2.7
Foamed phenolic	770	1.6	13.7	17.8

The thermoset structural foams described above have considerable potential in applications requiring stability from distortion at elevated temperatures, or an inherent flammability resistance (particularly with phenolic resin compositions), which is less readily achieved in thermoplastics structural foam counterparts, without substantial modification using costly additives.

4.5.2   Controlled Internal Pressure Molding

A new injection-molding technology was recently introduced. It involves controlled injection of a gas (generally nitrogen) into the center of molten polymer as it fills the mold cavity. This process can ameliorate many of the limitations inherent in conventional structural foam and bulk molding and is therefore creating extensive commercial interest as an alternative means for producing large thermoplastics moldings with combined thin and thick geometries [43].

Unlike the usual approach for producing structural foam moldings requiring dissolution of gas into the melt, in the controlled internal-pressure molding the gas does not mix with the polymer, but instead forms continuous channels through less viscous thicker sections of melt in the mold cavity. Hence, although the products are not strictly structural foams containing a cellular interior, they do possess many of the characteristics of integral skin-foamed core materials, such as an increased stiffness-to-weight ratio.

In one commercially proven variant of this process, the injected gas is delivered via the melt injection nozzle, such that optimum control of gas pressure is maintained at the start and during the injection phase of the cycle (Fig. 4.37). Plant required to achieve this operation comprises a means for accurate volumetric dosing of gas into a precisely metered volume of melt, commonly using a gas accumulator unit contain a hydraulic ram. This conversion facility can be readily fitted to both conventional high-pressure and structural foam injection-molding machines.

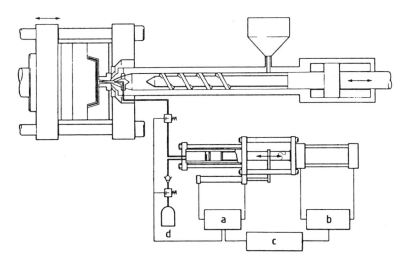

Figure 4.37 Adaptation of a conventional injection molding machine for controlled internal pressure molding: a) electrical system; b) hydraulic system; c) control panel; and d) gas cylinder.

Typically, a short shot of polymer is delivered to partially fill the mold, followed by injected gas to push the melt to the extremities of the cavity, whilst also maintaining internal (holding) pressure as it cools. This ensures positive contact between the polymer and cavity surface, creating a surface quality comparable to high-pressure injection molding and overcoming the limitation of a rough texture evident in short-shot structural foam moldings, or sink marks sometimes found in bulk, thick-section parts. However, since the internal pressure within the molding can be as low as 15% of that needed for an equivalent solid injection-molded component, clamping forces can be substantially reduced, allowing the use of low-cost tooling, e.g., aluminium.

A further benefit of low internal cavity pressures, particularly in thin-walled moldings is reduced frozen-in molding stress and greater freedom from distortion in flat or straight-edged pieces.

Experience has shown that cycle times for parts made in the controlled internal pressure injection molding can be up to 50% lower than those needed in structural foam technology. This is due to more efficient cooling and release of internal gas pressure before the mold is opened, thereby eliminating the possibility of post-blowing, which can occur in prematurely ejected structural foam moldings.

This technology can be readily applied to most thermoplastics materials, including filled and reinforced grades, in both thin-walled moldings (4 mm or less) and thicker sections in excess of 8 mm, or in combinations of both. However, since the injected gas will always take the route of least resistance, considerable attention must be given to component design in order to avoid problems of non-uniform gas distribution or the possibility of gas penetrating through the molding surface, as can occur when materials of low viscosity are injected into a higher viscosity skin medium [44]. In particular, specific use of ribs, box sections, or molding contours is needed to provide guidance for direction of gas flow, together with optimization of gate location.

Further developments of this technology include the use of multiple gating systems in hot-runner molds, where the polymer enters the cavity independently of the gas, introduced selectively via retractable gas-injection nozzles into thicker sections of the molding.

Very little detailed structural analysis has been reported on the parts produced by controlled internal pressure molding, although it is known that with fiber-reinforced thermoplastics moldings, a high level of preferred fiber alignment can be introduced in regions influenced by the gas flow.

4.5.3 Multiple-Component Structural Foams

The concept of introducing two materials into a mold cavity, such that one forms the skin and the other the foamed core, is well established and may be undertaken by sequential injection of the two materials or by co-injection through a single multichannel nozzle.

The principle of this simultaneous operation mentioned in Section 4.2, was schematically shown in Fig. 4.3 for two polymers A and B, the former being unfoamed and ultimately producing the molding skin, whilst the latter polymer contained chemical blowing agent which decomposes and develops a cellular core structure expanding until the mold cavity is completely filled. Thick-walled moldings can be produced by this method using various combinations of skin and core polymer giving density reductions of up to about 30% depending on molding geometry and polymer type.

A further refinement of this technology, giving additional scope for structural control within the molding, is to combine two-component (or sandwich) molding with a form of controlled gas-injection technique discussed previously.

It has been shown that, while in conventional two-component structural foam complete filling of the mold is achieved by expansion of the core material, it is possible to fill the mold using injected gas, as discussed previously for single-component moldings [45]. In this way, smaller quantities of plastic are needed to fill the cavity and the internal pressure of the gas ensures close contact of the outer molding skin and cavity wall, resulting in an outstanding surface quality.

Subsequent to mold filling, the foamable core material is allowed to expand inwards, creating a void-free uniform low-density foam, with a density below that which would be tolerable by the conventional low-pressure structural foam molding process, due to the unacceptable surface quality which would be evident.

Hence, this alternative technique offers opportunities for the manufacture of thick-section two-component moldings (with a wall thickness of at least 10 mm), combining excellent surface finish, low density, and high rigidity. Furthermore, using two-component molding technology, different polymer combinations are possible for skin and core regions allowing creation of special effects, such as a soft outer skin surrounding a hard low-density cellular core, thus extending possibilities for original component design.

4.6   CONCLUSIONS

The controlled formation of polymers into integral skin-foamed core structures with load-bearing capabilities, has become increasingly important, largely due to advances in process technology and, in particular, developments with thermoplastics injection molding. Various forms of moldings may be produced, as single- or multicomponent structural foams, with additional filled and fiber-reinforced compositions also possible.

More recent processes offering commercial potential include injection molding of thermosetting structural foams and controlled internal pressure injection molding, using gas-injection techniques.

Successful exploitation of structural foam products has also resulted from judicious design with these materials, combined with a greater understanding of the interrelationship between physical properties and structural variables. These depend strongly on manufacturing route and process conditions. Mechanical properties, in particular, are influenced by density changes within the material, the extent of surface skin formation, and fiber orientation in reinforced systems. Material toughness may also depend on surface texture, internal cell structure, and molecular morphology in moldings made from semicrystalline polymers, such as polypropylene.

Increased confidence in the design and application of structural foam products has taken place through procedures for predicting mechanical properties, notably flexural stiffness. This can be achieved from a knowledge of structural parameters and properties of the base polymer in the unfoamed state. Prediction of impact performance remains more tentative, however, relying on empirical models.

ACKNOWLEDGMENTS

Figure 4.2 reproduced with permission from ICI Chemicals and Polymers Ltd. Figure 4.3 reproduced with permission from Battenfeld Machinenfabriken GmbH. Figures 4.5, 4.6, 4.8, 4.15, 4.17, 4.18, 4.19, 4.20, 4.21, 4.22, 4.23, 4.26, 4.27, and 4.28 reproduced with permission from Elsevier Applied Science Publishers Ltd., and Fig. 4.37 reproduced with permission from Cinpres Ltd.

NOTATION

b	Beam width
b_c	Reduced beam width $= b(E_c/E_s)$
C	Shape factor
e	Ratio of skin thickness to total beam thickness
E_c	Elastic modulus of foamed core
E_s	Elastic modulus of solid beam
h	Solid beam or molding thickness
h_c	Thickness of foamed core
h_s	Skin thickness
h_{sf}	Overall structural foam thickness
I_e	Equivalent moment of inertia of foamed beam
I_s	Moment of inertia of solid beam
IS_c	Impact strength of foam core
IS_s	Impact strength of solid material
IS_{sf}	Impact strength of structural foam (combined skin and core)
L	Beam span
P	Load at center of beam
R	Ratio of core-to-solid density
Z	Ratio of distance away from center-line to bar thickness
δ	Beam deflection
Σ	Ratio of skin thickness to half-bar thickness
ρ_f	Average foam-core density
ρ_s	Solid polymer density
ρ_{sf}	Apparent density of structural foam (combined skin and core)
$\bar{\rho}$	Local foam density

REFERENCES

1. Becker, W. E., Ed., *Reaction Injection Molding*, Van Nostrand Reinhold, New York (1979).
2. Oertel, G., *Polyurethane Handbook*, Hanser Publishers, Munich (1985).
3. Shutov, F. A., *Integral/Structural Polymer Foams: Technology, Properties and Applications*, Springer-Verlag, Berlin (1985).
4. Throne, J. L., *J. Cellular Plast.*, **12**, 161 (1976).
5. Hornsby, P. R., *Mater. Eng.*, **3**, 354 (1982).
6. Anonymous, *Technology of "Celogen" blowing agents*, Uniroyal Chemical, Naugatuck, USA.
7. Anonymous, *Genitron blowing agent*, Product data sheets, FBC (plc), Cambridge, England (1979).
8. Reed, R. A., *Brit. Plast.*, **33**, 469 (1969).
9. Bathgate, R. J., Collington, K. T., *SPE Techn. Pap.*, **27**, 856 (1981).
10. Eckardt, H., 14th Annual SPI Structural Foam Conference, Boston, April 22 (1986).
11. Ugine Kuhlmann, *Br. Pat.* 1,184,688 (1970); *Br. Pat.* 1,223,968 (1971); *Br. Pat.* 1,318,608 (1973).
12. Shelly, R., Han, C. D., *J. Appl. Polym. Sci.*, **22**, 2573 (1978).
13. Malpass, V. E., *Plast. Rubber: Mat. Appl.*, November 149 (1978).

14. Tadmor, Z., Gogos, C. G., *Principles of Polymer Processing*, John Wiley & Sons, New York (1979).
15. Bowman, J., Harris, N., Bevis, M. J., *J. Mat. Sci.*, **10**, 63 (1975).
16. Katti, S. S., Schultz, J. M., *Polym. Eng. Sci.*, **22**, 1001 (1982).
17. Murphy, M. W., Thomas, K., Bevis, M. J., *Plast. Rubber Process. Appl.*, **9**, 3 (1988).
18. Folkes, M. J., *Short Fiber Reinforced Thermoplastics*, Research Studies Press, Chichester (1982).
19. Hubeny, H., Weiss, E., Dragaun, H., *J. Cellular Plast.*, **11**, 256 (1975).
20. Ahmadi, A. A., Hornsby, P. R., *Plast. Rubber Process. Appl.*, **5**, 35 (1985).
21. Throne, J. L., *J. Cellular Plast.*, **14**, 21 (1978).
22. Ahmadi, A. A., *PhD thesis*, Brunel University, England (1983).
23. Throne, J. L., *J. Cellular Plast.*, **12**, 264 (1976).
24. Ahmadi, A. A., Hornsby, P. R., *Plast. Rubber Process. Appl.*, **5**, 51 (1985).
25. Burton, R. H., Hornsby, P. R., *J. Mat. Sci., Lett.*, **2**, 195 (1983).
26. Hornsby, P. R., Russell, D. A. M., *J. Mat. Sci., Lett.*, **3**, 1061 (1984).
27. Throne, J. L., *J. Cellular Plast.*, **14**, 21 (1978).
28. Hornsby, P. R., Head, I. R., Russell, D. A. M., *J. Mat. Sci.*, **21**, 3279 (1986).
29. Throne, J. L., in *Mechanics of Cellular Plastics*, Hilyard, N. C., Ed., Applied Science, London (1982).
30. Hobbs, S. Y., *J. Cellular Plast.*, **12**, 258 (1976).
31. Moore, D. R., Iremonger, M. J., *J. Cellular Plast.*, **10**, 230 (1974).
32. Orgorkiewicz, R. M., Sayigh, A. A., *Plast. Polym.*, **40**, 64 (1972).
33. Moore, D. R., Couzens, K. H., Iremonger, M. J., *J. Cellular Plast.*, **10**, 135 (1974).
34. Progelhof, R. C., Throne, J. L., *Polym. Eng. Sci.*, **19**, 493 (1979).
35. Gibson, L. J., Ashby, M. F., in *Cellular Solids– Structure and Properties*, Pergamon Press, Oxford (1988).
36. Progelhof, R. C., Throne, J. L., *SPE Techn. Pap.*, **27**, 683 (1981).
37. Ackhurst, S. R., Bucknell, C. B., *Plastics and Rubber Institute Conference on Rigid Structural Plastics Foams*, Bradford, England (1978).
38. Hobbs, S. Y., *J. Appl. Phys.*, **48**, 4052 (1977).
39. Cotton, J. S., Suh, N. P., *Polym. Eng. Sci.*, **27**, 485 (1987).
40. Frisch, K. C., Saunders, J. H., Ed., *Plastics Foams*, Part II, Marcel Dekker, New York (1973).
41. Hornsby, P. R., Brenner, M. J., *UK Pat. Appl.*, GB 2,188,636A (1987).
42. Brenner, M., Hornsby, P. R., *Cell. Polym.*, **7**, 451 (1988).
43. Pearson, T., *Mats. Des.*, **7**, 315 (1986).
44. White, J. L., Lee, B.-L., *Polym. Eng. Sci.*, **15**, 481 (1975).
45. Eckardt, H. J., *Plastics and Rubber Institute International Conference on Polymer Processing Machinery*, University of Bradford, England, July (1987).

CHAPTER 5

RHEOLOGICAL MEASUREMENTS OF RIGID FOAM EXTRUSION

by Kwang Ung Kim and Byoung Chul Kim

Polymer Processing Laboratory
Korea Institute of Science and Technology
P.O. Box 131, Cheongryang, Seoul
KOREA

This chapter reviews the rheology of the rigid foam extrusion. After a brief introduction to the fundamental aspects of bubble dynamics, the rheology associated with the extrusion of gas-charged molten polymer systems is presented. The framework for the discussion of the foam rheology is based mainly on the foam extrusion of rigid polyvinylchloride (PVC) using two chemical blowing agents, azodicarbonamide (AZ) and sodium bicarbonate (SC). A consensus is that inclusion of the blowing agent reduces both normal stress and viscosity. It may also significantly change the pressure profile, particularly in the vicinity of the die exit. Some researchers have reported that the incorporation of blowing agent results in a small difference from the system with no blowing agent. For a chemical foaming process of rigid PVC it was found that an increase of temperature considerably increases the power-law index. Of two chemical blowing agents, AZ gives greater increase in power-law index than SC at the same usage level. Furthermore, all the viscosity data of various expandable PVC formulations appear to fall onto a single viscosity curve of the unexpanded PVC formulation above a critical shear rate. Recently, the increasing commercial importance of rigid structural foam processing requires a better understanding of the foam rheology of high-viscosity materials. Thus, a precise measuring technique is important for designing and utilization of structural foam products.

5.1   INTRODUCTION

Incorporation of gases into polymeric materials has long been used to produce synthetic polymeric foams [1-6]. The first commercial foam, a sponge rubber, appeared between 1910 and 1920 [2]. Since then, synthetic foams have found a variety of applications in comfort cushioning, thermal and electrical insulations, domestic and commercial refrigeration, residential and commercial construction, packaging, structural components, buoyancy, space filling, and seals [7-11]. Rigid foams are of special interest because of the high strength-to-weight ratio. At the same time, understanding the rheological behavior of gas-charged polymers is essential to processing of expandable formulations for appropriately foamed products with desired properties. A number of experimental and theoretical results have been reported but definite conclusions and generalization in respect to foam rheology have seldom been provided. However, in spite of little progress in theoretical analysis of foaming systems, considerable progress has been made in processing techniques to meet the diverse requirements of practical applications [9,12-26].

The most recent development in rigid foams is the emergence of the so-called structural foam, a foamed product having an unexpanded integral skin. Structural foams are prepared by molding or extrusion. There are several known fabrication methods including the Ugine-Kuhlmann structural foam process [27], the Union Carbide [28], the USM [29], the Kraus-Maffei [30], the Mitsubishi MJC [31], the TAF [32], and the ICI sandwich process [33]. Each process has a significant commercial advantage by virtue of its own characteristic features. Among them, the Ugine-Kuhlmann structural foam process (the so-called Celuka process) is a variation of conventional extrusion. It requires an extruder, a specially designed die, cooling, sizing, and take-up devices, as illustrated in Fig. 5.1. As with a conventional foam extrusion, the resin containing blowing agent is fed into an extruder. However, the Celuka process differs from conventional extrusion in the gas expansion method. A mandrel, centered in the die, blocks the center cone at the entrance of the shaper and thereby forces the foam to expand toward the core area in the shaper [27]. The unique feature of this process is that the expandable melt is forced through die jaws with the same configuration and dimensions as those desired in the final products. At the same time, the surface of the extrudate is quickly cooled forming a solid integral skin.

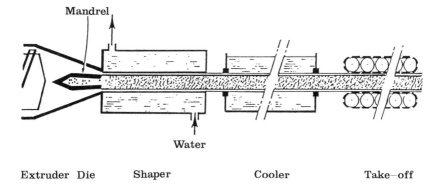

Figure 5.1 *Schematic representation of the Celuka structural foam process[27].*

The synthetic polymeric foams are produced by either a physical or chemical blowing agent. In the physical foaming process, the property of the charged gas is a dominant factor governing the foaming characteristics. In the chemical foaming process, however, the method of gas formation, hence the thermal decomposition mode of the blowing agent, also plays a significant role. In the chemical foaming process, the foaming characteristics are demonstrated rather well with a relatively high viscosity system. That is, high viscosity may suppress the extra bubble growth and retard the abrupt bubble formation in the extruder barrel and die sections. In consequence, the characteristic bubble-expanding behavior of a chemical blowing agent is easily observed in the resultant foamed products [34,35]. However, in the chemical foaming process with low viscosity, it is not easy to observe the foaming characteristics because of rather explosive bubble expansion and their rapid coalescence. This is ascribed to the fact that the blowing pressure of the chemical blowing agent exceeds the melt strength of the base resin. The rigid polyvinylchloride (PVC) foam extrusion is normally carried out at relatively low temperatures (poor thermal stability of PVC), frequently below the melting temperature of the syndiotactic polymer units. Therefore, the polymer phase is more viscous than usual polymer melts [34,35]. In addition to a useful correlation to the foam rheology, these foaming characteristics result in significant modification of the foam morphology [15,35-39]. Therefore, an understanding of the foaming characteristics greatly helps controlling the foam properties as well.

In spite of the increasing need for fundamental knowledge of foam processing, only a few publications or review articles are available [40-43]. This chapter emphasizes the rheology of rigid foam extrusion, particularly the rigid PVC foam process. The following sections briefly discuss bubble dynamics in a polymer phase and follow with phenomenological aspects of rigid foam extrusion considering the chemical blowing agents and processing parameters. The first two sections are intended to provide backgrounds of foam processing. The next section contains selective reviews of foam rheology, which owes much to the authors' own results on the rigid foam extrusion of PVC using azodicarbonamide (AZ) and sodium bicarbonate (SC) as two chemical blowing agents. The last part focuses on the potential of rheological properties for the practical foam processing. Further information on this topic can be found in Chapter 4.

5.2   BUBBLE DYNAMICS IN GAS–CHARGED SYSTEMS

An understanding of bubble dynamics is crucial for foam processors to manipulate foamed products for desired properties. Two fundamental steps are thought to be involved in the bubble-formation process: nucleation and growth. If the gas concentration exceeds a critical value, a bubble nucleation may be started from the supersaturated polymer phase. When the nucleated bubble has reached a certain minimum size, it continues to grow until the inside pressure, P_g, is in equilibrium with the external pressure exerted by the polymer phase, P_p, as a result of blowing agent from the molten polymer to the gas bubble. Then the growing bubble radius in a fluid, r_b, may be given by:

$$\Delta P = P_g - P_p = 2\nu/r_b \tag{5.1}$$

in which ΔP is the pressure difference across the interface, and ν is the interfacial tension.

In 1950, Epstein and Plesset [44] developed a pioneering theory on bubble-growth kinetics defined by:

$$r_b(t) = kt^{1/2} \tag{5.2}$$

where $r_b(t)$ is the bubble radius, growing with the time t, and k is a rate constant related to diffusion coefficient, solubility of the gas to the liquid, and the initial concentration of the gas in the liquid. The authors considered a diffusion-controlled growth-collapse mechanics of a stationary gas bubble suspended in Newtonian medium. Further, they assumed the medium to be inviscid and thus did not consider hydrodynamic resistance of the liquid to bubble growth. This expression, consequently, demonstrated its validity only in a narrow range of materials, particularly with several elastic media [45,46]. Later, Hobbs [47] combined the bubble coalescence concept with the Epstein-Plesset theory, deriving the following expression for the bubble growth kinetics:

$$r_b(t) = r_0 + ABt^{1/2} + (1 - A)B^2t/(2r_0 + Bt^{1/2}) \tag{5.3}$$

where r_0 is the initial bubble radius, and A and B are constants which depend on the diffusion coefficient, gas solubility, temperature, and pressure difference in bubbles of different sizes.

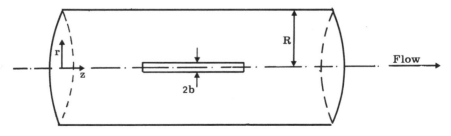

Figure 5.2 *Schematic representation of capillary flow with a bubble.*

Villamizar and Han [48] reformed the Epstein-Plesset solution for bubble growth kinetics, and gave a power-law type expression:

$$r_b(t) = k't^n \tag{5.4}$$

The authors reported that the exponent, n, was not constant but varied with both temperature and number of bubbles. They attributed discrepancies between their solution and the Epstein-Plesset's to an elastic effect on the bubble growth.

The effect of medium elasticity on bubble growth was first proposed by Street [49], who reported that the initial growth rate in a viscoelastic medium was greater than that in a Newtonian fluid of the same viscosity. Han and Yoo [50] formulated equations of bubble dynamics in the foam injection molding process by use of the Zaremba-Fromm-DeWitt model [51], which could describe such viscoelastic properties of medium as shear-dependent viscosity, normal stress effect, and stress relaxation. More recently, we reexamined the bubble dynamics in the foam extrusion process [52]. When a threadlike thin gas bubble of radius, r_b, travels along the centerline of a capillary die of radius, R, as described in Fig. 5.2, the equation of motion in cylindrical coordinates (r, θ, z) yields the following expression:

$$\rho \left[(\dot{r}_b{}^2 + r_b \ddot{r}_b) \ln \frac{R}{r_b} + \frac{r_b^2 \dot{r}_b^2}{2} \left[\frac{1}{R^2} - \frac{1}{r_b^2} \right] \right] = P_g - P(R) + \sigma_{rr}(R) - \frac{\nu}{r_b}$$

$$- \frac{4\eta_0}{\lambda} \int_0^t e^{(s-t)/\lambda} \left[\frac{r_b(s) \ \dot{r}_b(s)}{r_b^2(t) - r_b^2(s)} \right] \ln \frac{r_b(t)}{r_b(s)} ds - 2\sigma_{rr,0} e^{-t/\lambda} \ln \frac{R}{r_b} \tag{5.5}$$

In Eq (5.5), ρ is the medium density; P, the pressure in the medium; P_g, the pressure inside the bubble; σ_{ij}, the component of stress tensor; ν, the interfacial tension; η_0, the zero shear viscosity; λ, the relaxation time; t, time; r_b, $\dot{r}_b$ and $\ddot{r}_b$ are the bubble radius and its first and second derivative with respect to time t, respectively. Further, the equation of mass balance may be written as:

$$d(\rho_g r_b^2)/dt = 8 r_b^2 D_f \rho^2 (C_0 - C_w)^2/3(r_b^2 \rho_g - r_0^2 \rho_{g0}) \tag{5.6}$$

where ρ_g is the density of gas in the bubble; D_f, the diffusion coefficient; C_0, the solute concentration far from the bubble wall; and C_w, the solute concentration just outside the bubble wall. We have computed Eqs (5.5) and (5.6) simultaneously with regard to appropriate boundary conditions. The numerical appraisal of these equations reveals several important results: 1) the bubble radius and bubble growth rate are increased as the concentration of blowing agent increases, 2) an increase in medium viscosity suppresses bubble growth while an increase in medium elasticity promotes bubble growth, and 3) the bubble radius and the bubble growth rate are increased as the diffusion coefficient increases. However, the numerical solution of Eqs (5.5) and (5.6) leaves much room for studies on a general rheological model for bubble dynamics.

An understanding of the nucleation phenomena has been of great interest. Despite many pioneering works, it is still hard to determine the exact bubble initiation point [48,53-61]. On an empirical basis, it is known that the critical blowing pressure of a blowing agent is closely related to such complex factors as temperature, pressure, the mechanism of gas release by the blowing agent, and physical properties of the evolved gas. Visual experimentation by Han and Villamizar [40,48] showed that the critical pressure for bubble inflation decreased with increasing melt extrusion temperature, and increased with increasing concentration of the blowing agent. Furthermore, the authors reported that the critical pressure was independent of flow rate. Throne [62] obtained similar results by correlating the critical pressure to temperature, blowing agent level, solubility, and the initial bubble radius. Although there have been some attempts to determine the critical blowing pressure of a gas in polymeric medium, much of nucleation phenomena remains to be studied including size of initial microbubbles, dependency on flow geometry, and different combinations of the blowing gas and polymer system.

5.3   PHENOMENOLOGICAL ASPECTS OF RIGID FOAM EXTRUSION

Under pressures higher than the critical pressure of evolved gas from the blowing agent, the nucleated bubble may dissolve in the molten polymer phase. The exact mechanism of bubble nucleation from a supersaturated polymer phase is not well understood. Only limited information has been disclosed [44,63,64]. However, it is generally accepted that simultaneous nucleation of many bubbles produces a foamed product with uniform cell structure owing to the number of bubbles determined at the nucleation stage. Hansen and Martin [63] reported that nucleating agents let the bubble form at lower concentrations than in the case of a homogeneous nucleation. They further found that a tenfold increase in nucleating-particle concentration resulted in a thousandfold increase in the number of bubbles and a fiftyfold decrease in the bubble size, approximately at the same level of supersaturation. The nucleating agent consisted of well-dispersed metal particles, sodium aluminum silicate, silica, and iron(III) oxide, which could afford nucleating sites for the formation of bubbles from the dissolved gas. More recently, Yang and Han [64] investigated the effect of the type and concentration of nucleating agent on the foam extrusion characteristics of low-density polyethylene; they found that surface characteristics of a nucleating agent were of paramount importance in controlling the foam quality. Furthermore, these authors reported that as the nucleating agent concentration increased, the average cell size decreased and the fraction of open cells increased. However, unlike to foaming of polyethylene, polypropylene, and polystyrene, additional nucleating agents had no significant effect on the size or size distribution of cells in rigid PVC foam extrusion [35]. The finely dispersed metallic particles, a barium-cadmium type added as thermal stabilizer, in the polymer phase played a role of an auxiliary nucleating agent.

Variation of screw torque with volumetric flow rate at two temperatures during the foam extrusion of rigid PVC is shown in Figs. 5.3 and 5.4 [35]. The unexpanded formulation P-base in Figs. 5.3 and 5.4 contains thermal stabilizer, processing aid, and lubricant (see the Appendix). Measurement of screw torque was carried out with a Brabender Plasti-Corder ($D = 19$ mm, $L/D = 25$).

Two expandable PVC formulations exhibit a reduced torque comparing to the norms established for the unexpanded PVC formulation. The torque reduction may be a direct indication of bubble nucleation in the barrel. That is, the lubricating or plasticizing action of resultant bubbles reduced the torque [41]. Therefore, the amount of torque reduction at the same conditions may indicate that the nucleation occurs in the barrel. At the same usage level, AZ causes greater reduction of torque than SC. This means that nucleation of bubbles by SC is greatly affected by medium viscosity, hence by the external pressure. In other words, most

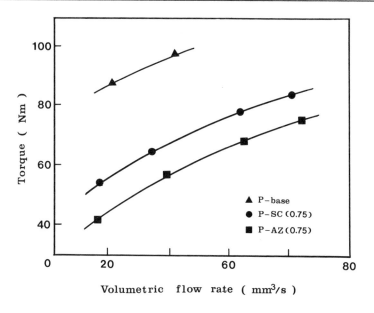

Figure 5.3 *Screw torque vs. volumetric flow rate for three PVC formulations with and without the blowing agent at 170°C (number in parenthesis gives the level of the blowing agent in phr).*

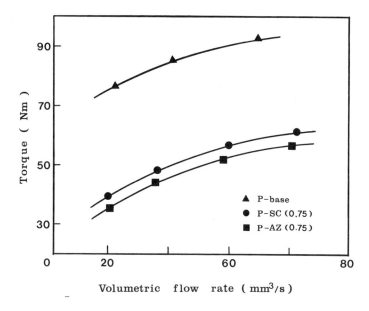

Figure 5.4 *Screw torque vs. volumetric flow rate for three PVC formulations with and without blowing agent at 180°C.*

bubbles nucleated by AZ may exist in the extruder barrel at high pressure as an irreversible dispersion state of microbubbles, whereas a considerable portion of bubbles nucleated from SC is dissolved at equilibrium in the polymer phase. This leads to questioning the conventional wisdom that, at elevated pressures, molten polymer can be supersaturated with a gas–forming component. Apparently, bubbles can be nucleated in the extruder barrel even in highly viscous PVC melt. Therefore, a blowing agent with high and invariant blowing pressure like AZ may nucleate more bubbles than a blowing agent with low and pressure-dependent blowing pressure such as SC.

Two chemical blowing agents (AZ and SC) yielded different axial bubble expanding profiles after leaving the short cylindrical die (D = 7.5 mm, L/D = 4, and converging angle of die entrance = 60°). A single-screw extruder (D = 25.4 mm, L/D = 25) was used for the experiment. Figures 5.5 and 5.6 show additional extrudate swell in terms of flow rate and temperature, respectively [35]. Naturally, a blowing agent with higher blowing pressure gives higher initial bubble growth rate. The bubble growth is enhanced as the volumetric flow rate increases due to increased medium elasticity and decreased viscosity. One can also note in Figs. 5.5 and 5.6 that AZ expands abruptly while SC appears to exhibit multi-step expansions. Obviously, the initial bubble growth rate is proportional to the number of bubbles nucleated. Therefore, AZ exhibits higher initial bubble growth rate, which shows a relatively weak temperature dependency as seen in Fig. 5.5. On the other hand, the initial bubble growth rate by SC exhibits a temperature dependence. An increase of temperature notably increases the initial bubble growth rate of SC. Consequently, this phenomenon brings about a significant difference in resultant cellular foam morphology.

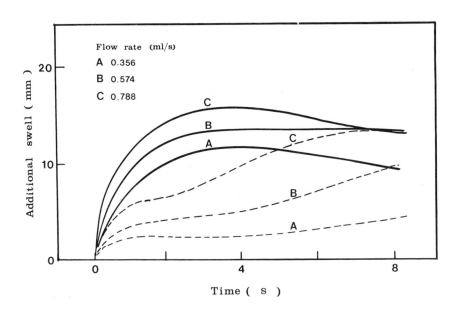

Figure 5.5 *Axial extrudate swell profiles of P-AZ (0.75 phr) (solid line) and P-SC (0.75 phr) (broken line) for three volumetric flow rates at 170°C.*

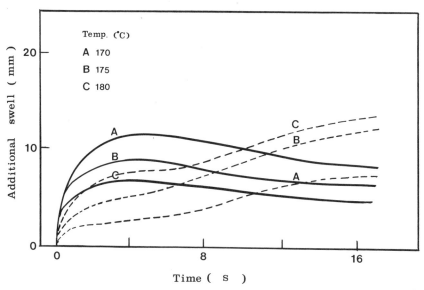

Figure 5.6 *Axial extrudate swell profiles of P-AZ (0.75 phr) (solid line) and P-SC (0.75 phr) (broken line) at three extrusion temperatures.*

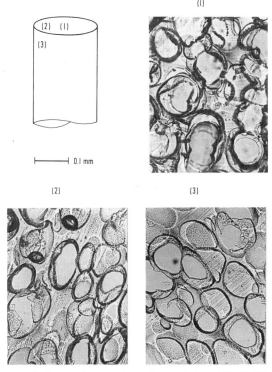

Figure 5.7 *Cellular morphology of extrudate foamed by AZ (0.75 phr) with density of 400 kg/m³: A, core, B, middle, and C, shell part of a rodlike foam product.*

Figures 5.7 and 5.8 represent photomicrographs of the foamed cellular morphology of rod-shaped extrudates by two chemical blowing agents [35]. For a fair comparison, foams with similar density (390 and 400 Kg/m^3) were chosen. AZ apparently yields foamed product with more uniform cells than SC. The result indicated that any processing condition leading to higher initial bubble growth rate produces more uniform cells. However, from a practical point of view, a blowing agent causing slow initial bubble growth rate (e.g., SC) may be advantageous for structural foam processing due to ease of controlling the foam morphology and foaming rate.

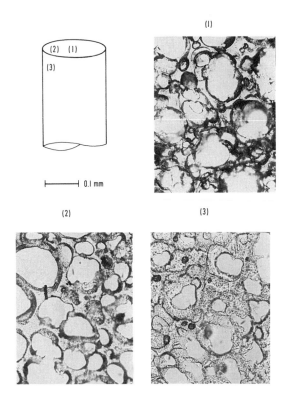

Figure 5.8 *Cellular morphology of extrudate foamed by SC (0.75 phr) with density of 390 kg/m^3: A, core, B, middle, and C, shell part of a rodlike foam product.*

As one may expect, foamed cellular morphology influences the mechanical properties of the product. A foamed product with spherical closed cells possesses greater load carrying capability because of more uniform stress distribution in the configuration than that with flattened or open ones [19,65]. Therefore, AZ produces foams with better mechanical properties than SC. This is shown in Figs. 5.9 and 5.10 [35].

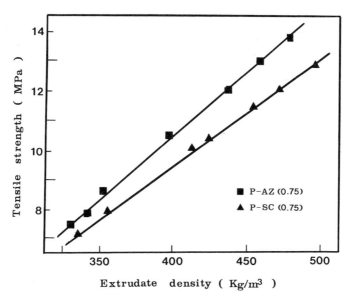

Figure 5.9 *Tensile strength vs. extrudate density for foamed product (formulation contained 0.75 phr boric acid as nucleating agent).*

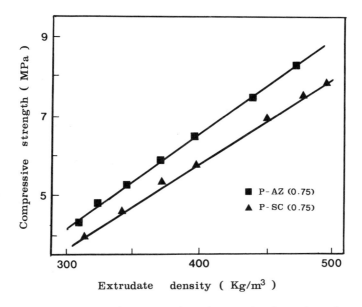

Figure 5.10 *Compressive strength vs. extrudate density for foamed product (formulation contained 0.75 phr boric acid as nucleating agent).*

When a mixture of two chemical blowing agents is used, intermediate morphology and mechanical properties are obtained [34,35] (See the material consideration section in the Appendix). The thermal decomposition mode of AZ and SC, and the physical properties of gaseous products evolved from them are given in detail [66-69]). For further discussion on the interrelationships between morphology and properties see Chapter 4.

5.4   RHEOLOGY OF GAS-CHARGED POLYMER SYSTEMS

5.4.1   General Background

Rheological properties of gas-charged polymer systems may provide useful information on bubble dynamics during processing. This information is important for better quality control of foamed products and for development of new foaming processes. However, rheological measurement of foaming systems was not feasible until a closed-type rheometer came into use in the early 1970s. Due to gas escape, the conventional rotational-type rheometers, such as cone-and-plate and parallel-plate, are unable to measure the rheological properties of gas-charged polymer systems. Only closed-type instruments, such as capillary and slit rheometers, are suitable. Some have questioned the validity of these rheometer types due to inherent pressure-hole error [70-74]. However, Han and Kim [75,76] have argued that the pressure hole-error is negligible, especially for flow of molten polymers. For gas-charged polymer systems, Bigg et al. [42] reported that the measured viscosity varied with the die geometry, but this effect may have originated in the bubble formation during the rheological measurement. In other words, the dynamic state of bubble forming may prevent the achievement of steady state for rheological measurements. The measuring technique can easily lead to an erroneous result if validity of the technique and state of the sample are not critically controlled. Several precautions are advisable to obtain a pressure profile of mixtures of molten polymer and blowing agent: 1) use a capillary or slit die with sufficiently large L/D or L/h ratio (at least over 12) [77], 2) employ a sufficiently large reservoir (reservoir diameter to capillary ratio should be at least over 12) [77], 3) allow sufficient time for steady-state flow before the measurements, and 4) opt for operation conditions that effectively suppress the bubble growth. Although there has been little theoretical analysis of the foam extrusion process, it has been recognized that the bubble dynamics is influenced by the type and amount of blowing agent, the type and amount of additives, and the processing conditions.

5.4.2   Pressure Profile

Several phenomena have been reported for flow of molten polymers containing blowing agent along a long capillary die: 1) wall normal stress is substantially reduced in comparison to that of polymer without a blowing agent [40-43,78-80], 2) deviation of axial pressure profiles from linearity was noticed, particularly near the die exit [40,78-80], and 3) a greater entrance pressure drop than that of the melt without a blowing agent was observed [40,78-80]. All these phenomena can be related to bubble nucleation and growth. Blyler and Kwei [41] suggested that lubrication by bubbles may reduce the wall normal stress. Han et al. [78] attributed the greater entrance pressure drop to the possible evolution of the dissolved gas from the molten polymer. Relatively large bubbles may be formed in the reservoir rather than in the die, particularly at low shear rates and at high temperatures. Presumably, these bubbles may considerably disrupt the converging flow at the die entrance, which results in excessive entrance-pressure losses.

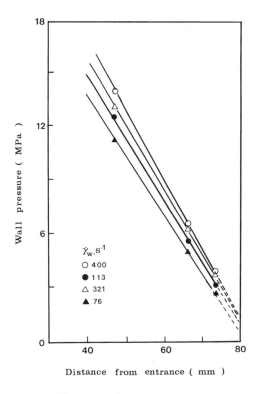

Figure 5.11 *Axial pressure profiles for P-base at various shear rates at 160°C (circle) and 170°C (triangle).*

Figures 5.11 through 5.13 give axial wall normal stress of PVC (P-base), PVC containing 0.75 phr AZ, and PVC containing 0.75 phr SC in a long cylindrical die (D = 4 mm, L/D = 20, and entrance angle of die = 60°) [35]. Figure 5.14 depicts the detailed layouts of die for the experiment. A careful examination of Figs. 5.11 to 5.13 reveals that two expandable formulations show deviation from a straight line in the vicinity of the die exit, whereas the unexpanded formulation gives a linear profile over the entire range of measured pressures. However, the curvature in the pressure profile decreases as shear rate increases. The shear rate at which the curvature disappears is much lower than the value reported by Han et al. [78] who used high-density polyethylene as polymeric medium. This implies that the bubble growth is effectively suppressed in high-viscosity medium even at low shear rates. Of the two chemical blowing agents tested, AZ gave greater reduction in wall normal stress and started to deviate from a straight line at higher shear rate than SC at the same concentration. Thus, a blowing agent exerting higher blowing pressure forms bubbles at a shorter distance in a die from the entrance at the same operating conditions. As one may surmise, the rheological implication of both reduction in wall normal stress and deviation of pressure profile from the linearity are evidence of phase-separated state, i.e., presence of the growing bubbles.

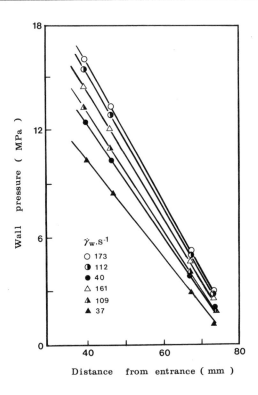

Figure 5.12 *Axial pressure profiles for P-AZ (0.75 phr) at various shear rates at 160°C (circle) and 170°C (triangle).*

Referring to the rheometry, for the steady simple shearing flow, the wall shear stress, σ_w, is calculated from the following expressions [81]:

$$\text{for a capillary flow:} \quad \sigma_w = (-\partial P/\partial z)\, D/4 \qquad\qquad (5.7)$$

and

$$\text{for a slit flow:} \quad \sigma_w = (-\partial P/\partial z)\, h/2 \qquad\qquad (5.8)$$

in which $(-\partial P/\partial z)$ is the pressure gradient along the capillary or slit in the fully developed region; D, the capillary diameter; and h, the slit thickness. The curvature of the pressure profile means that the pressure gradient $(-\partial P/\partial z)$ is not constant. Therefore, one may expect that the flow is no longer steady-state where the curvature exists. Nevertheless, Eqs (5.7) and (5.8) are still applicable.

5.4.3   Viscosity Behavior

In constructing the viscosity curves, two assumptions are inevitable; 1) a complete dissolution of the blowing agent in molten polymer, and 2) no change of polymer density by the dissolved blowing agent. The former can be fulfilled by taking the range of pressures where the pressure

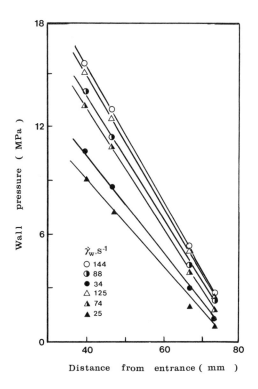

Figure 5.13 *Axial pressure profiles for P-SC (0.75 phr) at various shear rates at 160°C (circle) and 170°C (triangle).*

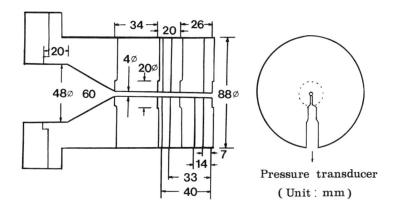

Figure 5.14 *Detailed layouts of reservoir and capillary die used for the rheological measurement.*

gradient $(-\partial P/\partial z)$ is constant. There are many reports that inclusion of a blowing agent obviously decreases both wall shear stress and viscosity [40-43,78-80]. Han and Villamizar [40] found that the dissolved gas decreased the viscosity of the system. They tried to explain the effect by means of the Doolittle's viscosity free volume equation [82]. They also reported that there was a temperature range where the viscosity was independent of the extrusion temperature, and that the shear sensitivity of viscosity increased with the chemical blowing agent content.

For a steady simple shearing flow, the power-law index (n) is defined by

$$n = d\ln \sigma_w/d\ln \dot{Q} = d\ln \sigma_w/d\ln \dot{\gamma}_a = d\ln \sigma_w/d\ln \dot{\gamma}_w \qquad (5.9)$$

where $\dot{Q}$ is the volumetric flow rate; $\dot{\gamma}_a$, the apparent shear rate; and $\dot{\gamma}_w$, the wall shear rate. The power-law index of unexpanded and various expandable PVC formulations at three different temperatures is shown in Table 5.1. Several facts are worth noting in Table 5.1: 1) addition of chemical blowing agent increases the power-law index of the PVC formulations (of the two agents, AZ gives greater increase in power-law index than SC at the same usage level), 2) with expandable formulations, the power-law index increases as the chemical blowing agent level increases, 3) an increase in temperature further increases the power-law index at the same chemical blowing agent level, and 4) the power-law index of expandable formulations containing AZ rapidly increases with temperature while the effect of temperature is less prominent in the SC system.

Table 5.1 *Power-law Index of PVC and Various Expandable PVC Formulations*

Formulation	Temperature (°C):	160	170	180
PVC		0.05	0.06	0.08
PVC with 0.5phr ADCA		0.19	0.24	0.31
PVC with 0.75phr ADCA		0.20	0.26	0.33
PVC with 1.0phr ADCA		0.22	0.29	----
PVC with 0.5phr SBC		0.14	0.16	0.20
PVC with 0.75phr SBC		0.14	0.16	0.21
PVC with 1.0phr SBC		0.15	0.17	0.23

The power-law describes the viscosity as [81]:

$$\eta = K \dot{\gamma}^{n-1} \qquad (5.10)$$

Where η is viscosity and K is a constant. Most polymeric materials have a power-law index less than unity, indicating pseudoplastic or shear thinning behavior.

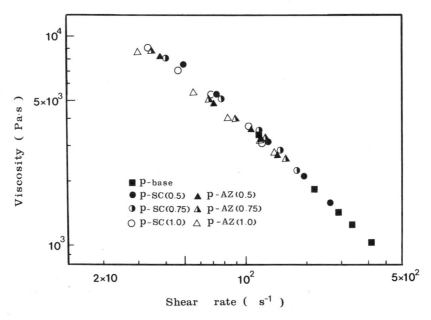

Figure 5.15 *Viscosity vs. shear rate for various PVC formulations measured by a long capillary die (L/D = 20) at 160°C.*

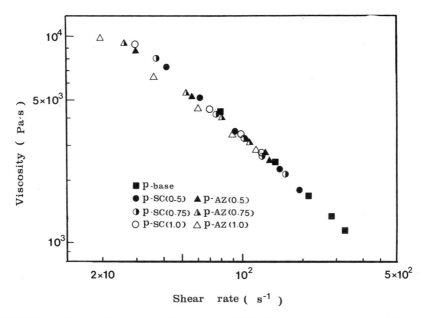

Figure 5.16 *Viscosity vs. shear rate for various PVC formulations measured by a long capillary die (L/D = 20) at 170°C.*

Figures 5.15 and 5.16 show viscosity curves of unexpanded and various expandable PVC formulations at two temperatures [35]. In determining the viscosity of expandable formulations, only the linear portion of the pressure profile was taken. The viscosity was calculated by [81]:

$$\eta = \sigma_w / \dot{\gamma}_w \qquad (5.11)$$

As may be seen from Figs. 5.15 and 5.16, inclusion of chemical blowing agent lowers the viscosity at shear rate below the critical value. This tendency was also observed with changes in the level of chemical blowing agent. Above the critical shear rate (ca. 100 sec^{-1} in this case) all the viscosity data appear to fall onto a single viscosity curve where the external pressure may be in equilibrium with the critical blowing pressure of the blowing agent. These results are somewhat different from those obtained for polyethylene and polystyrene foams [83].

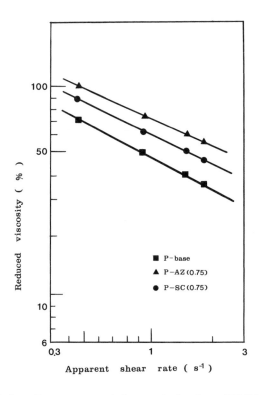

Figure 5.17 *Reduced viscosity vs. apparent shear rate for three PVC formulations measured by a short capillary die (L/D = 4) at 170° C.*

Han and Ma [83] reported that substantial viscosity reduction was invariant over the observed range of shear rates at a given temperature. The origin of this disagreement is still ambiguous. However, it can be imagined that medium viscosity may be a decisive factor in this discrepancy. Polyethylene or polystyrene melt, due to relatively low melt viscosity, may not produce external pressure high enough to suppress the bubble nucleation. Therefore, the result of Han and Ma [83] may be phenomenologically interpreted as follows. The resultant bubble size may remain almost constant over the shear rate range because the bubble growth is counterbalanced by the decrease of medium viscosity. Thus, on increasing shear rate the bubble growth is suppressed by the increase of external pressure. On the other hand, at the normal processing temperature range, PVC has much higher viscosity than polyethylene or polystyrene. Therefore, the external pressure exerted by PVC in the capillary exceeds the critical blowing pressure of two chemical blowing agents above the shear rate of 100 sec^{-1}. Consequently, the bubble nucleation may be completely suppressed above the critical shear rate. In addition, of the two chemical blowing agents, AZ produces a greater number of bubbles than SC due to higher blowing pressure. Therefore, at the same concentration AZ reduces viscosity more than SC.

The bubble formation process depends on the die geometry [42]. If bubbles are formed in the die, the rheological properties of gas-charged polymer are expected to exhibit a geometry dependence. The apparent viscosity, measured by using a short cylindrical die with a large opening (D = 7.5 mm, L = 30 mm, and converging angle of die = 60°), is shown in Figs. 5.17 and 5.18 [35].

From Figs. 5.17 and 5.18, two facts are worth mentioning: 1) reduction of viscosity by bubble formation in the short die is far greater than in the long capillary die, and 2) of the two chemical blowing agents, the effect of extrusion temperature was more prominent with AZ than with SC.

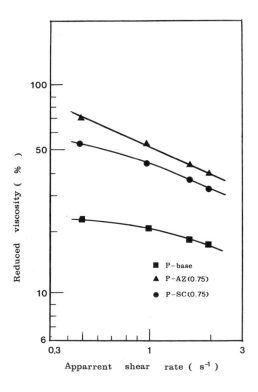

Figure 5.18 *Reduced viscosity vs. apparent shear rate for three PVC formulations measured by a short capillary die (L/D = 4) at 180°C.*

To correlate this viscosity with the screw torque, two parameters, torque reduction factor, $\Delta\tau_f$, and viscosity reduction factor, $\Delta\eta_f$, are defined as follows:

$$\Delta\tau_f = (\tau_f/\tau - 1)\,100 \tag{5.12}$$

$$\Delta\eta_f = (\eta_f/\eta - 1)\,100 \tag{5.13}$$

where τ_f and τ are torque of PVC with and without the chemical blowing agent, respectively, and η_f and η, viscosity of PVC with and without the chemical blowing agent, respectively. The results are given in Table 5.2. There is no doubt that $\Delta\tau_f$ is a measure of bubble nucleation in the extruder barrel and $\Delta\eta_f$ represents the bubble growth in the die. Table 5.2 also indicates that the torque of the expandable formulation containing SC approaches that of AZ. On the other hand, the viscosity of the expandable formulation containing SC is nearer to that of the unexpanded formulation. This indicates that the bubble formation process in a given die geometry greatly depends on the type of the chemical blowing agent and the processing conditions. AZ decomposes explosively at a narrow range of temperatures and produces relatively inert gases while the decomposition of SC is pressure dependent and yields gases partially soluble in PVC [66-69]. Consequently, AZ gives more abrupt phase separation by higher blowing pressure than SC.

Table 5.2 *Torque Reduction Factor (TRF) and Viscosity Reduction Factor (VRF) of Two Expandable PVC Formulations Containing 0.75phr Chemical Blowing Agent*

| | Parameter: | | TRF(%) | | | VRF(%) | |
	Temp. (°C):	170	175	180	170	175	180
Formulation	Q^a						
	2	63	66	72	28	66	69
PVC containing	4	51	54	61	28	53	61
0.75phr ADCA	6	50	49	54	30	48	58
	8	--	--	--	33	49	60
	2	46	56	65	11	18	22
PVC containing	4	38	46	54	9	13	18
0.75phr SBC	6	33	41	49	10	15	16
	8	--	--	--	12	15	18

a *Volumetric flow rate (ml/min.).*

Comparison of the viscosity measured in a die of small L/D ratio (Figs. 5.17 and 5.18) with that in one of large L/D ratio (Figs. 5.15 and 5.16) reveals that the effect of blowing agent type gradually vanishes with increasing L/D ratio. This may indicate that the critical shear rate of bubble inflation decreases with increasing L/D. Therefore, in order to eliminate the geometry dependency, longer capillary or slit die is preferred for the rheological measurements of gas-charged polymer systems.

5.4.4   Elastic Properties

There are many theoretical and experimental reports on elastic properties of gas-charged systems [44-50,53-62]. A general agreement is that an increase of medium elasticity enhances the bubble growth in the foaming processes. Therefore, a better understanding of elastic behavior will undoubtedly lead to a better operation of the foaming processing. For steady simple shearing flow along the capillary or slit rheometer, two elastic material functions, first normal stress difference (N_1) and, second normal stress difference (N_2), may be given by [81]:

$$N_1 = \sigma_{11} - \sigma_{22} = P_{exit} + \sigma_w \, dP_{exit}/d\sigma_w \qquad (5.14)$$

$$N_2 = \sigma_{22} - \sigma_{33} = -\sigma_w \, dP_{exit}/d\sigma_w \qquad (5.15)$$

where σ_{ij} represents the components of the deviatoric stress tensor; P_{exit} the exit-pressure; and σ_w, the wall shear stress. From Eqs (5.14) and (5.15), one can see that precise measurement of the exit pressure is a prerequisite for determining elastic properties accurately. However, owing to the pressure dependence of viscosity, the pressure profile of gas-charged polymer systems may deviate from a straight line as the melt approaches the die exit. Extrapolation of this curved pressure profile to the die exit often leads to negative values of exit pressure. According to Eqs (5.14) and (5.15), negative exit pressure does not have any meaning at all. Besides pressure reduction by bubble formation, one may suspect a pressure-hole error during measurement. However, if there is any effect of pressure-hole error, it must be a minor one because the slit die with flush-mounted transducers (without pressure-hole) gives essentially the same result [75,76]. It is our opinion that unsteady dynamic state exists near the die exit so that exact pressure measurement is almost impossible. Therefore, the negative exit pressure must be due to compressibility of the foamed melt. Sizable extrudate swell was observed even with these negative exit pressure systems.

Figure 5.19 shows N_1 for the unexpanded PVC formulation (P-base) at temperatures of 160°C to 190°C [35]. Although there is some data scattering, one may conclude that N_1 is not affected by temperature as demonstrated by others [81,84,85].

With the expandable formulations, however, the unsteady-state near die exit (hence nonlinear pressure profile) prevents use of Eqs (5.14) and (5.15) for the elastic property. Furthermore, as already mentioned, the conventional open-type rheometer is not suitable for the measurement. Thus, the elasticity of foaming systems should be estimated by other means. One such method can be, at least qualitatively, the extrudate swell as illustrated by Figs. 5.5 and 5.6.

5.5   RHEOLOGICAL POTENTIAL FOR PRODUCT DESIGN

An understanding of rheological behavior of polymeric material offers an essential clue to equipment design and product quality control [23,83]. In other words, rheological characterization of the foaming process provides not only valuable information on the bubble dynamics during processing, but also on the effect of the type and amount of blowing agent, type and amount of additives, and on the optimum processing conditions. However, up to now, the foam rheology has not fully contributed to the progress of foam technology, failing to give an adequate answer to problems encountered in actual processing. Recent attempts of rheological studies combined with bubble dynamics established a basic principle for foam rheology. This is mainly due to the work of Han and Villamizar [40,48]. Other attempts to measure the criteria for growth/collapse of bubbles and the critical blowing pressure of blowing agent have also been performed [55,56,80,86]. So far, some qualitative results have been reported, but little quantitative correlation was given to bubble dynamics.

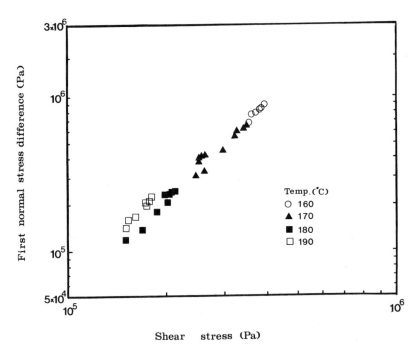

Figure 5.19 *First normal stress difference vs. shear stress for P-base at various temperatures measured by a long capillary die (L/D = 20).*

Han and Ma [83] suggest that a judicious selection of die geometry, processing conditions, and type and concentration of blowing agent is prerequisite for the production of good quality foams. Furthermore, they have pointed out that either a decrease in the die pressure or an increase of its temperature gives rise to premature foaming (inside the die). Consequently, extruded foams contain a high fraction of open cells. When the die temperature is increased above a critical value, the cells collapse and the foam density dramatically increases. The authors also proposed an experimental technique for selection of resins for foam extrusion and the guidelines for selecting optimum die geometry.

For the sandwich foam coextrusion process, Shetty and Han [79] reported that the size and size distribution of cells in the foamed core and the mechanical properties of the sandwich foam product are controlled by such factors as core-to-skin thickness ratio, melt extrusion temperature, and concentration of chemical blowing agent. As it may be surmised, the problem at hand is more complex in the coextrusion process. Additional study is needed to clarify the relation between processing, morphology, and property.

PVC structural foam has been studied with a conventional single-screw extruder by use of chemical blowing recipes [34,35,87,88]. Kim et al. [88] found that simultaneous incorporation of dioctylphthalate (DOP; 3 to 8 phr) and terpolymer of methylmethacrylate-butadiene-styrene (MBS 50/25/25; 3 to 8 phr) into the expandable PVC formulation promoted the integral skin formation. They attributed this to the retarded bubble growth. That is, DOP reduces the initial bubble growth rate by decreasing the medium elasticity and the terpolymer further diminishes it by increasing medium viscosity. An antiplasticization followed by the migration of less viscous materials such as DOP to the shell part is also considered to be helpful for the skin formation [88-92]. Therefore, a blowing agent showing lower initial bubble growth rate may produce PVC structural foam with thicker glossy skin than a blowing agent exhibiting higher initial bubble growth rate. This structural foam process possesses some merits over conventional structural foaming either by coextrusion or by a specially designed (split-line) die to give the skin layer. For example, this process requires neither specially designed die parts nor modification of the conventional single-screw extruder. In addition, the skin thickness is easily controllable by varying either processing parameters or formulations.

5.6 RECOMMENDATION FOR FURTHER STUDY

The literature on foam rheology agrees that inclusion of blowing agents substantially decreases both wall normal stress and the steady state shear viscosity. Most results indicate that the viscosity reduction by blowing agent is almost invariant over the examined range of shear rates at a given temperature. However, somewhat different results were reported for the viscosity behavior of rigid foam extrusion of highly viscous PVC. The degree of viscosity reduction was notably decreased as shear rate increased. Above a critical shear rate, all the viscosity curves of PVC with a chemical blowing agent converged to the viscosity curve of PVC at a given extrusion temperature.

In the foam extrusion process, it is generally accepted that fewer bubbles in the die enhance uniformity of resultant cells. For this, the foam processor may have to choose as low a die temperature as possible as well as high extrusion rates. This rule cannot account for the frequent formation of large voids at the center of the foamed product, as in the chemical foaming process of rigid PVC with SC at low die temperature and at high extrusion rate. Thus, the die temperature should be carefully selected in relation to melt viscosity, type of blowing agent, diffusivity and solubility of gas, die geometry, and extrusion rate.

Many foam processors suffer from lack of knowledge of the rheological principles of foam processing. First of all, reproducibility and reliability of the rheological properties remain controversial. To rule out this controversy, accurate rheological measurement and theoretical modeling of foam extrusion should be carried out. However, the accuracy of measurements suffers from equipment limitation. In practical terms, rheological equipment may not always reach sufficiently high pressure to suppress the bubble nucleation. Furthermore, knowledge of thermodynamic and transport properties of molten polymer and blowing agent mixtures, the phenomena of bubble nucleation and bubble growth, and the mass- and heat-transfer operations involving gas-liquid and gas-solid systems are still lacking. Therefore, further rheological study should be performed on a basis of bubble dynamics in order to develop firm and reliable principles for solving practical problems in rigid foam processing.

APPENDIX: MATERIAL CONSIDERATION

A suspension grade polyvinylchloride (PVC) has a number average molecular weight $M_n = 67 \pm 3.5$ kg/mol and a barium-cadmium stearate as thermal stabilizer. The basic unexpanded PVC formulation (P-base) was (in parts per hundred resin):

PVC (DP = $1,000 \pm 50$)	100
Ba-Cd type thermal stabilizer	2
Acryloid K-125 (Rohm & Haas Co., U.S.A.)	15
Stearic acid	0.5

Two chemical blowing agents, azodicarbonamide (AZ) and sodium bicarbonate (SC), showed almost opposite decomposition mode by heating [66,67]. Some distinctive features of AZ are: temperature sensitivity and explosive, irreversible, and exothermic decomposition. On the other hand, the important features of SC are pressure-dependent, slow, erratic, reversible, and endothermic decomposition. In addition, these two chemical blowing agents yield gases with very different physical properties. The gases evolved from AZ are reported to be: nitrogen (56 vol.%), ammonia (21 vol.%), carbon monoxide (15 vol.%), and carbon dioxide (8 vol.%), while SC generates carbon dioxide and steam. The nitrogen gas, a major product of AZ, has a relatively low critical temperature and low solubility in PVC, while the major gas from SC, carbon dioxide, has a relatively high critical temperature and is more soluble in PVC than nitrogen as shown below.

The inherently poor thermal stability and the relatively high viscosity of PVC during processing require various additives. These include plasticizer, stabilizer (thermal and light), lubricant, impact modifier, and processing aid. The expandable formulation can also contain blowing agent, nucleating agent, and filler. These additives undoubtedly affect the rheological properties, both in processing as well as of foamed products. This chapter discussed only the blowing agent on processing conditions. Understanding the effects of other additives on rheological and physical properties of PVC products is at present far from complete since only limited information is available [15,38,39]. Recently, Kim et al. [88] examined the effects of a processing aid (Acryloid K-125), a plasticizer (DOP), and an impact modifier (a terpolymer of methylmethacrylate-butadiene-styrene) on rheological and mechanical properties of PVC formulations with and without the chemical blowing agent. In an unexpanded formulation, the plasticizer DOP reduced the viscosity but no significant change in N_1 was noticed. The impact modifier increased the viscosity with no noticeable change in N_1. However, both viscosity and N_1 were increased by adding the processing aid (up to 15 phr).

Table A. *Physical Properties of Gases Evolved from AZ and SC [68,69]*

Gas	Critical temperature (°C)	Critical pressure (MPa)	Solubility in PVC[a]
N_2	−147.1	3.28	0.24
NH_3	132.4	10.93	−
CO	−139.1	3.43	−
CO_2	31.3	7.15	0.48
H_2O	374.2	21.37	−

[a] *Gas solubility at 298 K in m^3 (STP)/$m^3 MPa$.*

NOTATION

AZ	Azodicarbonamide or azobisformamide (ABFA)
DOP	Dioctyl phthalate (di-2-ethylhexyl phthalate)
phr	Part per hundred resin
PVC	Polyvinylchloride
SC	Sodium bicarbonate
A,B	Constants in Eq (5.3)
C_0	Solute concentration far from the bubble wall
C_w	Solute concentration outside the bubble wall
D	Diameter of capillary die or extruder screw
D_f	Diffusion coefficient in Eq (5.6)
h	Slit thickness
K	Constant in the power-law equation
k	Constant in Eq (5.2)
k'	Constant in Eq (5.4)
L	Length of die or extruder screw
N_1	First normal stress difference ($= \sigma_{11} - \sigma_{22}$)
N_2	Second normal stress difference ($= \sigma_{22} - \sigma_{33}$)
P	Isotropic pressure
P_{exit}	Exit pressure
P_g	Pressure inside bubble
P_p	Pressure of polymer phase
$\dot{Q}$	Volumetric flow rate
R	Radius of capillary die
r_b	Growing bubble radius
r_0	Initial bubble radius at nucleation
t	Time
$\dot{\gamma}$	Shear rate
$\dot{\gamma}_a$	Apparent shear rate
$\dot{\gamma}_w$	Wall shear rate
η	Shear viscosity
η_f	Shear viscosity of foamed polymer
η_0	Zero-shear viscosity
$\Delta\eta_f$	Viscosity reduction factor
λ	Relaxation time
ν	Surface tension
ρ	Density of molten polymer
ρ_g	Density of gas in the bubble
σ_{ij}	ij-th component of stress tensor
σ_w	Wall shear stress
τ	Torque of polymer phase
τ_f	Torque with blowing agent
$\Delta\tau_f$	Torque reduction factor

REFERENCES

1.   Goggin, W. C., McIntire, O. R., *Br. Plast.*, **19**(223), 528 (1947).
2.   Madge, E. W.: *Latex Foam Rubber*, John Wiley & Sons, New York (1962).
3.   Cooper, A., *Plast. Inst. Trans. J.*, **29**, 39 (1961).
4.   Ingram, A. R., *J. Cellular Plast.*, **1**, 69 (1965).
5.   Fields, R. T., *U.S. Pat.*, 3,058,166 (1962).
6.   Skochdopale, R. E., Rubens, L. C., Jones, G. D., *U.S. Pat.*, 3,062,729 (1962).
7.   Saunders, J. H., Frisch, K. C., *Polyurethanes, Chemistry and Technology*, Vol.II., Interscience Publishers, New York (1964).
8.   Lever, A. E., *Plastics* (London), **18**, 274 (1953).
9.   Frisch, K. C., Patel, K. J., Marsh, R. D., *J. Cellular Plast.*, **6**, 203 (1970).
10. Nicholas, L., Gmitter, G. T., *J. Cellular Plast.*, **1**, 85 (1965).
11. Beyer, C. E., Dahl, R. B., *U.S. Pat.*, 3,058,161 (1962).
12. Hansen, R. H., *SPE J.*, **18**, 77 (1962).
13. Lasman, H. R., *Mod. Plast.*, **42**, 314 (1964).
14. Weissenfeld, H., *Kunststoffe*, **51**, 698 (1961).
15. Shutov, F. A., *Integral/Structural Polymer Foams*, Springer Verlag, Berlin (1986).
16. Valgin, V. C., *Europlast. Mon.*, **46**(7), 57 (1973).
17. Throne, J. L., *J. Cellular Plast.*, **12**, 264 (1976).
18. Kolb, J. J., in *Engineering Guide to Structural Foam*, Wendle, B. C., Ed., Technomic Publishing Co., Westport, CT (1976).
19. Frisch, K. C., Saunders, J. H., *Plastic Foams*, Marcel Dekker, New York (1972, 1973).
20. Thomas, J. R., *Plast. Eng.*, **37**(1), 33 (1981).
21. Ahnemiller, J., *J. Plast. Eng.*, **39**(2), 21 (1983).
22. Peach, N., *Plast. Eng.*, **40**(8), 19 (1984).
23. Yang, H. H., Han, C. D., *J. Appl. Polym. Sci.*, **30**, 3297 (1985).
24. Zizlsperger, J., *French Pat.*, 1,446,187 (1966).
25. Zweigle, M. L., Humbert, W. E., *U.S. Pat.*, 3,066,382 (1962).
26. Toray Industries, Inc., *Br. Pat.*, 1,333,392 (1973).
27. Boutiller, P. E., *French Pat.*, 1,498,620 (1967).
28. Tann, D., *U.S. Pat.*, 3,268,638(1966); *U.S. Pat.*, 3,436,446 (1969).
29. Annis, Jr., R. E., Salem, W. J., Kyritsis, W. T., *U.S. Pat.*, 3,674,410 (1973); Kyritsis, W. T., Farms, B., *U.S. Pat.*, 3,697,204 (1973); Annis, Jr., R. E., Salem, W. J., Kyritsis, W. T., *U.S. Pat.*, 3,776,989 (1974).
30. Berlin, A. A., Shutov, F. A., *Strengthened Gas−Filled Polymers*, Khimia, Moscow (1980).
31. Mitsubishi Petrochemical Company, Ltd., *Mitsubishi Injection Compression System*, Techn. Bull. (1969).
32. Tunbridge, T., *Europlast. Mon.*, **46**(7), 81 (1973).
33. Garner, P. J., *U.S. Pat.*, 3,690,797 (1973); Nyquist, S. E., *U.S. Pat.*, 3,773,156 (1974).
34. Kim, K. U., Park, T. S., Kim, B. C., *J. Polym. Eng.*, **7**, 1 (1986).
35. Kim, B. C., Kim, K. U., Hong, S. I., *Polymer* (Korea), **10**, 143 (1986); ibid, **10**, 215 (1986); ibid, **10**, 324 (1986).
36. Berens, A. R., Folt, V. L., *Trans. Soc. Rheol.*, **11**, 95 (1967); *Polym. Eng. Sci.*, **8**, 5 (1968).
37. Collins, E. A., Metzger, A. P., *Polym. Eng. Sci.*, **10**, 64 (1970).
38. Batiuk, M., in *Encyclopedia of PVC*, Vol. 3, Nass. L. I., Ed., Marcel Dekker, New York (1977).

39. Shutov, F. A., *Adv. Polym. Sci.*, **51**, 155 (1983).
40. Han, C. D., Villamizar, C. A., *Polym. Eng. Sci.*, **18**, 587 (1978).
41. Blyler, L. L., Kwei, T. K., *J. Polym. Sci.*, Part C, **35**, 165 (1971).
42. Bigg, D. M., Preston, J. R., Brenner, D., *Polym. Eng. Sci.*, **16**, 706 (1976).
43. Oyanagi, Y., White, J. L., *J. Appl. Polym. Sci.*, **23**, 1913 (1979).
44. Epstein, P. S., Plesset, M. S., *J. Chem. Phys.*, **18**, 1505 (1950).
45. Stewart, C. W., *J. Polym. Sci.*, Part A-2, **8**, 937 (1979).
46. Gent, A. N., Tompkins, D. A., *J. Appl. Phys.*, **40**, 2520 (1969).
47. Hobbs, S. Y., *Polym. Eng. Sci.*, **16**, 270 (1976).
48. Villamizar, C. A., Han, C. D., *Polym. Eng. Sci.*, **18**, 699 (1978).
49. Street, J. R., *Trans. Soc. Rheol.*, **12**, 103 (1968).
50. Han, C. D., Yoo, H. J., *Polym. Eng. Sci.*, **21**, 518 (1981).
51. Bird, R. B., Armstrong, R. C., Hassager, O., *Dynamics of Polymeric Liquids*, Vol. 1, John Wiley & Sons, New York (1977).
52. Kim, K. U., Kim, B. C., Hwang, E. J., *Kor. J. Rheol.*, submitted
53. Lundberg, J. L., Wilk, M. B., Huyett, J., *J. Polym. Sci.*, **57**, 275 (1962).
54. Lundberg, J. L. Mooney, E. J., Rogers, C. E., *J. Polym. Sci.*, Part A-2, **7**, 947 (1969).
55. Durrill, P. L., Griskey, R. G., *AIChE J.*, **12**, 1147 (1966).
56. Durrill, P. L., Griskey, R. G., *AIChE J.*, **15**, 106 (1969).
57. Maloney, D. P., Prausnitz, J. M., *AIChE J.*, **22**, 74 (1976).
58. Stiel, L. I., Harnish, D. F., *AIChE J.*, **22**, 117 (1976).
59. Cheng, Y. L, Bonner, D. C., *J. Polym. Sci., Polym. Phys. Ed.*, **15**, 593 (1977).
60. Bonner, D. C., *Polym. Eng. Sci.*, **17**, 65 (1977).
61. Duda, J. L., Vrentas, J. S., *J. Polym. Sci.*, Part A-2, **6**, 675 (1968).
62. Throne, J. L., *J. Cellular Plast.*, **12**, 161 (1976); ibid, **12**, 264 (1976).
63. Hansen, R. H., Martin, W. M., *Ind. Eng. Chem. Res. Div.*, **3**, 137 (1964); *J. Polym. Sci.*, Part B, **3**, 325 (1965).
64. Yang, H. H., Han, C. D., *J. Appl. Polym. Sci.*, **29**, 4465 (1984).
65. Gemeinhardt, P. G., Backus, J. K., Darr, W. C., Szabat, J. F., *J. Cellular Plast.*, **3**, 210 (1967).
66. Throne, J. L., *Polym. Eng. Sci.*, **15**, 747 (1975).
67. Thomas, D. C., *Foamed Plastics*, Stanford Research Institute, Menlo Park, CA, p. 57-64 (1975).
68. Van Krevelen, D. W., Hoftyzer, P. J., *Properties of Polymers: Their Estimation and Correlation with Chemical Structure*, Elsevier/North-Holland, New York (1976).
69. Windholz, M., Budavari, S., Blumetti, R. F., Otterbein, E. S., *The Merck Index*, Tenth Ed., Merck & Co., Rahway (1983).
70. Tanner, R. I., Pipkin, A. C., *Trans. Soc. Rheol.*, **13**, 471 (1969).
71. Lodge, A. S., Higashitani, K., *Trans. Soc. Rheol.*, **19**, 307 (1975).
72. Higashitani, K., Pritchard, W. G., *Trans. Soc. Rheol.*, **16**, 687 (1972).
73. Baird, D. G., *J. Appl. Polym. Sci.*, **20**, 3155 (1976).
74. Lodge, A. S., *Polym. News*, **9**, 242 (1984).
75. Han, C. D., *AIChE J.*, **18**, 116 (1972).
76. Han, C. D., Kim, K. U., *Trans. Soc. Rheol.*, **17**, 151 (1973).
77. Han, C. D., Kim, K. U., *Polym. Eng. Sci.*, **11**, 395 (1971).
78. Han, C. D., Kim, Y. W., Malhotra, K. D., *J. Appl. Polym. Sci.*, **20**, 1583 (1976).
79. Shetty, R., Han, C. D., *J. Appl. Polym. Sci.*, **22**, 2573 (1978).
80. Kraynik, A. M., *Polym. Eng. Sci.*, **21**, 80 (1981).
81. Han, C. D., in *Rheology in Polymer Processing*, Academic Press, New York (1976).

82.   Doolittle, A. K., *J. Appl. Phys.*, **22**, 1471 (1951).
83.   Han, C. D., Ma, C. Y., *J. Appl. Polym. Sci.*, **28**, 831 (1983); ibid, **28**, 851 (1983); ibid,
      28, 2961 (1983); ibid, **28**, 2983 (1983).
84.   Han, C. D., Jhon, M. S., *J. Appl. Polym. Sci.*, **32**, 3809 (1986).
85.   Min, K., White, J. L., Fellers, J. E., *J. Appl. Polym. Sci.*, **29**, 2117 (1984).
86.   Throne, J. L., *Polym. Eng. Sci.*, **23**, 354 (1983).
87.   Gotoh, M., *Japan Kokai*, **48**-29, 777 (1973).
88.   Kim, K. U., Kim, B. C., Hong, S. M., Park, S. K., *Int. Polym. Proc.*, **4**, 225 (1989).
89.   Kim, K. U., Hong, S. M., Kim, B.C ., *J. Kor. Soc. Text. Eng. Chem.*, **24**, 29 (1987).
90.   Han, C. D., *J. Appl. Polym. Sci.*, **17**, 1289 (1973).
91.   Lee, B. L., White, J. L., *Trans. Soc. Rheol.*, **18**, 467 (1974).
92.   Han, C. D., *J. Appl. Polym. Sci.*, **19**, 1875 (1975).

CHAPTER 6

FUNDAMENTALS OF MORPHOLOGY FORMATION IN POLYMER BLENDING

by J. J. Elmendorp

Koninklijke/Shell Laboratorium, Amsterdam
Badhuisweg 3
1031 CM Amsterdam
THE NETHERLANDS

and

by A. K. Van der Vegt

Laboratory of Polymer Technology
University of Technology, Delft
Julianalaan 136
2628 BL, Delft
THE NETHERLANDS

In this chapter, we touch upon phenomena that occur on a microscale during polymer blending, discuss fundamentals thereof, and investigate their influence on the morphology resulting from a blending operation. The flow—induced deformation and break—up of droplets in both shear and elongational flow fields, the growth of capillary instabilities in highly deformed dispersed—phase domains, and the phenomenon of flow—induced coalescence will be treated.

It is shown that for low concentration of the dispersed phase and simple flow fields it is possible to predict morphologies using theoretical approaches of the behavior of isolated droplets.

6.1   INTRODUCTION

Recently, a wealth of papers have been published on the subject of polymer blending - the reason being that polymer blends can show properties that significantly exceed those expected from the rule of mixtures. Synergism arises from the fact that these materials often show two-phase behavior, which can result in behavior similar to an in-situ composite [1,2] leading to enhancement of impact and shock resistance. The charm of these materials is that the second phase need not be incorporated in its desired geometrical shape (as does a laborious composite preparation). Its structure originates during the blending step. Therefore, a proper control of the latter can be utilized to tailor-make the morphology and performance of the material for a specific purpose.

A prerequisite is that sufficient insight be available as far as the relation between the original morphology (domain size, domain shape, regions of co-continuity), the process variables (flow fields applied during blending and effect heat treatments), and material parameters (rheological behavior and miscibility) are concerned. This explains why a considerable effort is spent to apply the fundamental knowledge on the micro-behavior of flowing two-phase liquid systems from emulsion science to the complexity of the polymer blends. In this respect, the work of Utracki [3,4], Meijer [5], White [6], Karger-Kocsis [7], Kuleznev [8], and Plochocki [9,10] should be mentioned.

In this chapter we review our earlier work in the field of polymer blending [11-15]. We touch upon the main phenomena that act on a microscale during a polymer-blending operation, discuss some fundamentals thereof, and investigate their influence on the resulting morphology.

6.2   MISCIBILITY

The miscibility of components is perhaps the most important parameter in determining the morphology of a polymer blend. In principle, an unambiguous criterion of the miscibility can be assessed. Miscible systems show unlimited solubility and a zero interfacial tension, therefore a mixture of the components is expected to result in a homogeneous blend. Immiscible systems show a limited mutual solubility and a finite interfacial tension, resulting in a two-phase structure.

In the mechanical blending of molten polymers, however, there are several mechanisms that may disguise the effect of miscibility on the morphology of the blend.

6.2.1 Effect of Process Conditions

Recently, it was shown that the miscibility of components can be enhanced by severe shearing of the system [16,17]. This implies that a blend which would, on the basis of static measurements (interfacial tension, light scattering), appear as an immiscible system, behaves as a miscible one during the blending operation. The morphology will therefore be determined by the balance of solidification rate and phase separation after cessation of blending, rather then by the two-phase hydrodynamics.

6.2.2. Effect of the Length Scale of the Blend

In the initial stage of a blending the components are commonly brought together in the form of pellets. Heating the system and subjecting it to a mechanical deformation in the mixer reduces the length scale of the blend. In immiscible systems, the forces (due to the interfacial tension) being reciprocal to the local curvature of the interfaces, are increased. In miscible blends, on the other hand, mutual diffusion will occur and aid homogenization of the blend.

If the length scale is such that the forces due to interfacial tension are negligible with respect to the hydrodynamic forces (which is usually the case if the domains are above 10 μm), they will hardly influence the flow pattern on a microscale. In miscible blends, however, diffusion is slow due to the length of the polymer chains, and noticeable homogenization during the blending is expected only for the submicron length scale [18].

It can therefore be inferred that in the initial period of blending, or for short term blending, there will be little difference in the morphology of miscible and immiscible blends. Both will appear as two-phase blends with a morphology consisting of the initial pellets that are deformed along the flow lines of the blending device.

Another interesting phenomena that can arise from the slow diffusion is the occurrence of transient interfacial tensions. It was shown by Wu [19] that interfacial tensions in molten polymers are inversely proportional to the thickness of the interphasial region. Therefore, freshly created interfaces will exhibit a higher interfacial tension than those having an equilibrated interface of finite thickness. Care should especially be taken in applying the equilibrium interfacial tension in the initial period of the blending, where interfacial area is created at an enormous rate.

The equilibration period will depend on the thickness of the interphasial region and the rate of diffusion at the interface. In immiscible systems only segments of the polymer chains will rapidly diffuse across the interface. It can be shown [18] that for an immiscible system, having an interfacial thickness below 5 nm [19], the equilibration time is typically less then a second. In miscible systems bulk diffusion will have to take place to reach the equilibrium, homogeneous molecular mixture. Since this is a slow process, transient interfacial tensions can be present.

The existence of such a transient interfacial tension was shown experimentally by Smith et al. [20]. In their experiments the interfacial tension between two miscible silicon oils decayed to zero only after several minutes.

6.3 MICRORHEOLOGICAL PHENOMENA

For sufficiently intensive blending operations, the domains in miscible systems will be reduced to such a size that diffusion is able to homogenize the system and a single phase is obtained.

Immiscible systems will show interfacial-tension-induced phenomena below a certain length scale. In this section some of these phenomena will be briefly mentioned.

6.3.1   Capillary Instabilities

Considering the initial domains, the nearly spherical pellets deformation will result in ribbon like entities which will convert rapidly into cylinder like bodies under the action of the interfacial tension (this reduces the interfacial area). A further reduction of interfacial area can be obtained by the growth of distortions on the liquid cylinder. This is shown in Fig. 6.1 where the interfacial area per unit length is calculated as a function of amplitude of three sinusoidal distortions with different wavelengths, Λ. For $\Lambda < 2\pi R_0$ the distortion increases the interfacial area, meaning that these distortions will be damped, while for $\Lambda > 2\pi R_0$ the growth of the distortion decreases the interfacial area and will finally convert the thread into a row of droplets.

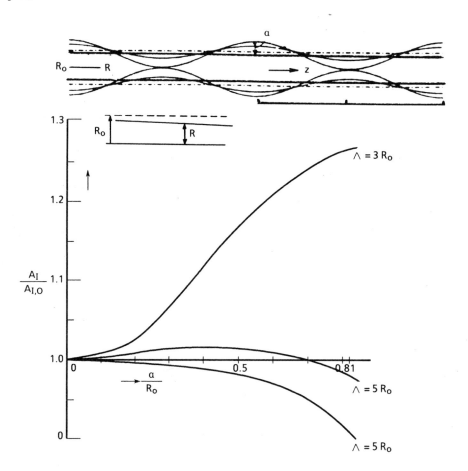

Figure 6.1 *Interfacial area A_I over undisturbed interfacial area $A_{I,0}$ vs. distortion magnitude for various values of the wavelength of distortion.*

The kinetics of the process have received considerable attention after the pioneering work of Rayleigh [21]. For Newtonian systems it can be shown that a sinusoidal distortion (see Fig. 6.1) grows exponentially with time:

$$\alpha = \alpha_0 \exp\{qt\} \tag{6.1}$$

with α_0 the distortion at $t = 0$ and the exponent q:

$$q = (\nu/2\eta_m R_0)\ \Omega(\Lambda, \lambda) \tag{6.2}$$

with ν, the interfacial tension, η_m, the matrix viscosity, η_d, the thread phase viscosity, λ, the viscosity ratio: $\lambda = \eta_d/\eta_m$, and $\Omega(\Lambda, \lambda)$, a tabulated function [22]. The value of Ω exhibits a maximum in Λ, meaning that when initially a spectrum of distortions is present (thermal noise) the dominant distortion will cause break-up. The wavelength of this dominant distortion and its growth rate depend on the viscosity ratio; a graphical representation can be found in [12].

In agreement with the theoretical expectations [23], the break-up of polymer-solution threads is delayed dramatically by strain-hardening effects due to the high elongation that accompanies thread break-up. Figure 6.2 shows an example of the growth of a distortion on a thread of polyacrylamide solution embedded in castor oil. The solid lines are calculated from Eqs (6.1) and (6.2) using the zero-shear rate value for the thread-phase viscosity. It is seen that, above a certain value of the distortion amplitude, the growth rate decreased. This implies that the time to reach break-up (at $\alpha/R_0 = 0.81$) is much longer than expected on the basis of

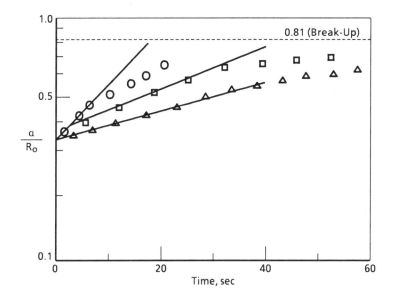

Figure 6.2 *Time-dependent distortion magnitude of a capillary instability growing on threads of a 2% polyacrylamide solution submerged in castor oil [12]. The solid lines were calculated from Eqs (6.2) and (6.3) using the zero shear viscosities of the thread phase. The dotted line represents the distortion magnitude at break-up.*

the Newtonian theory. It is relevant to notice that the onset of these deviations from the Eqs (6.1) and (6.2) can be monitored from the shape of the breaking thread. A distinct bead-string shape is then observed; see for example [24].

When the thread phase exhibits a yield stress exceeding the stresses due to the interfacial tension, the distortions will be unable to grow. A quantitative criterion on the value of the yield stress needed to arrest the growth of distortion was derived and verified experimentally by Elmendorp [12]. This might explain the remarkable ability of some block-copolymers to form continuous phases even in concentrations as low as 5% [25]. These materials are known to exhibit yield stresses in the melt [26].

In a recent paper [12], we reported measurements on the growth rate of capillary instabilities in molten polymer systems. In contrast with the break-up of high polymer solution threads, an exponential growth of the distortions was observed. The stresses induced by the interfacial tension during break-up were apparently too low to induce strain hardening in these medium-molecular-weight polymers. This latter feature was also evident from absence of bead-string shapes during break-up (see Fig. 6.3). Figure 6.4 gives the correlation between experimentally determined and calculated growth rates. The values of interfacial tension required to calculate the theoretical values of the growth rate were determined by means of a modified spinning drop tensiometer [15].

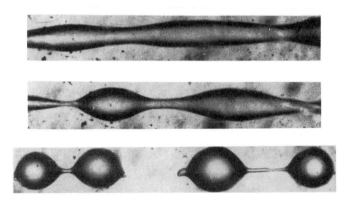

Figure 6.3 *Capillary instability growing on a polyamide-6 thread embedded in polyethylene matrix at 200 °C (after [12]).*

The above theoretical and experimental results are obtained for stationary systems, i.e., no motion of the outer fluid was assumed. Since under blending conditions the system is in motion, it is relevant to investigate the capillary break-up in flowing systems. The early work by Grace [27] revealed that in simple shear flow both dominant wavelength and break-up time are much larger than predicted from the Rayleigh-type considerations. Recently, Plochocki [28] carried out a stability analysis for liquid threads under simple shearing, and accounted for these longer break-up times.

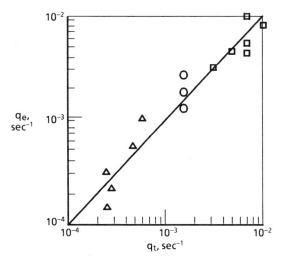

Figure 6.4 *Experimental growth rate, q_e, against its predicted value, q_t, for various molten polymer systems [12].*

The influence of extensional flow on the break-up of fluid threads was described as early as 1936 by Tomotika [22]. Again, an increase in break-up time and dominant wavelength was predicted.

6.3.2 Droplet Deformation and Break-Up

Simple shear flow. Droplets formed by the break-up of liquid cylinders will be deformed into liquid threads and again broken, thereby reducing the domain size. Since the instabilities that cause the break-up of the extended bodies grow faster with smaller domains [see Eq (6.2)], the number of droplets generated from a deformation and break-up cycle will decrease with decreasing domain size until just classical droplet-burst into two halves is feasible. An example of this is shown in Fig. 6.5. Further reduction of domain size will yield droplets that are able to withstand the hydrodynamic stresses and retain their size.

Since rather extensive reviews of the flow-induced deformation and break-up with respect to polymer blending have appeared in literature [18,29], we will limit ourselves to discussing the basic parameters involved, and treating some extensions of the literature toward polymer blending conditions.

The deformation and critical flow conditions of droplets in steady-flowing matrix liquids are determined by the rheological (viscosities, normal stresses) and interfacial (droplet size, interfacial tension) properties of the two-phase system. For the Newtonian system these influences can be expressed by two dimensionless parameters: the viscosity ratio, λ, and the capillary number, κ, a ratio shear stress, $\eta_m \dot{\gamma}$, and interfacial stress ν/R:

$$\lambda = \eta_d / \eta_m; \quad \kappa = \eta_m \dot{\gamma} R / \nu \tag{6.3}$$

a)

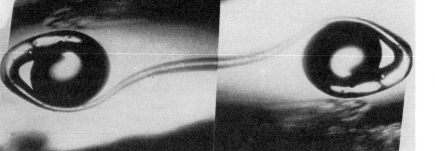

b)

Figure 6.5 *(a) Photograph of a deformed polystyrene droplet in a polyethylene matrix at 200 °C, below the critical shear rate [11]. (b) The same droplet at a shear rate slightly above the critical value [11].*

with R, the droplet radius. According to Cox [30], the deformation of a liquid droplet in simple shear flow, can be expressed in these two basic quantities as:

$$D = (L - B)/(L + B) = 5(19\lambda + 16)/4(\lambda + 1)/[(19\lambda)^2 + (20/\kappa)^2]^{1/2} \qquad (6.4)$$

with L and B the length and width of the deformed prolate droplet, respectively.

Eq (6.4) was derived on the basis of a first order perturbation theory, therefore its validity is limited to small deformations. Extension of these classical relations was made to second-order deformations [31], to systems containing surfactants [32,33], and using numerical approaches [34-36]. Results of the latter works are applicable to high deformations (which compensate for the large calculation times involved).

The break-up of droplets, by nature being an instability phenomenon, is more difficult to model. The only analytical result obtained so far stems from the classical work of Taylor [37]. This author calculated a stability criterion by equating the tensile stress on the droplet, calculated for an undeformed droplet, to its Laplace pressure:

$$4\eta_m\dot{\gamma}(19\lambda + 16)/(16\lambda + 16) = 2\nu/R \qquad (6.5)$$

therefore the critical shear rate and stress ratio κ_C are given by:

$$\dot{\gamma}_c = (\nu/2\eta_m)\,(19\lambda + 16)/(16\lambda + 16); \quad \kappa_c = (8\lambda + 8)/(19\lambda + 16) \tag{6.6}$$

In spite of its artificiality, Taylors's criterion has proven quite accurate for systems of viscosity ratio not too far from unity. For viscosity ratios in excess of 3.8 no break-up can be achieved in simple shear flow, regardless of the value of the interfacial tension! This experimental observation can be explained from Cox's relation, since for $\lambda > 3.8$ the deformation is predicted to be ≤ 0.32 for infinitely high values of κ. Such a deformation is insufficient for break-up. For very small viscosity ratios ($\lambda < 0.1$) numerical calculations [34] show that the critical κ_c is expected to follow a power-law with the viscosity ratio λ:

$$\kappa_c = 0.2\lambda^{-2/3}$$

Experimentally (see Fig. 6.6) these trends were verified by Grace [27].

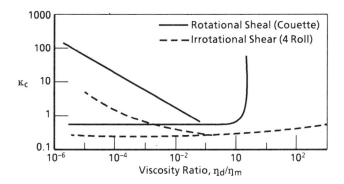

Figure 6.6 κ_c, ratio at burst vs. viscosity ratio, λ, according to Grace [27].

Non−Newtonian systems in simple shear flow. In contrast to the Newtonian case, little has been published on droplet behavior in non−Newtonian systems. Van Oene [38] derived on thermodynamical grounds a relation that incorporated the effect of fluid elasticity by altering the value of interfacial tension according to the relation:

$$\nu_{app} = \nu + [(\sigma_{11} - \sigma_{22})_d - (\sigma_{11} - \sigma_{22})_m]\,R/6 \tag{6.7}$$

with $\sigma_{11} - \sigma_{22}$ the first normal stress difference. Apparently, the droplet elasticity is expected to reduce the deformation and increase the critical shear rate of droplets, while matrix elasticity should increase the deformation and decrease critical shear rates. On the basis of a first-order perturbation calculation, van Dam [39] recently derived relations for the deformation of droplets in simple shearing matrices, assuming second−order fluid behavior.

According to this work, the influence of fluid elasticity is in qualitative agreement with van Oene's result.

In our early experiments [11] these trends were confirmed experimentally for both deformation and break-up droplets in simple shear flow. One example is given in Fig. 6.7 where the shear rate at burst is plotted against the inverse droplet size. The two sets of points represent the break-up of silicon oil droplets ($\eta = 1$ Pas) in two different polymer solutions. The shear thinning behavior of the matrices were comparable in the applied range of shear rates. The second matrix exhibited higher normal stresses than the first matrix. The curvature of the lines is due to the shear thinning behavior of the matrix. From the relative position of the lines it can be concluded that elasticity of the matrix enhances its ability to break-up droplets.

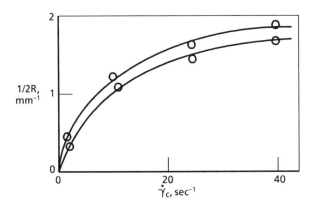

Figure 6.7 Break-up shear rates plotted against the inverse droplet size for silicon oil ($\eta = 1$ Pas) droplets dispersed in aqueous solutions of polyacrylamide. Top curve represents a 1.2% Separan AP30 (high-$\overline{M}_w$) solution, bottom curve represents a 0.7% Separan AP273 (ultra-high-$\overline{M}_w$) solution.

Droplet behavior in extensional flows, Newtonian systems. According to Cox, the deformation of droplets suspended in a matrix exerting an extensional flow at a constant extensional rate can be described by:

$$L/L_0 - 1 = A(\eta_m \dot{\epsilon} R/\nu)\,(19\lambda + 16)/(16\lambda + 16) \qquad (6.8)$$

with constant A having a value of 1.5 in axisymmetric extensional flow and 2 in hyperbolic flow. L_0 and L are, respectively, the length of the undeformed and deformed droplet with $\dot{\epsilon}$ being the rate of elongation.

However, an extensional flow with constant extensional rate, as experienced by the traveling droplet, can only be generated in singularities of the flow field, such as observed in hyperbolic flows [13]. Therefore, it is relevant to discuss the deformation of droplets in non-stationary extensional flows. Cox [30] gave an expression for the response of a droplet to a sudden exposure to a constant extensional flow field:

$$L/L_0 - 1 = A(\eta_m \dot{\varepsilon} R/\nu) \ (1 - \exp\{-20\nu t/19\eta_d R\}) \ (19\lambda + 16)/(16\lambda + 16) \qquad (6.9)$$

Any time dependent extensional flow field may be approximated by a large number of successive small step functions. Thus, one may postulate that the deformation of droplets in these flow fields is given by the superposition of Eq (6.9) [15]:

$$L/L_0 - 1 = [20A(19\lambda + 16)/19\eta_d(16\lambda + 16)] \int_0^t \eta_m \dot{\gamma}(t') \exp\{-20\nu(t - t')/19\eta_d R\}dt' \qquad (6.10)$$

A practical example is the case of a droplet flowing through a conical contraction shown in Fig. 6.8. It can be shown [13,15] that the time-dependent extensional rate as experienced by the droplet can be approximated by:

$$\dot{\varepsilon} = 2/3 \ (t_0 - t) \qquad (6.11)$$

with t_0 the time required for the droplet to travel through the cone:

$$t_0 = \pi \ (z_0 \tan \alpha_c)^2/6\Phi$$

where α_c, z_0 and Φ are defined as in Fig. 6.8.

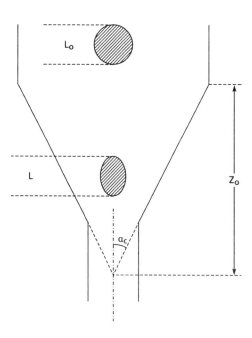

Figure 6.8 *Definition of* α_c, z_0, L, L_0, *and the flow rate* Φ.

Combining Eqs (6.10) and (6.11) yields time-dependent deformation of the droplet flowing through the cone:

$$L/L_0 - 1 = [5(19\lambda + 16)/57\lambda(\lambda + 1)] \exp\{Q(t)\} \int_{Q(t)}^{Q(0)} \exp\{-y\}/y \, dy \qquad (6.12)$$

with $Q(t) = 40\nu/57\eta_d \dot{\epsilon} R$.

Figure 6.9 gives the experimentally determined deformation of droplets of various silicon oils with different viscosities submerged in castor oil flowing through a conical contraction. The solid lines were predicted by Eq (6.12). Fair agreement is observed. It was shown by Van der Reijden and Sara [13], that a more accurate representation of the flow field in the cone considerably improves the agreement.

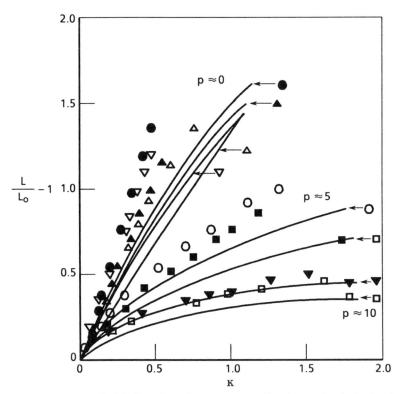

Figure 6.9 *Predicted (solid lines) and experimentally determined (points) stretches of droplets flowing through a conical contraction vs. κ for various viscosity ratios. For detailed description of the experimental conditions see [13].*

It can be seen in the graph that, although the deformation of droplets in a steady state extensional flow field is nearly independent of the viscosity ratio [see Eq (6.9)], in time-dependent flow fields the deformation of droplets with viscosity ratio above unity lags behind its equilibrium value. Indeed, it is known [40] that the formation of fibrillar dispersed phases due to elongation flow of immiscible blends is less successful with large viscosity ratios.

Break—up of droplets in extensional flow. The above relations are concerned with the deformation of droplets in extensional flow. Break-up of droplets in these types of nonstationary flows, however, is not to be expected. The continuous elongation of the droplet does not allow for the instability to grow to such an extent that actual burst into two halves can be completed. As time-independent extensional flow fields are not likely to be encountered in actual polymer blending devices, the break-up of droplets due to extensional flow fields will primarily occur by capillary break-up of droplets that are extended in conical or wedge-shaped contractions.

Elastic spheres in extensional flow. Although some work was done on viscoelastic systems [41,42], no experimental or theoretical work has been reported on the time-dependent deformation of droplets in elongating viscoelastic systems. The mathematical complexity involved in time-dependent two-phase non-Newtonian flow will hamper exact solutions. Here we will discuss the deformation of droplets behaving according to the other limit of the Deborah scale, viz. solid elastic spheres.

From a recent paper of Brunn [43], it can be shown that the deformation of an elastic sphere responds to a step in extensional rate as:

$$L/L_0 - 1 = 7.5\,\kappa_E\,(1 - \exp\{-2Et/9\eta_m\}) \tag{6.13}$$

with $\kappa_E = \eta_m\dot{\epsilon}/E$, where E is the Youngs modulus of the elastic sphere.

Applying the same superposition principle as above, gives a relation for the time-dependent deformation of these spheres flowing through a conical contraction:

$$L/L_0 - 1 = 1.11\exp\{Q_E(t)\}\int_{Q_E(t)}^{Q_E(0)} \exp\{-y\}/y\;dy \tag{6.14}$$

with $Q_E = 4E/27\eta_m\dot{\epsilon}$.

To investigate the validity of Eq (6.14), elastic spheres were prepared from a polyvinylalcohol solution physically crosslinked with Congo red dye. A mixture of silicon oil and elastic spheres were then guided through a tube with a conical contraction and the deformation of the spheres measured. Figure 6.10 gives the deformation of the three sets of elastic spheres as a function of the local value of κ_E. The points are experimental, the curve is calculated from Eq (6.14). Good agreement was observed for stretches smaller then 30%. For higher stretches nonlinear elastic effects such as strain hardening of the spheres occur.

These results might be applicable to the in-situ preparation of composite materials by flow-induced fibrillation of the dispersed polymer [1,2,44,45]. To obtain high-modulus fibrils, elastic deformation of the dispersed phase is required. When the Youngs modulus of this phase is known, the necessary flow conditions can be estimated.

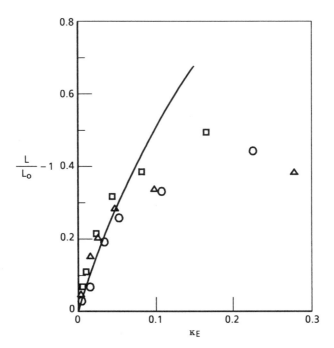

Figure 6.10 *Theoretical (solid line) and experimentally determined stretches points of elastic spheres flowing through a cone vs. κ_E. Three different batches of spheres are used.*

6.4   FLOW–INDUCED COALESCENCE

6.4.1   General Considerations

In the preceding section the behavior of isolated droplets was briefly discussed. It should, however, be realized, that collision of droplets can only be avoided at extremely low dispersed phase fractions. Supposing for a start that the break–up of droplets whose size are above the maximal stable droplet size requires a matrix shear strain $\gamma_c = \dot{\gamma}t$. The number of droplet bursts per unit volume and time is then given by:

$$N_b = (\dot{\gamma}/\gamma_c)\, 3C/4\pi R^3 \qquad (6.15)$$

with C being the disperse phase fraction.

From Smoluchowski's work [46], one can obtain the number of collisions per unit volume and time:

$$N_c = 6\dot{\gamma}C^2/\pi^2 R^3 \qquad (6.16)$$

From Eqs (6.15) and (6.16) the disperse phase fraction at which these number rates are equal can be expressed as:

$$C_{c=b} = \pi/8\gamma_c \qquad (6.17)$$

According to Grace [27] the critical shear strain is at least 30, yielding a value for $C_{c=b}$ as low as 1.2%.

The above relations are extremely simplified since they take into account neither deformation of the droplets nor their mutual hydrodynamic interactions. They show nevertheless, that if the fraction of collisions that results in an actual recombination of the disperse phase is substantial, the influence of coalescence on blend morphology cannot be neglected.

Previously [14] we proposed a route to estimate the probability of collision of two droplets resulting in a coalescence. In the model that was used the colliding droplets were assumed to be spherical. Upon collision the colliding faces of the droplets deformed and the remainder of the matrix liquid was squeezed out from the film, separating the colliding droplets. The closest approach of the droplets was calculated and compared to the critical thickness below which the liquid film separating the colliding droplets was expected to rupture due to Van der Waals interactions. In our calculations we used a value for the critical film thickness of 50 nm as proposed by Vrij [47]. Since the closest approach of the colliding droplets appeared to depend on the spatial separation of their streamlines, only a fraction of the collisions was predicted to result in coalescence. It is this fraction that is defined as the coalescence probability.

An important factor influencing the magnitude of this coalescence probability is the mobility of the interface, i.e., the ability of the interface to deform with the liquid it is bounding without exerting a tangential stress on the fluid system. A possible origin of this tangential stress is a gradient in interfacial tension along the interface due to flow-induced concentration variations of surface-active substances. From emulsion science it is known that small concentrations of impurities can considerably reduce the interfacial tension of oil-water mixtures, thereby inducing interfacial immobility.

In the polymer-polymer systems polarity differences are usually small, leading to a positive, but small, enthalpic contribution to the free energy of mixing. The fact that the negative entropic contribution is even smaller explains the immiscibility of these systems. It can therefore be expected that additives or impurities present in a polymer blend have a small tendency to adhere to the interface and reduce its interfacial tension. Therefore, a high degree of interfacial mobility can be assumed. Substances that do preferentially adsorb at the interface, such as block copolymers, will induce interfacial immobility. This would imply that the emulsifying effect of block copolymers [48-50] is not only due to the reduction of interfacial tension, but also to the reduction of coalescence probability.

Attempts to determine the magnitude of interfacial mobility by studying the gravity-induced coalescence in molten polymer systems yielded, in spite of promising initial experiments [14], inconclusive results. Further work is required.

6.4.2 Influence on Domain Size

It was shown [14] that the coalescence probability, Pc, depends on domain size, R, and the ratio, κ. Pc was shown to be zero above a certain threshold value for R, depending on κ. This renders coalescence to be self limiting.

The solid line in Fig. 6.11 represents the droplet size above which the coalescence probability is zero according to our model. The dashed line represents the droplet size below which no droplet burst is expected, according to Taylor's Eq (6.6). Full interfacial mobility was assumed. The area bound by these two criteria is the region in which the droplet sizes are expected. Droplets larger than the coalescence limit can only break up to reduce their size,

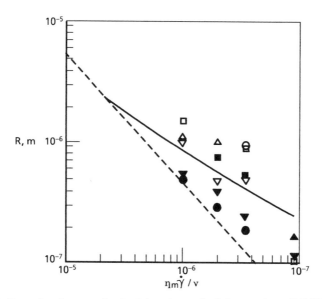

Figure 6.11 *Domain size vs. the "mixing intensity" for various PS/PP blends in various disperse phase fraction: • = 0.25%; ▼ = 0.5%;∇ = 1%;+ = 2%;□ = 4%;∆ = 8%; ○ = 16%. The various data points were obtained by varying the matrix polymer and the blending conditions.*

while droplets smaller than Taylor's limit can only coalesce to increase their size. Droplets in the region between the two criteria can both coalesce and break up. In this region the average domain size and distribution will depend on the complex balance between the break-up and coalescence phenomena. It can be expected that, with increasing concentration, the number of collisions per unit volume and time increase leading to an increase of the average droplet size.

6.4.3   Experimental Verification

To investigate the influence of coalescence on the domain size as generated in an actual polymer mixer, blends were prepared using a single-screw extruder. The back flow was minimized by removing the die-head to facilitate a proper modeling of the flow field. Since this results in an inefficient mixing apparatus, the thus-prepared blend was granulated and reextruded until an equilibrium morphology was attained. Brittle fracture surfaces were examined using the scanning electron microscope (SEM) and the domain size was determined. The domain-size distribution proved to be rather narrow. The average values of the domain sizes are plotted against the "mixing intensity" κ/R in Fig. 6.11. It can be seen that the domain size is strongly concentration dependent, indicating the presence of flow-induced coalescence. Quantitatively, both the calculated upper and lower limit of the domain size were in moderate agreement with the experiments.

6.4.4 Deficiencies of the Model

The predictions on coalescence probabilities were obtained assuming the colliding droplets to be initially spherical, which is unrealistic at $\kappa > 0.1$. Furthermore, the thinning of the film separating the colliding droplets was described with Newtonian hydrodynamics which may lead to serious errors. A more realistic modeling of flow-induced coalescence, by taking into account the deformation of the droplets due to the external shear field and non-Newtonian flow in the draining of the film is being investigated.

A factor that may significantly enhance the coalescence probability relative to those in classical emulsions is the fact that the viscosity of thin polymer layers is considerably lower than their bulk viscosity. This phenomenon was explained in terms of depletion [51] or a reduction in entanglement density due to the geometrical constraint [52]. Experimental evidence of this viscosity reduction was provided by Burton et al. [53] using small gap rheometry.

It may be just this phenomenon that explains the fact that droplet sizes in molten polymer systems are far more dependent on disperse-phase fraction than in low-molecular-weight emulsions. This may even account for the fact that polymer blends are the only class of liquid-liquid dispersions that can exhibit co-continuous structures.

6.5 CONCLUSIONS

In this chapter we have discussed the microrheological phenomena that can contribute to the formation of a polymer-blend morphology (or in fact any other fluid dispersion).

The first part dealt with the behavior of isolated disperse domains, applicable only to blends with an extremely low disperse-phase fraction. It appeared that capillary instabilities growing on highly extended domains reduce the domain size to such a level where the interfacial forces become large enough to withstand the viscous forces and size reduction is ceased. In industrial compounding of polymer blends the dispersed-phase fraction usually is so large that collisions of the dispersed domains cannot be neglected. If a considerable number of collisions results in a coalescence, the average droplet size is increased above the stable droplet size and a complex balance of break-up and coalescence events occurs. When the droplet size only slightly increased, break-up in two halves will take place. If, on the other hand, a considerable increase in size occurs the break-up mode will be via the growth of capillary instabilities of highly deformed domains.

At yet higher concentrations the extended bodies can coalesce before break-up has set in, thus yielding a co-continuous morphology. The processing of the blend after the actual blending will determine whether the co-continuous structure is preserved or converted into a dispersion-type morphology by interfacial tension driven processes. An example of the latter is shown in Fig. 6.12.

Quantification of these break-up and coalescence phenomena is only feasible under the conditions where the rheological nature of the fluid involved and the present flow fields agree with the model assumptions. As a result, the domain size and its concentration dependence can only be approximated in a highly modeled blending device such as the open-ended single-screw extruder. Considering the complexity of an actual polymer-blending process, involving inhomogeneous and time-dependent flow fields, complex rheological properties of the fluids, and high disperse-phase fractions, it is unrealistic to expect a full quantification of polymer-blending operations along these lines.

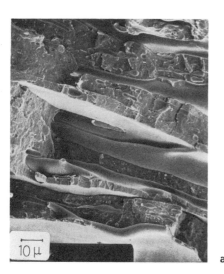 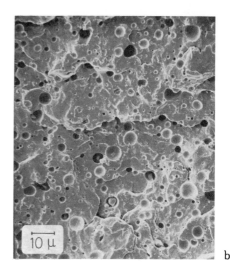

a)   b)

Figure 6.12 (a) Micrographs of a polypropylene/polystyrene blend, prepared on a two-roller mill at 200 °C and quenched during blending. (b) Micrographs of the blend after annealing for 10 min at 200 °C.

In view of these complexities, it is worthwhile to consider polymer blends as ordinary liquid dispersions. The role of surfactants being fulfilled by their polymeric analogue, e.g., block copolymers. Complementary information with possible routes to a better quantification might then be obtained from stochastic approaches that are routine in emulsion science [54]. The spreading of dyes or deuterated disperse material [55] or the time-dependent response of the blend morphology to the stepwise changes in mixing intensity might give information on the break-up and coalescence rates. These latter can be properly modeled and substituted into a "population balance equation" to yield predictions on the concentration dependence of the domain size, the onset of co-continuity, and the concentration at which phase inversion takes place.

ACKNOWLEDGMENTS

One of the authors (J. J. Elmendorp) acknowledges the support of his present employer, Shell Development Company, during the preparation of this paper. Dr. J. L. Thomason is gratefully acknowledged for reading the manuscript.

NOTATION

B	Width of droplet
C	Disperse phase concentration
$C_{c=b}$	Concentration of equal number of break-ups and coalescences
D	Droplet deformation

E	Elasticity modulus
L	Length of deformed droplet
L_0	Initial length of droplet
R	Droplet or cylinder radius
R_0	Initial radius
z_0	Length of conical contraction
α	Distortion amplitude
α_0	Initial distortion amplitude
α_c	Cone angle
γ	Shear deformation at burst
$\dot{\gamma}$	Shear rate
$\dot{\gamma}_c$	Shear rate at burst
$\dot{\epsilon}$	Rate of elongation
η	Viscosity
κ	Ratio of viscous forces to interfacial forces
κ_c	Critical ratio
κ_E	Ratio of viscous forces to elastic forces in disperse phase
λ	Viscosity ratio
$\sigma_{11} - \sigma_{22}$	First normal stress difference
ν	Interfacial tension
Φ	Flow rate
Λ	Wavelength

subscripts d and m stand for disperse and matrix phase, respectively.

REFERENCES

1. Van der Vegt, A. K., Verbraak, C. L. J. A., in *Interrelations between Processing, Structure and Properties of Polymeric Materials*, Seferis, J. C., Theocaris, P. S., Eds., Elsevier (1984).
2. Verbraak, C. L. J. A., van Dam, J., Van der Vegt, A. K., *Polym. Eng. Sci.*, **25**, 431 (1985).
3. Utracki, L. A., *Polym. Eng. Sci.*, **22**, 1166 (1981); *SPE Techn. Pap.*, **33**, 1339 (1987).
4. Utracki, L. A., *Polymer Alloys and Blends*, Hanser Publishers, Munich (1989).
5. Meijer, H. E. H., Lemstra, P. J., Elemans, P. H. M., *Macromol. Chem., Macromol. Symp.*, **16**, 19 (1988).
6. White, J. L., Min, K., *Macromol. Chem., Macromol. Symp.*, **16**, 19 (1988).
7. Karger-Kocsis, J., Kallo, A., Kuleznev, V. N., *Polymer*, **25**, 279 (1984).
8. Kuleznev, V. N., *Int. J. Polym. Mat.*, **8**, 295 (1980).
9. Plochocki, A. P., *Polym. Eng. Sci.*, **22**, 1153 (1982).
10. Plochocki, A. P., *Polym. Eng. Sci.*, **25**, 82 (1985).
11. Elmendorp, J. J., Maalcki, R. J., *Polym. Eng. Sci.*, **25**, 1041 (1985).
12. Elmendorp, J. J., *Polym. Eng. Sci.*, **26**, 418 (1986).
13. Van der Reijden, C., Sara, A., *Polym. Eng. Sci.*, **26**, 1229 (1986).
14. Elmendorp, J. J., Van der Vegt, A. K., *Polym. Eng. Sci.*, **26**, 332 (1986).
15. Elmendorp, J. J., *PhD thesis*, Delft Univ. Technology, Netherlands (1986).
16. Lyngaae-Jorgensen, J., Sondergaard, K., *Polym. Eng. Sci.*, **27**, 334, 359 (1987).
17. Lyngaae-Jorgensen, J., in *Multiphase Polymers: Blends and Ionomers*, L. A. Utracki, R. A. Weiss, Eds., Amer. Chem. Soc., Symp. Ser. Vol. 395, Washington, D.C. (1989).

18.   Elmendorp, J. J., in *Mixing in Polymer Processing*, Raauwendaal, C., Ed., Marcel Dekker, New York (1990).

19.   Wu, S., *Polymer Interfaces and Adhesion*, Marcel Dekker, New York (1983).

20.   Smith, P. G., Van der Ven T. G. M., Mason, S. G., *J. Colloid. Interface Sci.*, **80**, 332 (1981).

21.   Rayleigh, J. W. S., *Phil. Mag.*, **34**, 145 (1892).

22.   Tomotika, S., *Proc. Roy. Soc.* (London), **A153**, 302 (1936).

23.   Schumer, P., Tebel, K., *J. Fluid Mech.*, **93**, 231 (1983).

24.   Gordon, M., Yerushelmi, J., Shinnar, R., *J. Fluid Mech.*, **38**, 689 (1969).

25.   Kraus, G., Rollmann, K. W., *Kautchuk Gummi*, **34**, 648 (1981).

26.   Ghysels, A., Raadsen, J., *Pure Appl. Chem.*, **52**, 1339 (1980).

27.   Grace, H. P., *Chem. Eng. Comm.*, **14**, 225 (1982).

28.   Plochocki, A. P., *Polym. Eng. Sci.*, **28**, 62 (1986).

29.   Han, C. D., *Multiphase Flow in Polymer Processing*, Academic Press, New York (1981).

30.   Cox, R. G., *J. Fluid Mech.*, **37**, 601 (1969).

31.   Chaffey, C. E., Brenner, H., *J. Colloid Interface Sci.*, **24**, 258 (1967).

32.   Flumerfelt, R. W., *J. Colloid Interface Sci.*, **76**, 330 (1980).

33.   Phillips, W. J., Graves, R. W., Flumerfelt, R. W., *J. Colloid Interface Sci.*, **76**, 350 (1980).

34.   Acrivos, A., Lo, T. S., *J. Fluid Mech.*, **86**, 18 (1978).

35.   Barthes-Biesel, D., Acrivos, A., *J. Fluid Mech.*, **86**, 641 (1987).

36.   Rallison, J. M., Acrivos, A., *J. Fluid Mech.*, **89**, 191 (1978).

37.   Taylor, G. I., *Proc. Roy. Soc.* (London), **A138**, 41 (1932).

38.   Van Oene, H., in *Polymer Blends*, Vol. 1, Paul D. R., Newman, S., Eds., Academic Press, New York (1978).

39.   Van Dam, J., Laboratory of Polymer Technology, Delft (to be published).

40.   White, J. L., Plochocki, A. P., Tanaka H., *Polym. Eng. Rev.*, **1**, 217 (1981).

41.   Chin, H. B., Han, C. D., *J. Rheol.*, **23**, 357 (1979).

42.   Chin, H. B., Han, C. D., *J. Rheol.*, **24**, 1 (1980).

43.   Brunn, P. O., *J. Fluid Mech.*, **126**, 553 (1983).

44.   Tsebrenko, M. V., Vinogradov, V. G., Ablazova, T. I., Yudin, A. V., *Kolloid Zh.*, **38**, 200 (1976).

45.   Tsebrenko, M. V., Ablazova, T. I., Yudin, A. V., Vinogradov, G. V., *Kolloid Zh.*, **38**, 204 (1976).

46.   Smoluchowski, M. Z., *Phys. Chem.*, **92**, 129 (1917).

47.   Vrij, A., *Disc. Farad. Soc.*, **42**, 2505 (1966).

48.   Barentzen, W. M., Heikens, D., *Polymer*, **14**, 579 (1973).

49.   Barentzen, W. M., Heikens, D., Piet, P., *Polymer*, **18**, 69 (1974).

50.   Plochocki, A. P., Dagli, S. S., Mack, H. H., *Kunststoffe*, **78**, 237 (1988).

51.   Cohen, Y. Metzner, A. B., *J. Rheol.*, **29**, 67 (1985).

52.   de Gennes, P. G., *Rev. Mod. Phys.*, **57**, 827 (1985).

53.   Burton, R. H., Folkes, M. J., Narth, K. A., Keller, A., *J. Mater. Sci.*, **18**, 315 (1983).

54.   Coulaloglou, C. A., Tavlarides, L. L., *Chem. Eng. Sci.*, **32**, 1289 (1977).

55.   Roland, C. M., Bohm, G. G. A., *J. Polym. Sci., Polym. Phys. Ed.*, **22**, 74 (1984).

CHAPTER 7

MELT RHEOLOGY AND MORPHOLOGY OF LINEAR LOW DENSITY POLYETHYLENE/POLYPROPYLENE BLENDS

by M. M. Dumoulin, L. A. Utracki

Industrial Materials Institute
National Research Council Canada
Boucherville, QC, J4B 6Y4
CANADA

and

by P. J. Carreau

Chemical Engineering Department
Ecole Polytechnique
Montreal, QC, H3C 3A7
CANADA

This work examines the viscoelastic behavior of molten blends of polypropylene with two linear low density polyethylenes having different molecular weights. Several data treatments were used. The zero-shear viscosity, η_0, was evaluated through a curve-fitting procedure using a four-parameter equation that proved useful for polymers with wide relaxation spectrum. These η_0 values were in good agreement with those estimated from the Cole-Cole plot as well as from the crossover point of the dynamic moduli. It was found that η_0 is related to the location of the maximum in the relaxation time spectrum. Examination of the blends morphology indicated that the high strains and pressures encountered by the material during injection molding induce partial solubilization of the dispersed phase. On the basis of morphological evidence, it was also possible to define the concentration region for phase inversion. The presence of two co-continuous phases for those concentrations was reflected in the rheological properties through changes in the relaxation time spectrum. The influence of the long relaxation time components originating from this structure was responsible for an apparent yield stress. The changes in the relaxation spectra can also be discerned in the Cole-Cole plots.

7.1.   INTRODUCTION

Plastics are maintaining a fast pace in displacing more conventional materials and creating new opportunities. Technical and financial incentives make polymeric materials an attractive alternative in an ever-increasing number of applications. The development of new resins, which, for several decades, attracted most of the research effort, now tends to be supplemented by new approaches. The strength of these new tendencies is clearly shown by the rapid emergence of polymer blends and composites as engineering materials. The transport industry has been the leader in the development and use of these new materials [1]. On the other hand, this technology has by no means been limited to specific fields. For example, the addition of a rubbery phase to impart impact resistance to a more brittle one has given birth to materials such as high-impact polystyrene and toughened engineering resins for which applications are countless and diversified [2,3].

Commercial polymer blends are often immiscible. Properties of multiphase materials are strongly influenced by the morphology, which, in turn, depends on the thermodynamic interactions between the two polymers, their rheological behavior, and processing conditions. Rheological studies are therefore essential since they give access to information pertaining to the structure, morphology, and processing of the materials [3].

The polyolefin family constitutes the most widely used group of plastics. It was estimated that 22 million metric tons (Mt) of polyolefins were produced in 1980, making them the undisputed leaders in the plastic materials world market with 39% of total consumption [4]. By 1986, the world consumption of polyolefins increased to 31 Mt (including 8.4 Mt of polypropylene, PP) or 57% of all plastics [5,6]. Polyolefin blends have attracted keen interest for technical as well as commercial reasons. For example, about 60% of the LLDPE production goes into blends with other polymers such as high-density and conventional high-pressure polyethylenes (HDPE and LDPE respectively) or ethylene-vinyl acetate copolymers [7].

Only a handful of technical papers deal with PP/LLDPE blends [3,8,9] even though LLDPE has been commercially manufactured for over two decades. The impact of LLDPE and its blends is particularly strong in film blowing technology where it tends to displace

LDPE. In terms of total consumption of PE of low density in the U.S.A., 13% of it was LLDPE in 1981 while this figure rose to 38% by 1985 [10]. In 1986, LLDPE represented 43% of the North American market while in Europe its share was about 10% [11].

Even though LLDPE's penetration of the commodity resins market has been rapid, it was delayed by processing difficulties. The change of resin from LDPE to LLDPE often resulted in a throughput decrease of 20 to 50%. This could be avoided by costly equipment modifications. However, the most widely used approach has been blending LDPE with LLDPE to take advantage of the good processing characteristics of the former and the final properties of the latter [12,13].

The capital expenditures and time required for bringing new materials on line are considerably less for blends and composites than through other routes. This makes these technologies particularly attractive to countries like Canada where markets and capital are limited. Since Canada is a major producer of polyethylene, especially the linear low-density type, it is certainly no coincidence that a number of research centers in Canada developed strong expertise in rheology and processing of polyolefin blends and composites.

Binary blends of PP with LLDPE are commercially attractive for their strength, modulus, and low-temperature impact performance. In spite of immiscibility, it is possible to select grades of these two polymers which will result in additivity or even synergism of properties. LLDPE, which can be manufactured with widely different molecular weights and molecular weight distributions as well as having different concentrations of comonomers, proves to be a more suitable blend ingredient than the historically older HDPE. The latter blends invariably required significant amounts (5 to 25 wt%) of a compatibilizer, e.g., EPR or EPDM.

The present communication is the fourth in the series in which the behavior of blends of PP with LLDPE is examined. The first two papers concentrated on the solid-state behavior [14,15]. The first reported on the temperature and composition dependence of low-strain tensile properties while in the second the mechanical properties were discussed in a more global fashion, relating them to morphology and crystallinity. In part three, the time-temperature superposition of dynamic tensile moduli in a wide range of temperature was examined [16]. This fourth part aims at discussing the melt rheology. The melt flow data will therefore be presented and discussed while relevant observations from the earlier work will be brought in whenever necessary.

7.2   EXPERIMENTAL

7.2.1   Materials

Properties of the PP and the two LLDPE commercial resins are shown in Table 7.1. All three polymers were used as obtained from the manufacturer. LLDPE-1's density, $\rho = 951 \text{ kg/m}^3$, obviously is in the high-density range, but it still bears the name of LLDPE because of its fluidized bed gas-phase manufacturing process. Small quantities (0.1% by weight) of a commercial antioxydant mixture (Ciba-Geigy Irganox 1010 and Irganox 1024) were added during the extrusion. Blends containing LLDPE-1 will be referred to as System-1 while those containing LLDPE-2 as System-2.

These materials were selected in order to obtain two chemically similar systems showing reversed viscosity ratios, $\lambda = \eta_{PP}/\eta_{PE}$. The zero-shear viscosity values in Table 7.1 show that for System-1, LLDPE is the high viscosity component while the situation is reversed for System-2.

Table 7.1 *Properties of Polymers*

Material	PP	LLDPE-1	LLDPE-2
Supplier	Hercules	Esso	Dow
Designation	6701	LPX-24	Dowlex 2517
Density at 23°C (kg/m^3)	900	951	917
η_0 at 190°C (kPas)	13.0	8,120	1.32
Melt index (dg/min)	0.8	0.3	25
M_w (kg/mole)[a]	392	216	108
M_w/M_n [a]	3.7	7.6	3.0

[a] *Obtained by SEC, in equivalent mass of PS.*

7.2.2   Preparation of Blends

Binary blends were prepared in a Werner & Pfleiderer co-rotating twin-screw extruder, model ZSK-30 using high dispersive mixing screw configuration. At screw speed of 350 rpm, a throughput of approximately 4 kg/hour was achieved. The temperature of the extruder ranged from 170°C (in the melting zone) to 240°C (in the metering zone and the die). The extruded filaments were quenched, dried, and granulated.

7.2.3   Size Exclusion Chromatography (SEC)

Molecular weights of all materials were measured using a Waters Scientific chromatograph, model GPC/LC 150-C with data module and computer data acquisition system. The samples were dissolved in 1,2,4-trichlorobenzene, (0.1% by weight) for 1.5 h at 165°C then injected at 140°C. Four PL-Gel columns, from Polymer Labs Inc., with pore sizes of 10^3, 10^4, 10^5, and 10^6 Å were calibrated with twelve polystyrene, PS, samples of narrow molecular-weight distribution ($M_w/M_n < 1.1$).

7.2.4   Morphology

The state of dispersion of the blends was examined using a Jeol scanning electron microscope, SEM, model JSM-35CF. The surfaces were prepared by fracturing under liquid nitrogen either extruded or injection-molded specimens. A gold/palladium coating was applied on the surfaces prior to their SEM examination.

7.2.5   Melt Rheology

The dynamic flow properties of the molten materials were measured with a Rheometrics Mechanical Spectrometer, RMS, Model 605. The shear complex moduli were measured using dia. = 50 mm parallel plates at 190°C under dry nitrogen. First, time sweeps at a frequency of $\omega = 1$ rad/s showed no apparent change within 30 min. The strain, γ, was then varied in order to define the linear viscoelastic region. Finally, the dynamic moduli were determined with $\gamma = 10$ or 15%, well below the linear limit of viscoelastic response, while ω was varied from 10^{-2} to 10^2 rad/s.

The Rheometrics Stress Rheometer, RSR, was used for the complementary tests carried out at 190°C under constant stress. During these tests, the strain was monitored for 200 s while the stress, adjusted to values between 0 and 3 kPa, was kept constant.

7.3   RESULTS

7.3.1   Size Exclusion Chromatography

The different molecular weight averages obtained from SEC are plotted in Fig. 7.1. For each data point, the standard deviation of the measurements is shown by the vertical error bar. Curves drawn through the points show the trends while the broken lines illustrate the computed averages:

$$M_n = (\Sigma w_i/M_{ni})^{-1} \tag{7.1}$$

$$M_w = \Sigma w_i M_{wi} \tag{7.2}$$

$$M_z = \Sigma w_i M_{zi} M_{wi} / \Sigma w_i M_{wi} \tag{7.3}$$

where subscript $i = 1, 2$ refers to the two components.

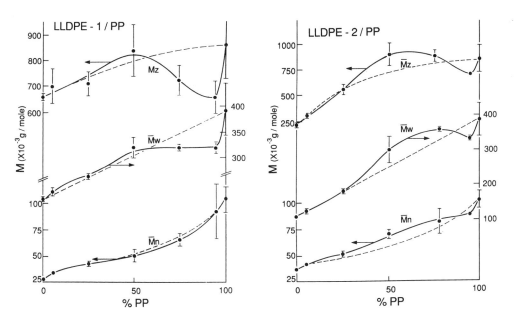

Figure 7.1 *Average molecular weights obtained with SEC for System-1 (left) and System-2 (right); points are experimental, broken lines are computed.*

7.3.2   Morphology

Results from the morphological investigation are presented in the form of SEM micrographs. Figure 7.2 shows the state of dispersion in extruded blends. The two phases can be distinguished in blends containing either 5% or 25% of LLDPE-1 (Figs. 7.2a,b). In both cases the particles of PE are dispersed in a PP matrix. Figure 7.2c shows the surface fracture of a blend containing 50% LLDPE-1. Immiscibility of the co-continuous polymers is again apparent but the two phases are not readily identifiable.

The subsequent micrographs (Fig. 7.3) were taken of samples that went through an additional processing step, either injection or compression molding. The effect of a gradual change in composition is illustrated by Figs. 7.2d to 7.3c showing fracture surfaces of blends containing 5 to 75% LLDPE-1. The amount of dispersed phase visible in the injection-molded samples generally seems to be at odds with the nominal concentration. Figure 7.2d, for example, is a representative illustration of the state of dispersion in blends containing 5% LLDPE-1. There appear to be very few PE particles on the fractured surface. The same comment can be made in regard to the blend containing 25% LLDPE-1; there is a clear difference between Fig. 7.2b (prior to the injection) and Fig. 7.3a (after). Thus processing reduced the apparent amount of dispersed phase. In addition, the visible particles have lost their spherical geometry.

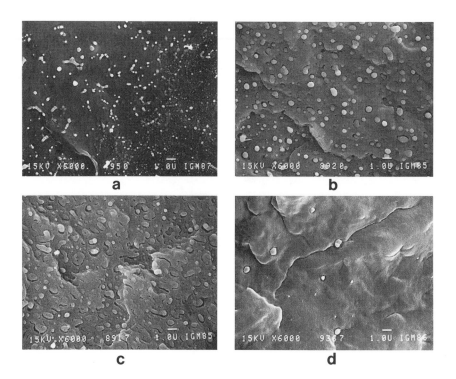

Figure 7.2 *Micrograph of the PP/LLDPE-1 blend at magnification 6000X; concentration of the second polymer in (a) and (d) was w = 5 wt%, while in (b) and (c) w = 25 and 50 wt%, respectively. Specimens (a–c) were extruded, while (d) was injection molded.*

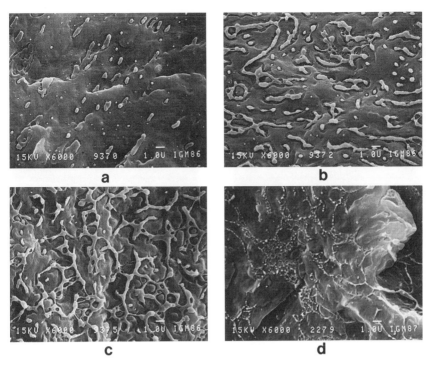

Figure 7.3 *Micrograph of injection-molded samples of the PP/LLDPE blend at magnification 6000X; concentration of LLDPE-1 in (a) to (c) was w = 25, 50, and 75 wt%, respectively, while in (d) that of LLDPE-2 was w = 50 wt%.*

Figure 7.3b shows the fracture surface of the injection-molded blend containing 50% of LLDPE-1. Comparing it with Fig. 7.2c, one sees again two phases, but a large portion of the PE exists in the form of a network immersed in a PP matrix. The same type of structure is observed in Fig. 7.3c for the blend containing 75% LLDPE-1. In this case, the network is more complete, with more points of contact. Figure 7.3d shows the fracture surface of an injected sample of the blend containing 50% of LLDPE-2. A network similar to the one observed in the two previous pictures is also seen.

In an attempt to elucidate the origin of a peculiar deviation of the maximum strain at break [15], a series of injection-molded samples of the blend containing 5% of LLDPE-1 were heated to 170°C and annealed in order to remove the effects of the injection. Figures 7.4a,b show two micrographs (different magnifications) of the same fracture surface of an annealed sample; two different types of particles are visible: 1) a few large particles with diameters of 0.5 μm or more (similar to those seen in Fig. 7.2a) show an irregular shape, and 2) there are a great number of small particles which were invisible in any unannealed sample. Figure 7.4b, taken at a magnification of 16,000, enables determination of the shape and size of these small particles, whose diameter is on the order of a fraction of a micron. They seem to originate from the breakage of fibrillas under forces of the Taylor capillary instability waves (see [3] and Section 6.3 in Chapter 6).

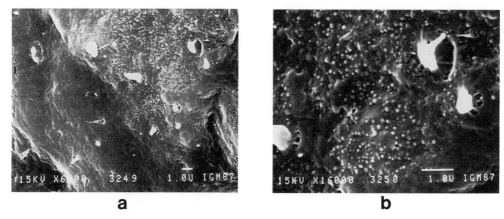

Figure 7.4 *Micrograph of remelted and annealed injection-molded sample of the blend containing 5% LLDPE-1 at magnification 6000X (a) and 16,000X (b).*

7.3.3   Melt Rheology

The dynamic moduli are the basic rheological functions obtained in dynamic experiments from which all others were computed. The storage, G', and loss, G", shear moduli at 190°C for Systems-1 and -2 are plotted respectively in Figs. 7.5 and 7.6. The curves are identified by their LLDPE content. For the sake of clarity, the curves corresponding to 5 and 95% compositions were omitted but their behavior will be described below.

It can be seen in Fig. 7.5 that the curves corresponding to the 50 and 75 wt% blends show an upward deviation in the low-frequency region. The deviation is particularly large for the 50% sample for which the slope of the G' curve decreases significantly when the frequency is decreased. The G" dependence shows a similar tendency.

In Fig. 7.6 the storage moduli for the same compositions of System-2, also show a peculiar behavior. An upward deviation can once again be observed for the curve corresponding to the blend containing 75% of LLDPE-2, but not for the one containing equal proportions of the polymers. In the latter case, the frequency dependence of G' seems slightly bimodal. On the other hand, G" exhibits normal behavior.

Figure 7.7 shows complementary results obtained in the Rheometrics Stress Rheometer, RSR. Plotted on this graph is the shear strain, γ, as a function of shear stress, σ_{12}, for blends containing 50% of each polymer for both systems.

7.4   DISCUSSION

7.4.1   Molecular Weights

Since the liquid state behavior depends strongly on the molecular weight, the SEC results will first be discussed before examining the melt rheology. The highlight of the molecular-weight measurements, as shown in Fig. 7.1, is the apparent degradation in the PP-rich region for both systems. For samples with LLDPE as the major component, the SEC results follow the broken

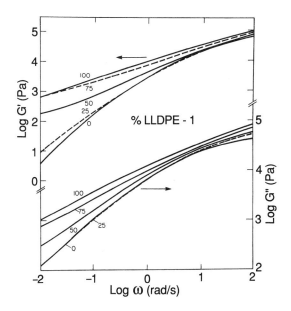

Figure 7.5 *Dynamic shear moduli as a function of frequency, ω, for System-1.*

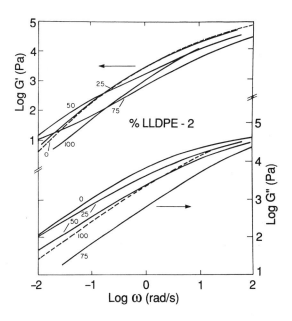

Figure 7.6 *Dynamic shear moduli as a function of frequency, ω, for System-2.*

lines representing the computed averages from Eqs (7.1) to (7.3). On the other hand, M_W and M_z for the blends containing 5 or 25% of LLDPE-1 and 5% of LLDPE-2 show a strong negative deviation while M_n for the same compositions approximately follows the broken lines. The presence of the PE seems to cause degradation of the PP matrix. Apparently, the longer macromolecular chains are primarily affected. The more viscous LLDPE-1 appears more effective in causing a molecular weight decrease. Impurities contained in the PE can possibly accentuate the mechanical and thermal degradation already taking place in the extruder. Because of this, the blends containing 5 or 25% of LLDPE-1 or 5% of LLDPE-2 must be considered different materials. They represent samples for which the molecular weights were each reduced by a different amount. This complicates the analysis of the composition dependence of the rheological functions.

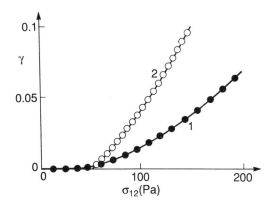

Figure 7.7 *Shear strain, γ, as a function of shear stress, σ_{12}, for blends containing 50% of each polymer; Curve 1: PP/LLDPE-1. Curve 2: PP/LLDPE-2.*

7.4.2   Yield Stress

In Figs. 7.5 and 7.6 deviations were observed in the low-frequency regions of the G' curves for the blends containing 75 and 50% of LLDPE. Such deviations have been related to the apparent yield stress, frequently reported for filled melts and composites [3,17]. An apparent yield stress has been observed in concentrated suspensions under conditions where the attractive forces between the suspended particles are such that they form a three-dimensional network [18]. For rubber modified styrene-acrylonitrile copolymers, Münstedt observed behavior similar to that reported in the present work [19]. He reported sharp increases of shear viscosity for decreasing shear rates. The shear and elongational flows as well as examination of morphology indicated that the rubber particles created a three-dimensional network which had to be broken before any flow could take place.

Similar deviations were also reported for acrylonitrile-butadiene-styrene (ABS) copolymers [20]. The extent of the deviations was directly related to the butadiene rubber phase content. It was demonstrated that the G' curves for these materials were similar to those obtained for an irradiated PS. In the latter case, the extent of the deviations was related to the irradiation dose or, in other words, to the crosslinking density. This analogy confirmed the presence of a physical network in the rubber modified samples, a network similarly postulated for our blends.

An apparent yield stress is a function of the test frequency or the time of the experiment, t:

$$\sigma_\gamma = \sigma_\gamma^0 [1 - \exp\{-t/\tau_y\}]^\alpha \tag{7.4}$$

where σ_γ^0 is the permanent yield stress observed when $t \to \infty$ or $\tau_y \to 0$ and $0 \le \alpha \le 1$, α being an exponent dependent on concentration of the temporary crosslinks and τ_y the network relaxation time [3]. Therefore an apparent yield stress can originate in the presence of long, but not infinite, relaxation times.

The rheological behavior can be more clearly presented by plotting the elastic component of the complex viscosity, η'', as a function of its viscous component, η', as in Fig. 7.8. This type of graph, often called the Cole-Cole plot, was first introduced for presentation of dielectric data [21]. When the relaxation mechanism is simple, e.g., for small molecule liquids, polymers with narrow molecular weight distribution, or some miscible blends, the dependence should be semi-circular [22,23]. Under these circumstances, the Cole-Cole plots can be used to evaluate the zero-shear viscosity [8,24].

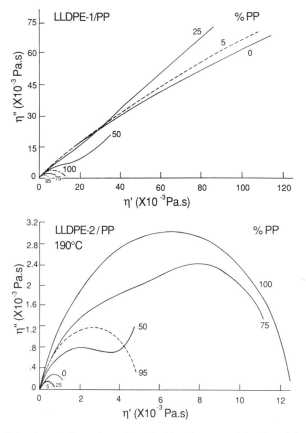

Figure 7.8 Cole-Cole plot of η'' as a function of η' for System-1 (top) and System-2.

For System-1, only the PP-rich compositions display this type of behavior. Owing to a broad relaxation times spectrum, only a small part of the curve can be seen for LLDPE-1-rich compositions. The relevance of these curves lies in the fact that they can demonstrate the presence of multiple relaxation mechanisms. For System-1, for example, the curves corresponding to blends containing 75 and 50% of LLDPE show two regions. At first, the curve seems to tend toward the semi-circular shape, then a second region shows a deviation due to a significant increase of η''. This deviation indicates the presence of the long relaxation times associated with a yield stress. The same phenomenon was observed for immiscible polystyrene/polycarbonate (PS/PC) blends while the curves for the pure polymers displayed a semi-circular shape [25].

It can be seen in Fig. 7.8 that the curves for the 50% blends also show two regions, the increase in η'' being due to the contribution of the three-dimensional structure of the material. The peculiar shape of the curve corresponding to the blend containing 25% of LLDPE-2 should be noted. Its semi-circular shape is slightly distorted, as if resulting from summation of two distinct contributions. Such bimodality was observed for homologous polystyrene blends by Montfort et al. [26]. The contribution of each component is related to its molecular weight ratio and concentration.

The presence of an apparent yield stress in the two blends containing 50% of each polymer was confirmed by an independent method. The results of the RSR experiments (see Fig. 7.7) illustrate that, within a time period of 200 s, for Systems-1 and -2 the stress had to reach a value of 36 and 52 Pa respectively, before any strain could be recorded. These results confirm the presence of an apparent yield stress postulated on the basis of the dynamic test results.

The apparent yield stress is a real physical phenomenon, especially strong in the blend containing 50% LLDPE-1 where, as shown in Figs. 7.2c and 7.3d, there is dual phase continuity. Since the 75% LLDPE-1 blend shows the same peculiar rheological behavior, one can postulate that the phase inversion region extends to this concentration. Indeed, the presence of two co-continuous phases in this blend can be seen in Fig. 7.3c. The same phenomenon was observed for System-2, both in the rheological properties as well as in the observed morphology. Thus, it is fair to postulate that the phase inversion region covers the same concentration range in both systems even though the viscosity ratio of the components is very different. This observation contradicts the semi-empirical rule proposed by Paul and Barlow [27] and found valid by Jordhamo et al. [28] for predicting the phase inversion region:

$$\phi_1/\phi_2 = \eta_1/\eta_2 \tag{7.5}$$

This expression implies that only the viscosity ratio of the two components determines the volume fraction at which the phase inversion takes place, i.e., an exchange of dispersed/matrix phase roles of the two polymers. The inversion leads to co-continuous phases. Several other systems did not follow this rule as well [29].

It is interesting to quantify these effects. It was shown that the yield stress of a composite can be determined using the modified Casson equation [3,30,31]:

$$\sigma^{1/2} = \sigma_y^{1/2} + a_0 \sigma_m^{1/2} \tag{7.6}$$

where σ is the shear stress for the composite, σ_y, the yield stress, and σ_m, the stress for the matrix. From dynamic test data, one can determine a yield stress, σ_y', corresponding to the elastic component, G', and σ_y'' for the loss component. Following the modified Casson equation approach, σ_y' is obtained by plotting $\sqrt{G'}$ for the blend against $\sqrt{G'_m}$ for the matrix. A similar procedure with the loss moduli yields σ_y''. This method is illustrated in Fig. 7.9. The LLDPE was taken as the matrix for the blends containing 25 wt% PP while PP was considered the matrix for the other compositions. The square root of the yield stress is given by the intercept of these curves. It can be seen that the curve corresponding to the blend containing 75% of LLDPE-2 extrapolates to zero, indicating absence of the yield stress. The same behavior was observed for the 5 and 95% compositions which, for sake of clarity, were not plotted. Similar curves with the loss modulus (not shown) enabled the determination of σ_y''. Table 7.2 gives the values of σ_y' and σ_y'' thus determined, as well as the material taken as the matrix for each blend. The small values of $\sigma_y'' \ll \sigma_y'$ are to be noted. This method gives a yield stress of 57.8 Pa for the blend containing 50% LLDPE-1 and 5.3 Pa for the one containing 50% LLDPE-2 while direct measurements in the RSR gave 36 and 52 Pa respectively. The two methods are not quantitatively equivalent because of different time scales. Nevertheless, they do confirm the presence of the yield stress.

Table 7.2 *Yield Stresses*

Composition (wt%)	σ_y' (Pa)	Matrix	σ_y'' (Pa)
95 LLDPE-1	0	LLDPE	0
75 LLDPE-1	39.6	LLDPE	0
50 LLDPE-1	57.8	PP	1.0
25 LLDPE-1	0.8	PP	0
5 LLDPE-1	0	PP	0
5 LLDPE-2	0	PP	0
25 LLDPE-2	1.8	PP	1.4
50 LLDPE-2	5.3	PP	0
75 LLDPE-2	0.5	LLDPE	0.2
95 LLDPE-2	0	LLDPE	0

The values for σ_y' and σ_y'' have been subtracted from G' and G'', respectively. The moduli thus corrected can then be used to calculate rheological functions which are free from the yield stress phenomenon. All the results used in the following discussion were corrected in this fashion.

7.4.3 Compositional Dependence of the Viscosity

The frequency dependence of the complex viscosity, η^*, for Systems-1 and -2 is presented in Fig. 7.10. The points are experimental while the curves illustrate calculations that will be discussed later. The behavior of the blend containing 95 wt% LLDPE-1 is close to that of the

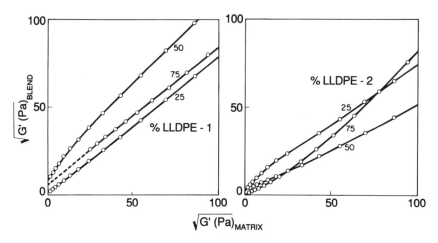

Figure 7.9 *Modified Casson plot: square root of the storage modulus of the blend as a function of the square root of the storage modulus of the matrix for System-1 (left) and System-2 (right).*

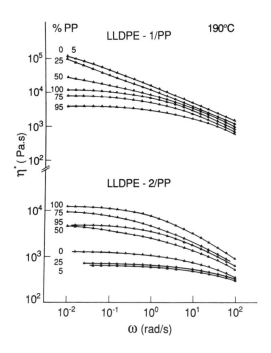

Figure 7.10 *Complex viscosity, η^*, vs. frequency for System-1 (top) and System-2 (bottom); points experimental, curves computed.*

pure PE (curve 0). The difference between the two is real but so small that only one curve could be drawn. The addition of 5% of a less viscous PP resin therefore produces little effect on the blend viscosity. The behavior is different at the other end of the concentration spectrum, where the relationship between composition and viscosity does not appear linear. The material showing the lowest viscosity is the blend containing 95% PP. The addition of up to 25% LLDPE-1 still produces a viscosity lower than that of the pure PP even though the LLDPE-1 viscosity is significantly higher.

For System-2 the behavior seems more complex. As for System-1, adding 5% of LLDPE-2 to PP induces a large reduction in viscosity. However, here the viscosity reduction is also observed at the other end of the composition range. On the other hand, the shape of the curves corresponding to the blends containing 50 and 75 wt% PP differs slightly from the others. These two curves seem to be further away from the Newtonian plateau than others, thus implying that the relaxation time spectrum is broader for these two blends.

The relationship between viscosity and composition is presented in Fig. 7.11 where the complex viscosity, η^*, measured at the lowest attainable frequency (0.01 rad/sec) is plotted as a function of PP content. The broken straight lines illustrate the logarithmic additivity rule while the solid ones were drawn through the experimental points only to illustrate trends. These curves underline the previously discussed decrease of viscosity for LLDPE-1 as the minor phase. Note that Fig. 7.1 showed a decrease of molecular weight for these compositions. Thus, these high PP-content blends are not directly comparable to the other compositions. However, it is possible to overcome this difficulty. For low polydispersity polymers, the zero-shear viscosity can be estimated from:

$$\eta_0 = KM_w^a \qquad (7.7)$$

where a $\approx$ 3.4 for linear polymers [32,33]. It has been shown that for polydisperse polymers with log-normal distribution of molecular weight, another dependence should be used [24]:

$$\eta_0 = KM_\eta^a \qquad (7.8)$$

where M_η is the melt-viscosity average molecular weight calculated from the other averages according to:

$$M_\eta = M_z (M_w/M_n)^{0.2} \qquad (7.9)$$

Thus, corrected values of the viscosity can be computed from Eq (7.8). These are shown in Fig. 7.11 as open circles and squares, representing the blends behavior without degradation. Now the η^* vs. PP content dependencies can be discussed.

For System-1 at low PP content, the LLDPE-1 seems to dominate the behavior, giving birth to a mild positive deviation. In the PP-rich region, a small negative deviation is observed, indicating that once again the matrix dominates the behavior. The shape of this curve is typical of an immiscible system where the reversal in behavior is due to a change in morphology.

The curve illustrating the behavior of System-2 is more complex. The corrected viscosity value for the blend containing 95% PP is close to the additivity rule. As also evident in Fig. 7.10, an addition of the more viscous PP to LLDPE-2 induced a decrease in viscosity. This phenomenon translates in Fig. 7.11 into a large negative deviation for blends rich in

LLDPE-2. There was no sign of any molecular weight change for these blends, i.e., degradation is not the source of this deviation. Such negative deviations have been observed for immiscible blends with high (positive) value of the interaction parameter as well as for systems approaching the phase separation [3]. A drop in viscosity can also be related to a change in morphology [34] or to an inter-layer slip [35] (see also Section 6.4.4 in Chapter 6). Absence of the negative deviation for high LLDPE-1 compositions in System-1 may suggest that the origin of such a variation in lower molecular weight System-2 may originate from partial miscibility at low PP content. The postulate needs further verification.

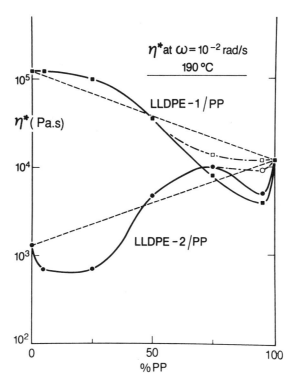

Figure 7.11 *Complex viscosity, η^*, at $\omega = 10^{-2}$ rad/sec vs. composition for Systems-1 and-2.*

7.4.4   Modeling

In the following paragraphs, other data treatments will be discussed and their advantages for data interpretation will be illustrated. It is useful to characterize the material behavior with quantifiable parameters. The viscosity at fixed stress or frequency can help represent the overall behavior. However, the zero-shear viscosity is a more important fundamental parameter. For high-molecular-weight polydisperse materials, it is often impossible to observe the Newtonian plateau. This is the case for several samples used in this work (see Fig. 7.10). In these cases, it is difficult to determine η_0. One approach is to use an appropriate extrapolative model. There are several models cited in the literature whose ability for describing the data is often proportional to their complexity. Carreau proposed a simple one for monodisperse polymer solutions or melts [36]. Elbirli and Shaw extended this approach to polydisperse systems [37]. Recently, their equation was re-derived by Utracki [38]:

$$\eta' = \eta_0 \left[1 + (\tau^* \omega)^{m_1} \right]^{-m_2}$$

(7.10)

where η_0 is one of the adjustable parameters, the other three being: τ^*, the mean relaxation time and the two exponents, m_1 and m_2. The experimental dynamic viscosity, $\eta' = G''/\omega$, from which the yield stress had already been subtracted, was fitted to this equation using a nonlinear regression technique. The parameters obtained from the regressions are listed in Table 7.3. An excellent description of the data was obtained, as shown in Fig. 7.10. Here, the curves are computed from Eq (7.10) using the parameters listed in Table 7.3 while the points are experimental.

Table 7.3 *Characteristic Rheological Parameters*[a]

Material	η_0 (kPas)	τ^* (ms)	m_1	m_2	η_{0c} (kPas)
LLDPE-1 : 100%	8120	541	0.107	9.79	-----
LLDPE-1 : 95%	6270	332	0.108	9.80	-----
LLDPE-1 : 75%	380	104	0.162	6.85	-----
LLDPE-1 : 50%	43.2	73.6	0.272	4.19	-----
LLDPE-1 : 25%	8.52	244	0.551	1.47	8.5
LLDPE-1 : 5%	3.97	165	0.741	1.03	4.4
PP	13.0	411	0.623	1.37	12.5
LLDPE-2 : 5%	5.33	209	0.587	1.33	4.95
LLDPE-2 : 25%	19.3	4.86	0.235	6.13	11.7
LLDPE-2 : 50%	6.09	4.80	0.279	4.62	-----
LLDPE-2 : 75%	0.706	14.2	0.586	1.12	0.72
LLDPE-2 : 95%	0.695	38.4	0.592	0.807	0.68
LLDPE-2 : 100%	1.32	321	0.680	0.597	1.35

[a] *Parameters for Eq (7.10).*

For $\tau^* \cdot \omega \gg 1$, the equation simplifies to:

$$\eta' \simeq \eta_0 [\tau^* \cdot \omega]^{-m_1 \cdot m_2}$$

(7.11)

Comparing this relation to the power law model:

$$\eta = k \dot{\gamma}^{n-1}$$

(7.12)

one can see that at high frequencies, $0 \leq m_1 m_2 \leq 1$. A value outside this range indicates either an experimental error (in the data or in the curve-fitting procedure) or significant changes in the melt structure.

The zero-shear viscosity, η_0, obtained by fitting the data to Eq (7.10) is plotted for both systems in Fig. 7.12 as a function of composition. The values corresponding to the compositions for which the degradation was observed can be corrected according to the procedure described above; these are indicated by open circles and squares. The shape of the curves in Fig. 7.12 is similar to those of Fig. 7.11 as at $\omega = 10^{-2}$ rad/s for several materials (especially those of System-2) the Newtonian plateau was nearly reached. On the other hand, η_0 for LLDPE-1 rich blends is much larger than the experimental viscosities. In these cases, the Newtonian plateau is far down from the accessible frequency range. The η_0 vs. composition curve for System-1 is therefore quite different from that shown in Fig. 7.11. There is no positive deviation noted in Fig. 7.12. It is worth noting that η_0 of the materials containing 95 and 100% LLDPE-1 is larger by two decades than η^* measured at $\omega = 10^{-2}$ rad/s. These values of η_0 correspond to frequencies of about 10^{-10} rad/sec.

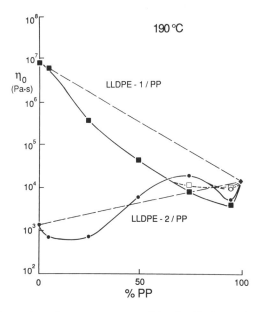

Figure 7.12 *Zero-shear viscosity, η_0, vs. composition for Systems-1 and -2. Open points were calculated for undegraded samples.*

It was mentioned above that the Cole-Cole plots can be used for estimating the zero-shear viscosity if the semi-circular shape could be obtained. By definition, η_0 equals η' when η'' and ω go to zero. Values of η_0 have been determined by extrapolating the curves in Fig. 7.8 (those that could be assumed to be symmetrical) to the abscissa. The values thus obtained are called η_{0c}, the subscript c indicating that the determination was done with the

Cole-Cole plot, and are listed in Table 7.3. In most cases, the agreement between these and the values obtained from Eq (7.10) is good. A significant discrepancy is observed only for the blend containing 25% LLDPE-2. The curve-fitting operations were performed using data corrected for the yield stress while the Cole-Cole plots used uncorrected data. The curve for the blend containing 25% LLDPE-2 clearly shows an irregular shape, indicating changes in the relaxation spectrum which question the accuracy of η_{0c}.

7.4.5 Crosspoint Coordinates

Since in the melt the storage shear modulus, G', shows a stronger frequency dependence than the loss shear modulus, G", the two functions are equal at one point, called the cross point. The coordinates of this point are the crosspoint modulus, $G_x \equiv G' = G''$, and crosspoint frequency, ω_x. It was shown for PP, that these coordinates were related to the molecular parameters; G_x was found to be proportional to $(M_n/M_w)^a$ and ω_x to $(1/\eta_0)^b$ where a and b are positive, empirical parameters [39].

The crosspoint coordinates for Systems-1 and -2 are plotted in Fig. 7.13 as functions of composition. It is to be noted that variations of ω_x qualitatively correspond to these of $1/\eta_0$ (viz. Fig. 7.12). For System-1, η_0 increases with the LLDPE-1 content while ω_x decreases. Similarly, reversed dependencies are observed for System-2. In addition, the negative deviation of viscosity in the PP-rich region is reflected here in the form of a local increase in ω_x. Therefore it seemed useful to check the relation between η_0 and ω_x. Figure 7.14 illustrates the dependence. The data for System-1 show a regular decrease of η_0 with increase of ω_x down to the blend containing 5% LLDPE-1. In effect, addition of LLDPE-1 to this blend generates a series of mixtures with additive properties. Note that the point corresponding to PP shows

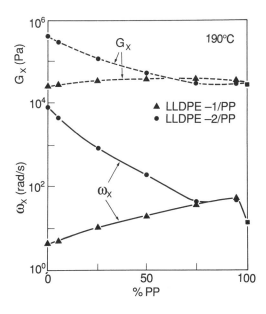

Figure 7.13 *Crosspoint coordinates, G_x and ω_x, vs. composition for Systems-1 and-2.*

quite different behavior. The data for System-2 behave in a more erratic way. The η_0 vs. ω_x dependence has a maximum and two local minima. Such behavior may suggest more dramatic changes of blend morphology with composition than those in System-1.

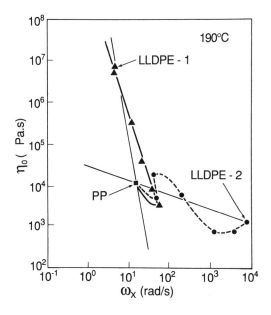

Figure 7.14 *Zero-shear viscosity, η_0, vs. crosspoint frequency, ω_x, for Systems-1 and -2.*

Both moduli G' and G" can be described in terms of the relaxation spectrum, $H(\tau)$:

$$G' = \int_0^\infty z^2 H(\tau) \, d\ln \tau / (1 + z^2) \tag{7.13}$$

$$G'' = \int_0^\infty z \, H(\tau) \, d\ln \tau / (1 + z^2) \tag{7.14}$$

where $z = f \cdot \tau$, f being the frequency and τ the relaxation time [40]. By definition, $G' \simeq G''$ at the crosspoint. Under this condition, Eqs (7.13) and (7.14) must be equal, which means that $z = 1$ or, in other words:

$$f_x = 1/\tau \tag{7.15}$$

For a single Maxwell element, G' is defined as:

$$G' = (\eta_0 \tau f^2) / (1 + \tau^2 f^2) \tag{7.16}$$

Combining Eqs (7.15) and (7.16), one obtains:

$$G_x = \eta_0 f_x/2 \qquad (7.17)$$

The zero-shear viscosity of a single Maxwell element, η_{0M}, is thus related to the ratio of the crosspoint coordinates by the expression:

$$\eta_{0M} = 4\pi G_x/\omega_x \qquad (7.18)$$

in which a factor of 2π accounts for the conversion to angular frequency [3]. A graph of η_{0M} as a function of η_0, which can be seen in Fig. 7.15, shows that the low viscosity data are concentrated around the straight line: $\eta_0 = \eta_{0M}$. The high viscosity points depart from this equality at about $\eta_0 = 20$ kPas. It should be noted that the same phenomenon was observed for blends of LLDPE with LDPE or LLDPE [24] for which the upper limit was 60 kPas.

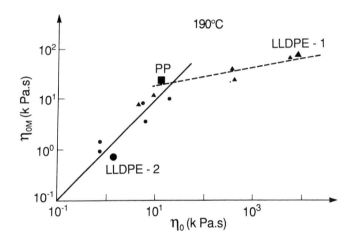

Figure 7.15 *Maxwellian viscosity, η_{0M}, vs. zero-shear viscosity, η_0.*

7.4.6 Relaxation Spectrum

The rheological functions are interrelated by the relaxation spectrum. The frequency relaxation spectrum was defined [40] as:

$$H_G(\omega) = (2/\omega\pi)\,\text{Re}\,G''\,(\omega\exp\{\pm i\pi/2\}) \qquad (7.19)$$

where Re represents the real part of the loss shear modulus written in terms of the complex frequency $\omega e^{\pm\pi/2}$. Since $G'' = \eta'\omega$, Eqs (7.10) and (7.19) can be combined to yield [24]:

$$H_G(\omega)/\eta_0 \equiv \tilde{H}_G(\omega) = (2/\pi)\sin(m_2\theta)/r^{m_2} \qquad (7.20)$$

with

$$r \equiv [1 + 2\,(wt^*)^{m_1}\cos(m_1\bar\eta/2) + (wt^*)^{2m_1}]^{1/2}, \text{ and}$$

$$\theta \equiv \arcsin\{(wt^*)^{m_1}r^{-1}\sin(m_1\bar\eta/2)\}$$

where $\tilde{H}(\omega)$ represents the reduced frequency relaxation spectrum. The spectra calculated with this expression are shown in Fig. 7.16. There is a significant variation in the shape of these curves with composition. For materials with large zero-shear viscosity, the spectrum is broader, spreading to lower frequencies. For System-1, the broadest spectra are observed for the blends with the largest η_0 values, i.e., containing 100 and 95% LLDPE-1. For System-2, the blends containing 25 and 50% LLDPE-2 have the broadest spectra. In Sections 7.4.2 and 7.4.3, describing the complex viscosities, a broad relaxation spectrum for these two materials was postulated. This premise is confirmed by Fig. 7.16.

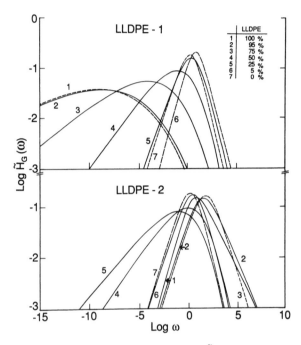

Figure 7.16 *Reduced frequency relaxation spectrum, $\tilde{H}(\omega)$, vs. frequency, ω, for System-1 (top) and System-2 (bottom).*

The shape of the relaxation spectra can be discussed quantitatively in terms of the coordinates of the maxima, ω_{max}, and $\tilde{H}_{G,max}$. The position of the maximum, ω_{max}, and η_0 is shown in Fig. 7.17. Since they are normalized, the area under the curve is equal to unity:

$$\int_{-\infty}^{+\infty} \tilde{H}_G(\omega) \, d\ln \omega = 1 \tag{7.21}$$

The height of the spectrum is therefore inversely proportional to its width. The relationship between the spectra breadth and the polydispersity of the materials can then be illustrated by plotting $\tilde{H}_{G,max}$ as a function of the inverse of the polydispersity index, M_n/M_w, as in Fig. 7.18.

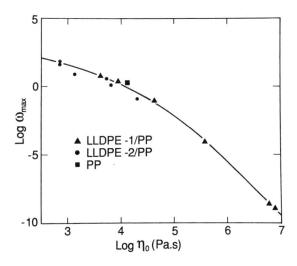

Figure 7.17 *Frequency at which* $\tilde{H}_G(\omega)$ *attains maximum*, ω_{max}, *vs. zero-shear viscosity*, η_0.

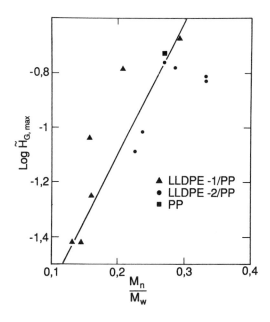

Figure 7.18 *Maximum of relaxation spectrum*, $\tilde{H}_{G,max}$, *vs.* M_n/M_w.

The data in Fig. 7.17 seem to follow one curve with the scatter well within the experimental error. However, the apparent regularity of the ω_{max} changes with η_0 is misleading. It should be noted that for both blend series, the data points for the blends containing small amounts of LLDPE are placed on the opposite side of the curve from PP than the rest. In short, the ω_{max} vs. η_0 relation indicates strong nonlinearity of the blend behavior, evidence of immiscibility [3]. For the same reason, the $\tilde{H}_{G,max}$ vs. M_n/M_w plot in Fig. 7.18 is a complex one.

While the spectrum was derived from the dynamic viscosity, it can also be used to calculate other rheological functions. For example, the storage shear modulus in Fig. 7.19 was computed from the $H_G(\omega)$. It should be mentioned that Eq (7.10) was the starting point for the calculation of the relaxation spectrum. Since the viscous component of the viscoelastic response is used in this expression, calculating the elastic one from the spectra and then comparing it with the experimental data constitutes a proper verification of the method. In Fig. 7.19, the points are experimental and the curves have been computed. The agreement is satisfactory. The small fluctuations observed at low frequencies are due to the fact that at low frequency, the measured torque is small and, therefore, less precise. In conclusion, the agreement between the experimental data and the computed curves confirms the validity of the method and indicates that the materials behave like a linear viscoelastic system.

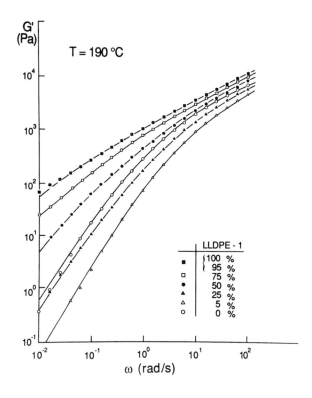

Figure 7.19 *Storage shear modulus, G', vs. frequency, ω. The points are experimental and the curves are computed for a linear viscoelastic material.*

Since this approach utilizes Eq (7.10) for the curve fitting, its applicability will depend on the feasibility of describing the frequency dependence of η' with the four-parameter expression. Therefore there may be some difficulties in handling complex rheological responses. Phenomena like yield stresses have to be corrected for before the method is to be applied. However, when curve fitting is feasible, the method is analytical and simple. An alternative is to use a two-step approach by fitting Eq (7.10) to data at low then at high frequencies. The advantage of this approach is the possibility of constructing a bimodal relaxation spectrum realistically describing the phase-separated rheologically complex systems [3].

7.5   CONCLUSIONS

The first observation to be made concerns the apparent degradation of PP during the blending step. The large decrease of the molecular weight reported for PP-rich compositions seems to be induced by the presence of PE. The effect is particularly large for the more viscous LLDPE-1. However, once this effect is compensated for, the composition dependencies of the viscosity are typical of two-phase systems. Each material dominates the behavior in the region in which it constitutes the major component.

Examination of morphology of the extruded and injection-molded samples has led to a number of observations, the changes due to the injection step being the most interesting. The high shear rates and pressures encountered during the injection molding induced a partial solubilization of the molten materials. The subsequent rapid cooling hindered the complete segregation of the domains, reducing the apparent concentration of the dispersed phase.

Blends containing 25 and 50% PP showed phase co-continuity; therefore the phase-inversion region covers this concentration range. It should be noted that this is the case for both systems despite the reversed viscosity ratios. This phase co-continuity has been linked to changes in the relaxation-time spectrum resulting in an apparent yield stress. This phenomenon was inferred from the shape of the frequency dependence of the dynamic moduli and confirmed by direct measurements.

Several data-treatment approaches were used. First, the crossover coordinates were found to be related to the material's molecular characteristics by means of the zero-shear viscosity, η_0. The use of the Cole-Cole plots was also shown to be beneficial, allowing determination of η_0 as well as making it possible to appreciate more fully the subtle changes in the material's behavior. Finally, the use of the four-parameter equation to model the frequency dependence of the dynamic viscosity was described. Determination of the parameters allowed calculation of the relaxation spectrum from which any linear viscoelastic function could be computed.

Thus, the PP/LLDPE blends were found to be immiscible and rheologically complex, the nonadditivity originating in the blend structure. The effect of the interphase was particularly strong within the phase-inversion region where a co-continuity of phases was observed. In the solid state, the two-phase nature of these materials strongly affected the ultimate properties, viz. the yield stress or the maximum strain at break [14,15]. On the other hand, the time-temperature superposition of low-strain dynamic tensile moduli was found to be valid at temperatures below the glass transition point [16]. Comparing the results of this work with an earlier study [41,42] one must conclude that the PP molecular weight plays a dominant role in determining the blend performance [43].

NOTATION

ABS	Acrylonitrile-butadiene-styrene
EVA	Ethylene-vinyl acetate copolymer
HDPE	High density polyethylene
HIPS	High impact polystyrene
LDPE	Low density polyethylene
LLDPE	Linear low density polyethylene
PB	Polybutadiene
PC	Polycarbonate
PE	Polyethylene
PP	Polypropylene
PS	Polystyrene
PVC	Polyvinylchloride
RMS	Rheometrics Mechanical Spectrometer
RSR	Rheometrics Stress Rheometer
SEC	Size exclusion chromatography
f	Frequency
G_x	Crossover shear modulus
G'	Storage shear modulus
G''	Loss shear modulus
$H_{G,max}$	Maximum of the relaxation spectrum
$H_G(\omega)$	Frequency relaxation spectrum
$\tilde{H}_G(\omega)$	Reduced frequency relaxation spectrum
$H(\tau)$	Time relaxation spectrum
M	Molecular weight
M_n	Number average molecular weight
M_w	Weight average molecular weight
M_z	Z-average molecular weight
Pa	Pascal
rad	Radian
R	Ideal gas constant
Re	Real part of a complex function
sec	Second
t	Time
T	Temperature
z	Reduced frequency, $f \cdot \tau$
a	Parameter
γ	Shear strain
η	Viscosity
η_0	Zero-shear viscosity
η'	Dynamic viscosity
η^*	Complex viscosity
ρ	Density
σ	Stress
σ_{12}	Shear stress
σ_y	Yield stress
σ_y^0	Permanent yield stress
σ_m	Stress for the matrix phase

τ	Relaxation time
τ^*	Mean relaxation time
ω	Angular frequency
ω_{max}	Frequency at which $\tilde{H}(\omega)$ is maximum
ω_x	Crossover frequency

REFERENCES

1. Wigotsky, V., *Plast. Eng.*, **43**, 21 (1987).
2. Manson, J. A., Sperling, L. H., *Polymer Blends and Composites*, Plenum Press, New York (1976).
3. Utracki, L. A., *Polymer Alloys and Blends*, Hanser Publishers, Munich (1989).
4. Romanini, D., *Polym. Plast. Technol. Eng.*, **19**, 201 (1983).
5. Anonymous, *Chemical Marketing Reporter*, Nov. 23, p. 327 (1987).
6. Strobel, W., *Kunststoffe*, **77**, 967 (1987).
7. Hamielec, L., *Polym. Eng. Sci.*, **26**, 111 (1986).
8. Dumoulin, M. M., Farha, C., Utracki, L. A., *Polym. Eng. Sci.*, **24**, 1319 (1984).
9. Yeh, P.-L., Birley, A. W., *Plast. Rubb. Proc. Appl.*, **5**, 249 (1985).
10. Ward, T. W., *Caoutch. Plast.*, **62**, 39 (1985).
11. Herner, M., Wolter, H.-J., *Kunststoffe*, **77**, 959 (1987).
12. Speed, C. S., *Plast. Eng.*, **38(7)**, 39 (1982).
13. Nancekivell, J., *Canad. Plast.*, **40(5)**, 27 (1982).
14. Dumoulin, M. M., Utracki, L. A., Carreau, P. J., Proc. *2nd Conference of European Rheologists*, Prague, Czechoslovakia, June 17-20, 1986; *Rheol. Acta*, **26**, 215 (1988).
15. Dumoulin, M. M., Carreau P. J., Utracki L. A., *Polym. Eng. Sci.*, **27**, 1627 (1987).
16. Dumoulin, M. M., Utracki, L. A., in *Polymer Rheology and Processing*, Collyer, A. A., Utracki, L. A., Eds., Elsevier Appl. Sci., London (1990).
17. Utracki, L. A., Fisa, B., *Polym. Comp.*, **3**, 193 (1982).
18. Dzuy, N. Q., Boger, D. V., *J. Rheol.*, **27**, 321 (1983).
19. Münstedt, H., *Polym. Eng. Sci.*, **21**, 259 (1981).
20. Zosel, V. A., *Rheol. Acta*, **11**, 229 (1972).
21. Cole, K. S., Cole, R. H., *J. Chem. Phys.*, **9**, 341 (1941).
22. Wisniewski, C., Marin, G., Monge, Ph., *Eur. Polym. J.*, **20**, 691 (1984).
23. Malik, T. M., Prud'homme, R. E., *Polym. Eng. Sci.*, **24**, 144 (1984).
24. Utracki, L. A., Schlund, B., *Polym. Eng. Sci.*, **27**, 367 (1987).
25. Wisniewski, C., Marin, G., Monge, Ph., *Eur. Polym. J.*, **21**, 479 (1985).
26. Montfort, J. P., Marin, G., Arman, J., Monge, Ph., *Polymer*, **19**, 277 (1978).
27. Paul, D. R., Barlow, J. W., in *Polymer Blends*, Paul, D. R., Newman, S., Eds., Academic Press, New York (1978).
28. Jordhamo, G. M., Manson, J. A., Sperling, L. H., *Polym. Eng. Sci.*, **26**, 517 (1986).
29. Favis, B. D., Chalifoux, J. P., Van Gheluwe, P., *SPE Techn. Pap.*, **33**, 1326 (1987).
30. Onogi, S., Matsumoto, T., *Polym. Eng. Rev.*, **1**, 45 (1981).
31. Utracki, L. A., *Rubber Chem. Technol.*, **57**, 507 (1984).
32. Cross, M. M., *J. Appl. Polym. Sci.*, **13**, 765 (1969).
33. Mills, N. J., *Eur. Polym. J.*, **5**, 675 (1969).
34. Han, C. D., *Multiphase Flow in Polymer Processing*, Academic Press, New York (1981).
35. Lin, C.-C., *Polym. J.*, **11**, 185 (1979).

36.   Carreau, P. J., *Trans. Soc. Rheol.*, **16**, 99 (1972).

37   Elbirli, B., Shaw, M. T., *J. Rheol.*, **22**, 561 (1978).

38.   Utracki, L. A., in *Current Topics in Polymer Science*, Ottenbrite, R. M., Utracki, L. A., Inoue, S., Eds., Hanser Publishers, Munich (1987).

39.   Zeichner, G. R., Patel, P. D., Proc. *2nd World Cong. Chem. Eng.*, **6**, 333 (1981).

40.   Gross, B., *Mathematical Structure of the Theories of Viscoelasticity*, Hermann, Paris (1968).

41.   Dumoulin, M. M., Utracki, L. A., Farha, C., *SPE Techn. Pap.*, **30**, 443 (1984).

42.   Dumoulin, M. M., Farha, C., Utracki, L. A., *Polym. Eng. Sci.*, **24**, 1319 (1984).

43.   Utracki, L. A., Chapter 7 in *Multiphase Polymers: Blends and Ionomers*, Utracki, L. A., Weiss, R. A., Eds., ACS Symp. series No. 395, Amer. Chem. Soc., Washington, D.C. (1989)

CHAPTER 8

CRYSTALLIZATION BEHAVIOR IN POLYMER BLENDS

by V. M. Nadkarni and J. P. Jog

National Chemical Laboratory
Pune 411008
INDIA

Thermoplastic blends and alloys form a commercially important class of materials. A majority of the engineering polymer blends contain at least one crystalline component. The properties of the finished articles are governed by the morphology created as a result of the interplay of processing conditions and the inherent polymer characteristics, including crystallizability. Therefore, a scientific understanding of the crystallization behavior in polymer blends is necessary for optimization of the compounding and processing conditions for effective manipulation and control of properties.

In a multicomponent system, the thermal and chemical environment under which a polymer crystallizes is modified as a result of the presence of the second component. Therefore, the investigations concerning crystallization in polymer blends are reviewed by categorizing them in terms of the physical conditions of crystallization as follows: 1) concurrent crystallization; 2) crystallization in the presence of molten phase of the second component; and 3) crystallization in the presence of solidified domains of the second component. The crystallization behavior and morphology of a polymer is modified significantly as a result of blending. There are a number of factors that govern the extent and direction of change. These include miscibility of the component polymers, their glass transition and melting temperatures, their relative melt viscosities at the temperatures of crystallization, chemical compatibility, relative crystalizabilities, and the as-compounded phase morphology. In view of the specificity and complex interdependence of these factors, this chapter summarizes the general trends.

8.1   INTRODUCTION

The technology of thermoplastic blends and alloys has led to the development of new thermoplastics applications, such as automotive bumpers and body panels, blow-molded "barrier" containers for oil products, electrical switchgear, and other engineering components. All such applications are functional and therefore require materials with specific balanced combinations of physical properties, processing characteristics, and cost. The main property improvements sought through polymer blending/alloying include impact strength, heat-distortion temperature, processability, flame retardancy, and controlled permeability [1]. The versatility of polymer blends and alloys in upgrading the performance of plastics is well illustrated by the fact that about four hundred different blends and alloys are being marketed commercially [2]. The study of polymer blends and alloys forms an active area of research in polymer science [3-16].

The properties of the molded and extruded products made from thermoplastic blends are governed by the morphological features such as domain size, shape, interfacial interactions, degree of crystallinity, crystallite size, and supermolecular morphology developed in processing as a result of the interplay of processing conditions and inherent polymer characteristics such as miscibility, melt flow, and crystallizability of the component polymers in the blend [17]. Therefore, a scientific understanding of the crystallization behavior of the component polymers in thermoplastic blends is necessary for optimizing processing conditions and properties.

8.2   CRYSTALLIZATION IN HOMOPOLYMERS

The crystallization in polymers is similar to that of low-molecular-weight compounds involving the formation of nuclei and their subsequent growth. In the case of polymer crystallization from a homogeneous melt, nucleation involves spontaneous aggregation of extended polymer–chain segments aligned to form ordered units that can lead to stable nuclei. The nucleation process is reversible up to a certain critical size of the nuclei. Once a critical-size nucleus is formed, the subsequent addition of chains is irreversible and the growth process is deemed to have commenced. The distribution of such crystalline embryos in the melt is assumed to be random. Usually the nucleation step is rate controlling. The critical nucleus size depends on both the temperature of crystallization (it increases with increasing temperature) and the molecular structure of the polymer. At temperatures close to the melting point, T_m, the critical nucleus size is large. It is, therefore, necessary to supercool the polymer melt to a temperature below T_m at which formation of stable nuclei is facilitated. However, the rate of nucleation goes through a maximum as the temperature decreases. This is because although the thermodynamic driving force increases with decreasing temperature, the mobility decreases, adversely affecting the "transport" of chain segments for the formation of stable nuclei. The reduced mobility at low temperatures also slows the rate of crystal growth. Thus, the lower temperature limit for polymer crystallization is the glass transition temperature of the polymer, T_g. Consequently the crystallization of polymers is possible only over a definite range of temperatures between T_m and T_g, and the maximum rate of crystallization is usually observed at temperatures close to $(T_g + T_m)/2$ [18]. For the theoretical details on the kinetics of polymer crystallization, the reader is referred to standard textbooks on the subject [19-21].

Investigations of the polymer crystallization may be carried out under both isothermal and nonisothermal conditions. The nonisothermal crystallization studies help to elucidate structure development in melt processing whereas the isothermal studies are useful for investigating the mechanisms of crystallization.

A variety of experimental techniques are used for carrying out the crystallization studies. The most common include optical microscopy [22-29], differential scanning calorimetry [30-39], and dilatometry [40-50]. These techniques involve measurements of a characteristic parameter of the crystallization process such as spherulitic growth rate, heat of crystallization, and specific volume, respectively. However, other techniques such as wide-angle X-ray diffractometry, infrared spectroscopy, and nuclear magnetic resonance are also used for investigating the crystallization behavior and crystalline morphology of polymers.

The general treatment of the kinetics of a phase change was developed by Avrami [51] and was applied to polymeric systems by Morgan [52] and Mandelkern et al. [53]. According to this theory, the weight fraction of crystallized polymer may be expressed as:

$$1 - x_t = \exp\{-Kt^n\} \tag{8.1}$$

where x_t is the weight fraction crystallized at time, t, K is the rate constant, and n is the Avrami exponent. The rate constant, K, includes the combined effects of nucleation and growth. The Avrami exponent provides qualitative information about the nature of the nucleation and growth processes on a relative basis. The plot of the extent of crystallization, x_t, versus the logarithm of time at a fixed temperature is termed a crystallization isotherm. The isotherms have characteristic sigmoidal shapes. There is an initial induction period followed by a period of accelerated crystallization. In a few cases, this accelerated crystallization is followed by a slower crystallization process which is known as secondary crystallization. The analysis of these sigmoidal curves leads to the estimation of different crystallization parameters such as induction time, crystallization halftime, rate constant, and Avrami exponent. These parameters are used to describe the isothermal crystallization behavior of polymers.

The isothermal crystallization parameters obtained from this analysis are useful in elucidating the inherent crystallizability of the polymer. However, in polymer processing operations such as extrusion, injection molding, and fiber spinning, the temperature varies with time and space, and the polymer crystallizes under nonisothermal conditions. The study of nonisothermal crystallization of polymers is thus relevant for optimizing processing conditions to obtain the required product properties. The investigations for quantitative evaluation of the nonisothermal crystallization process are rather limited [54-65].

Ziabicki [54,55] and Ozawa [56] have derived kinetic equations based on Avrami's theory. Ziabicki's analysis is based on the assumptions that the nonisothermal crystallization process may be treated as a sequence of isothermal crystallization steps and the importance of the secondary crystallization is negligible. Ozawa has also extended Avrami's theory to nonisothermal crystallization conditions. According to his theory the degree of crystallization at any temperature is given by:

$$-\ln\left[1 - x_t\right] = C(t)/q^{n'} \tag{8.2}$$

where q is the heating or cooling rate, C(t) is a cooling function of the process, and n' is the Ozawa's exponent.

Thus, the crystallization from the "solid" state occurring during heating and the crystallization from the molten phase taking place during cooling of the melt are studied to investigate the effect of external conditions on the crystallization behavior of polymers.

The crystallization process results in different morphologies depending on the crystallization conditions and the inherent crystallizability of the polymer. The main structural parameters used to describe the crystalline morphology include the degree of crystallinity, crystallite size and its distribution, as well as the supermolecular structure describing the geometrical arrangement of crystallites.

The crystallite size and size distribution change with crystallization conditions and can be studied using DSC, SAXS, SALS, and other techniques (the abbreviations are given in the NOMENCLATURE at the end of this chapter). The geometrical arrangement of these crystallites can lead to different morphological structures such as spherulites, axialites, shishkebab, or fibrils. Spherulites are the most commonly observed structures in polymers crystallized from melt. They consist of lamellae radiating from a central point. A spherulite becomes a three-dimensional spherical object by virtue of the fact that the lamellae occasionally branch. The structure of spherulites has been a subject of intensive investigations and it has been reviewed in detail by Keller [66] and by Geil [67].

Axialites consist of multilayered lamellas which are splayed about a common axis. The fibrils and shishkebab structures are often observed in flow-induced crystallization from melt- or strain-induced crystallization from the "solid" state. However, polymers crystallized from quiescent melts generally exhibit spherulitic morphology.

Recently, characterization of the polymer-chain conformations in the "interfacial region" between the crystalline and amorphous zones has been the subject of intensive research, employing sophisticated techniques such as broad-line proton NMR and Raman spectroscopy [68].

Various morphological parameters govern the physical properties of the molded or extruded parts of crystalline polymer blends. Thus, investigations on crystallization in polymer blends and alloys concern elucidating the effect of blending on the rate of nucleation and growth, degree of crystallinity, crystallite size, and size distribution, and on the interfacial and supermolecular structure.

8.3   CRYSTALLIZATION IN POLYMER BLENDS

The binary crystalline polymer blends and alloys may be categorized in terms of crystalline/crystalline systems, wherein both the component polymers are crystallizing, and crystalline/amorphous systems, wherein only one of the components is crystallizing [2]. The thermal and crystallization behavior of the component polymers in the blend are influenced by their relative amounts, chemical compatibility, and the level of dispersion achieved in the compounding process. Other factors such as the relative crystallizability, diffusion of the noncrystallizing component, and kinetic factors associated with the conditions of processing also influence the phase morphology [69].

The crystallization in binary polymer blends was discussed by Martuscelli [70] with special reference to the effect of composition and processing conditions on the nucleation, growth, overall rate of crystallization, the degree of crystallinity, and morphology. He analyzed the data in terms of primary nucleation, spherulitic growth, overall crystallization rates, and melting behavior. He noted that, in the case of miscible blends, generally, a depression in the melting point and the radial growth rate is observed.

The crystallization in miscible polymer blends has also been reviewed by Paul and Barlow [71]. The authors reported that the kinetics of crystallization and the crystalline morphology are affected as a result of blending. The observed changes in the crystallization kinetics were attributed to the changes in T_g caused by blending. Recently Runt and Martynowicz [72] have reviewed some aspects concerning crystallization in miscible blends with special reference to the estimation of polymer-polymer interaction using experimental data on melting-point depression.

The quantitative analysis of the spherulitic growth-rate equation in miscible polymer blends has been presented by Olabisi, Robeson, and Shaw [73]. The authors reviewed the crystallization behavior in miscible polymer blends, such as PVC/PCL, PPE/i-PS, PVDF/PEMA, PVDF/PMMA, PBT-THF/PVC, and PCL/CAB.

In the case of immiscible blends, the addition of a second noncrystallizing polymer results in significant changes in the spherulitic morphology. The effect of the second phase on the overall rate of crystallization was found to be either positive or negative depending on the conditions of crystallization, composition, molecular structure and molecular weight of the noncrystallizing component. The incorporation of one or more diluents in a crystallizable polymer may lead to the following modifications in its crystallization behavior: 1) no effect on crystallization rate or morphology; 2) retardation of crystallization with or without change in morphology; 3) prevention of crystallization at high loadings; 4) acceleration of crystallization with or without morphological change; and 5) crystallization of normally noncrystallizing polymer as a result of induced mobility.

The investigations concerning crystallization in polymer blends are summarized in Table 8.1. In a multicomponent system, the thermal and chemical environment under which a polymer crystallizes is modified as a result of the presence of the second component. The melting points and the normal ranges of crystallization temperatures of the component polymers largely determine the physical conditions of crystallization. Thus, if the melting points of the component polymers are comparable, as for example is the case for PP/PVDF blends, the component polymers may crystallize concurrently or crystallize sequentially depending upon the individual crystallizability of the polymers. On the other hand, if the difference in the melting points is significant, then the high-T_m polymer would crystallize in presence of the melt of the second component, whereas the low-melting second component would crystallize in presence of the solid phase of the other component. The presence of the second component either in the molten or solid state influences the relative rates of crystallization and the degree of crystallinity thereby modifying morphology of the blend. Tables 8.2 and 8.3 present the thermal data of principal polymers used in the various blend systems. These were taken from the references indicated in the table and from [18]. In order to elucidate the effect of the presence of the second component on the crystallization in polymer blends, the reported data may be categorized in terms of the physical conditions of crystallization, as follows: 1) concurrent crystallization; 2) crystallization in the presence of molten phase of the second component; and 3) crystallization in the presence of solidified domains of the second component.

Table 8.1 *Summary of Reported Work on Polymer Blends and Alloys*

No.	Base polymer	Modifying polymer
1.	Low-density polyethylene (LDPE)	Linear low-density polyethylene, low-density polyethylene, high-density polyethylene, elastomers, polyamide-6
2.	Linear low-density polyethylene (LLDPE)	Linear low-density polyethylene, ultrahigh-molecular-weight polyethylene, thermoplastic elastomers, polyamide-6
3.	High density polyethylene (HDPE)	Poly-1-butene, ethylene-propylene, copolymer, polystyrene, polycarbonate, polyamide-6, polycaprolactone
4.	Polypropylene (isotactic) (PP)	Poly-1-butene, high-density polyethylene, low-density polyethylene, polyisobutylene, polyamide-6, styrene-ethylene-butylene-styrene terpolymer, ABS, PC, PS, PVDF
5.	Polyvinylidenefluoride	Polyethylacrylate, PMMA, PEMA, PVMK
6.	Polybutyleneterephthalate (PBT)	Liquid crystalline polymers, PC, polyarylate, Phenoxy, EVA, PET
7.	Polyethyleneterephthalate (PET)	PC, PEC, polyamide-6, polyamide-66, liquid crystalline polymers, PP, HDPE, PMMA, Phenoxy
8.	Copolyester (PETG)	Polycarbonate, polyamide
9.	Polyphenylenesulfide (PPS)	HDPE, PET
10.	Polyamide-6	EVA, polyamide-11
11.	Polyamide-12	Liquid crystalline polymer
12.	Polyamide-66	EPR
13.	PEEK	PEEK, PEK

Table 8.2 Thermal Data on the Component Polymers in Crystalline/Crystalline Blends

No	Blend system		Glass Transition temperature T_g (°C)		Melting point T_m (°C)		Crystallization temperature T_c (°C) at 10°C/min		Physical conditions for crystallization of A	Physical conditions for crystallization of B	Refs.
	Component A	Component B	A	B	A	B					
1.	LDPE	LDPE	-123	-123	115-123	115-123	101	101	Cocrystallization		79
2.	LDPE	LLDPE	-123	-123	115-123	---	101	---	Concurrent crystallization		77,78
3.	LDPE	HDPE	-123	-123	115-123	132-140	101	118	Solidified HDPE	LDPE melt	81
4.	LLDPE	HDPE, UHMWPE	-123	-123	125	140	---	123			75,76
5.	HDPE	PB	-123	-45	132-140	114	118	---	PB melt	HDPE melt	83
6.	HDPE	PP	-123	-20	132-140	165	118	121	Crystallized PP	HDPE melt	90,92,94
7.	HDPE	PPS	-123	88	132-140	280	118	256	Crystallized PPS	HDPE melt	129,130
8.	HDPE	PET	-123	70	132-140	265	118	212	Crystallized PET	HDPE melt	131,132
9.	PP	HDPE	-20	-123	165	132-140	121	118	HDPE melt	Crystallized PP	90-93
10.	PP	PB	-20	-45	165	106, 142	121	---	PB melt	Crystallized PP	95-97
11.	PP	LDPE	-20	-123	165	115-123	121	101	LDPE melt	Crystallized PP	87,89
12.	PP	PA-6	-20	52	165	210	121	199	Crystallized PA-6	PP melt	150
13.	PP	PVDF	-20	-45	165	172	121	140	Crystallized PVDF	PP supercooled	131
14.	PP	PCL	-20	-70	165	68	121	---	PCL melt	Crystallized PP	115
15.	PBT	LCP	34	---	221	---	200	---			125,126
16.	PBT	PET	34	70	221	265	212	212	Crystallized PET	PBT melt	111
17.	PET	HDPE	70	-123	265	132-140	212	118	HDPE melt	Crystallized PET	110,131
18.	PET	PP	70	-20	265	165	212	121	PP melt	Crystallized	110
19.	PET	PA-66	70	46	265	258	212	236	Crystallized PA-66	PET supercooled	151-155
20.	PPS	HDPE	88	-123	280	132-140	256	118	HDPE melt	Crystallized PPS	129,130
21.	PPS	PET	88	70	280	265	256	212	PET supercooled	Crystallized PPS	131,132
22.	PEEK	PEK	141	152	335	365	316	---	Concurrent crystallization		80
23.	PEEK	PEEK	141	150	335	365	316	---	Cocrystallization		80
24.	PA-6	PA-11	52	47	225	182	199	---	PA-11 melt	Crystallized PA-6	156

Table 8.3 Thermal Data on the Component Polymers in Crystalline/Amorphous Blends

No	Blend system		Glass transition Temperature T_g (°C)		Melting or softening point of A, T_m or T_s (°C)	Crystallization temperature at 10°C/min T_c (°C)	Physical state of B during crystallization of A	References
	Component A	Component B	A	B				
1.	LDPE	EPDM, Elastomer	-123		115-123	101	Melt	74
2.	LLDPE	Elastomer	---		---	---	Melt	74
3.	HDPE	EPR	-123		132-140	118	Melt	84-86
4.	HDPE	PS	-123	100	132-140	118	Highly viscous melt	149
5.	PP	PIB	-20	---	165	121	Melt	99,103,108
6.	PP	SEBS	-20	---	165	121	Melt	106
7.	PP	EPDM, EPM	-20	100	165	121	Melt	98-104,107
8.	PVDF	Polyethylacrylate	-45	-20	172	140	Melt	145
9.	PVDF	PMMA	-45	105,126	172	140	Melt	134-136,138-144
10.	PVDF	PEMA, polyacrylates	-45	12, 65	172	140	Melt	133,137,145-148
11.	PBT	PC	34	147	221	200	Highly viscous melt	121
12.	PBT	Phenoxy	34	96	221	200	Melt	123
13.	PBT	Polyarylate	34	187	221	200	Highly viscous melt	122
14.	PBT	Amorphous polyamide	34	138	221	212	Highly viscous melt	120
15.	PET	PC	70	147	265	212	Melt	112-114,119
16.	PET	PEC	70	177	265	212	Melt	115
17.	PET	Phenoxy	70	96	265	212	Melt	116
18.	PET	Amorphous polyamide	70	138	265	212	Melt	120
19.	PETG	PC	87	147	277	217	Melt	127,128
20.	PETG	Amorphous polyamide	87	138	277	217	Melt	120
21.	PET	PMMA	70	100-125	260	---	Melt	119

8.4 CONCURRENT CRYSTALLIZATION

Concurrent or simultaneous crystallization of component polymers can occur only if the temperature ranges of crystallization of the polymers are overlapping and if their crystallizability is comparable.

Simultaneous crystallization does not necessarily mean cocrystallization, which occurs only when the component polymers are isomorphic or miscible in both the amorphous and crystalline phases. The cocrystallization requires chemical compatibility, close matching of chain conformations, lattice symmetry, and comparable lattice dimensions [73]. It has been observed only in a few blend systems consisting mainly of polyethylene and its copolymers. Starkweather [74] reported that in blends of LDPE with EPDM, cocrystallization occurred resulting in a gradual increase in the "a" parameter of the unit cell with increasing amount of EPDM in the blend. It is also reported that the development of spherulites of LDPE in the blend was inhibited resulting in improved transparency of LDPE films and unexpectedly high tensile strength.

Concurrent crystallization of the component polymers may certainly lead to mixed spherulites or axialites with a more coarse and open packing of lamellae. However, it may not induce changes in the crystal structure and unit cell parameters.

Kyu and Vadhar [75,76] have studied the blends of ultrahigh-molecular-weight polyethylene with HDPE, LLDPE, and LDPE. They have reported that the polymers crystallize concurrently in blends of UHMWPE with HDPE and LLDPE, whereas in the case of LDPE blends, separate crystals are formed. The linear variation of tensile properties with composition was attributed to cocrystallization in UHMWPE/HDPE and UHMWPE/LLDPE blends.

Recently Hu et al. [77] reported that blends of LLDPE with LDPE exhibit a single melting and crystallization peak and the peak temperatures varied linearly with composition. The WAXD studies and the density measurements also showed a linear dependence of the degree of crystallinity on the composition. These results indicate the possibility of concurrent crystallization.

Siegmann and Nir [78] also studied the melting and nonisothermal crystallization behavior in 50/50 wt% blends of LDPE and LLDPE (grades of identical density and comparable MFI). However, contrary to the results of Hu et al. [77], the authors reported occurrence of three melting peaks at 124 °C (corresponding to LLDPE), 108 °C (corresponding to LDPE), and an intermediate peak at 117 °C attributed to the cocrystallites of LDPE with LLDPE chains containing more densely distributed branches.

Cocrystallization takes place in binary mixtures of different molecular weight fractions of polyethylenes, as reported by Rego-Lopez and Gedde [79] for LLDPE. Three types of supermolecular structures in the blends, namely, banded spherulites, nonbanded spherulites, and axialites, depending upon the blend composition and the molecular weights of the components. The 80/20 wt% blend of the extreme-molecular-weight fractions of 2,500 and 66,000 exhibited axialites, whereas all other samples formed spherulites. The band spacing on the spherulites increased with increasing content of the low-molecular-weight fraction and also with decreasing molecular weight of the second component. Thus, the crystalline morphology in blends is affected by molecular size considerations.

Apart from the blends of polyethylene, concurrent crystallization was also reported for blends of polyaryletherketones. Sham et al. [80] reported that the blends of PEEK/PEK exhibited cocrystallization upon rapid quenching from melt but separate crystalline phases when the blends were isothermally crystallized or annealed at high temperatures. Thus, the "cocrystallite" morphology may not represent thermodynamically stable structures. This aspect must be considered when processing polymer blends that can cocrystallize; a "meta-stable" structure in the extruded or molded article could adversely affect its dimensional stability and integrity.

8.5   CRYSTALLIZATION IN THE PRESENCE OF MELT

For most polymer blends, whether crystalline/crystalline or crystalline/amorphous, the crystallization of one of the component polymers occurs in presence of the molten phase of the other component, as indicated in Tables 8.2 and 8.3. The blend systems discussed in the present review have therefore been classified on the basis of the major component polymer.

8.5.1   Polyethylene Blends

The different types of polyethylenes are distinguished in terms of their density and structure as low-density (LDPE), linear low-density (LLDPE), and high-density (HDPE). The different densities result from a variation in the crystalline packing ability of polyethylenes due to the difference in the level and length of branching. Thus, HDPE with a predominantly linear chain structure exhibits the highest crystallinity and melting point (132°-135°C) whereas LDPE, owing to its branched structure, is less crystalline, with a melting point of around 110°C.

In the case of HDPE/LDPE blends, the crystallization of HDPE takes place in the presence of LDPE melt. A considerable amount of work has been done on HDPE/LDPE blends. It was observed that blending does not significantly change crystallinity. However, factors such as molecular weight, molecular weight distribution, and the level of branching were found to affect the rate of crystallization and crystal perfection [81].

Bhateja and Andrews [82] reported on the blends of LDPE with UHMWPE. They found lower than expected values of the degree of crystallinity in the blends, which were attributed to the suppressed mobility of LDPE in the presence of UHMWPE.

A similar drop in the degree of crystallinity of LDPE and HDPE was observed in their blends with polybutene, PB [83]. In these blends, the polyethylenes crystallize in the presence of PB melt. The nonisothermal crystallization temperature for LDPE (in the DSC cooling scans) increased from 90° to 95°C, indicating enhanced nucleation of LDPE, whereas the degree of supercooling of HDPE was not affected. However, the spherulitic growth rate of both LDPE and HDPE were adversely affected leading to coarser spherulites.

Blends of HDPE with ethylene-propylene rubber, EPR, containing a different percentage of ethylene were reported to exhibit higher crystallization temperature (during cooling) and higher melting temperature (relative to virgin HDPE) indicating enhanced nucleation and better crystal perfection [84,85]. The overall crystallinity of the blends was found to be lower than that expected from the additivity rule.

Kalfoglou [86] investigated the blends of HDPE with EPDM reporting that the presence of EPDM melt enhances nucleation of HDPE leading to a reduced crystallite size and a decrease in the crystalline perfection. These observations have been attributed to the reduction of the crystal growth rate of HDPE by the high viscosity of EPDM.

8.5.2   Polypropylene Blends

The blends of polypropylene, PP, with polyethylene are immiscible and have been a subject of great interest owing to their commercial importance. In these blends crystallization of PP takes place in the presence of polyethylene melt.

    Teh [87] reported that the addition of LDPE had a dual effect on the morphology of PP. LDPE acted as a nucleating agent and reduced the average spherulite size of PP. In addition to the large number of smaller-size spherulites, the other effect concerned the formation of some large distinct spherulites which melted at 155°C corresponding to the β form of PP crystals.

    Polypropylene is known to crystallize in the monoclinic α form at temperatures above 132°C and in the hexagonal β form at temperatures below 132°C. However, it has been reported by Jacoby et al. [88] that addition of nucleating agents can result in different levels of the β form depending on the concentration and dispersion of the nucleating agents. The enhanced nucleation of PP in the presence of LDPE melt leads to initiation of the crystal growth of PP at temperatures below 132°C resulting in the formation of β crystallites.

    The spherulitic growth rate of PP in its blends with LDPE was found to remain unchanged [89]. Thus, the major changes observed in the PP morphology in PP/LDPE blends seem to result from the modification of the nucleation process.

    In the case of PP/HDPE blends, the addition of HDPE was found to decrease the nucleation density at high crystallization temperatures, $T_c \geq 127$°C, whereas it increased with increasing HDPE content at temperatures below 127°C, where the HDPE melt is in a "supercooled" state as reported by Bartczak et al. [90]. The spherulite size and distribution were found to be altered as a result of blending. However, the spherulitic growth rate was found to be constant indicating that diffusion of PP molecules is not affected by the presence of molten HDPE. Lovinger and Williams [91] also noticed that the average spherulite size of PP decreased as a result of blending at $T_c = 100$°C. These observations are supported by Gupta et al. [92] and Teh [87], who also observed an increase in nucleation density. The spherulitic structure of PP was found to become increasingly irregular and coarser with increasing PE content [93]. Bartczak and Galeski [94] reported that spherulitic crystallization of a polymer near the interface causes deformation of the interphase increasing the interfacial area. This may lead to improvement of toughness and impact properties. Note that the interfacial area is determined by the dispersion level governed by the compounding conditions, blend composition, and the relative melt viscosities of the polymers at the compounding temperature and stress.

    Thus, it was observed that, in general, blending PP with either LDPE or HDPE results in enhanced nucleation, reduced spherulite size, and a coarser irregular spherulitic structure of PP. However, the spherulitic growth rate remained unchanged.

    The blends of PP with PB have been reported by Siegmann [95,96]. He observed that although the onset temperature of crystallization of PP during cooling was not significantly affected by the presence of PB melt, the spherulitic morphology was. As PB concentration increased, the spherulites were found to change from coarser and fragmented to branched. The degree of crystallinity of these blends was found to be higher than the value calculated by the additivity rule. Gohil and Peterman [97] also reported that the PP in PP/PB blends exhibited reduced crystalline order as evidenced by broadening of the X-ray diffraction peaks.

    Blends of polypropylene with different elastomers have been investigated by many researchers [98-107]. Karger-Kocsis et al. [98] reported that incorporation of ethylene-propylene copolymer led to enhanced nucleation of PP, evidenced by the decrease in the degree of supercooling from 60 to 42°C. As a result, a decrease in the average spherulite size was observed. Martuscelli et al. [99] studied, under optical microscopy, the PP blends with EP-copolymer and reported that the nucleation density of PP increased as a result of blending and that the extent of change in nucleation density depended on the chemical structure and molecular weight of the elastomer.

Kalfoglou [100] also observed a reduction in the spherulite size of PP as the EP-copolymer content was increased. Recently Greco et al. [101,102] have confirmed these findings in blends of PP with EP-copolymers having different propylene content. The spherulitic texture was coarser in the PP/elastomer blends as compared to the compact and larger spherulites of virgin PP. However, contrary to the finding of Martuscelli [99], the degree of crystallinity of PP was found to be lower than that calculated from the additivity rule [101,102]. The discrepancy in these observations may be attributed to the differences in the ratios of ethylene to propylene and the molecular weights of the EP-copolymers used. Greco et al. used EP-copolymers containing 65 to 92 wt% of propylene, whereas Martuscelli et al. worked with 30 to 50 wt% propylene content.

Although the presence of EP-copolymer enhanced the nucleation process of PP, the limited compatibility of PP with the EP-copolymer may restrict the growth. The degree of crystallinity of PP in the blends could therefore vary depending upon the relative predominance of these two opposing effects on the crystallization process.

Martuscelli et al. [103] investigated blends of polypropylene with ethylene-propylene-diene terpolymer, EPDM. They observed that the nucleation density increased monotonically with EPDM content. However, the radial growth rate was found to decrease with increasing rubber content. Bartczak et al. [104] reported similar results attributing the depression in radial growth rate to the fine dispersion of EPDM which has to be circumvented by the growing PP crystallites. The elastomer particles were incorporated in the spherulites with minor changes in the structure, as shown in phase-contrast photographs by Karger-Kocsis and Kiss [105]. However, in the case of styrene-butadiene rubber, SBR, the presence of SBR caused fundamental changes in the fibrillar structure of the spherulites [105].

The block copolymer styrene-ethylene-butylene-styrene, SEBS, was investigated by Gupta and Purwar [106] for toughening PP. The crystallization studies indicated that the degree of crystallinity of PP was reduced and its nucleation and growth rates were adversely affected by blending. On the other hand, Danesi and Porter [107] reported marginal effect of blending EPDM and EPR with PP on its crystallization. The different observations concerning the effect on PP crystallization reported for its blends with EP-copolymers, EPDM and EPR [98-104,107] and those with styrenic copolymers [105,106] underscore the influence of chemical structure and compatibility on the crystallization behavior.

Blends of PP with polyisobutylene, PIB, were studied by Martuscelli et al. [99,103] and by Bianchi et al. [108]. Martuscelli used three samples of PIB with viscosity average molecular weights ranging from 66 to 3500 (kg/mol). It was observed that for all three PIB samples, the maximum nucleation density occurred at PP/PIB = 90/10 and the radial growth rate decreased with increasing PIB content. The extent of change depended on PIB molecular weight. The reduction in the radial growth rate was attributed to the diluent effect of PIB. PP/PIB blends were found to exhibit a lower degree of crystallinity and hence lower T_m relative to homopolymer PP. However, Bianchi et al. [108] reported that addition of PIB did not produce any significant changes in the nucleation density, thus contradicting the results of Martuscelli et al. [103]. The disagreement may be attributed to the different conditions of sample preparation. Bianchi et al. used the polymers in the as-received state; whereas Martuscelli et al. used them after dissolution in xylene and acid washing; followed by reprecipitation using methanol. There is a possibility that these latter operations changed the tacticity of PP. The differences in the level of isotactic index of PP used by the two investigating groups could lead to the observed discrepancy in results.

Kalfoglou [109] studied the crystallization of PP in its miscible blends with poly-ε-caprolactone, PCL, over the entire composition range. The degree of crystallinity of PP was found to be comparable to or higher than that in homopolymer PP, exhibiting a maximum at intermediate-blend compositions. The PP spherulite size decreased with increasing PCL content. The influence on PP crystallization rate was not investigated.

8.5.3 Polyethyleneterephthalate Blends

Polyethyleneterephthalate, PET, is a versatile polymer used extensively in synthetic fibers, films, and molded products. Considerable work has been reported on PET blends with polyolefins, polycarbonate, Phenoxy, and polyamides. The main PET property improvements sought by blending are heat-distortion temperature, toughness and improved moldability.

Toughening of PET using low concentrations (1-10 wt%) of polyolefins has been reported by Wilfong et al. [110]. The authors studied the blends of PET with LLDPE, HDPE, PP, and poly(4-methyl pentene-1). Due to the large difference in the melting points of PET and the polyolefins (Table 8.3), the crystallization of PET takes place in the presence of polyolefin melt. It was observed that polyolefin melts did not enhance the nucleation of PET, as evident from the large PET spherulites in the blend. The spherulites were found to be 2.5 to 3 times larger than those in neat PET with a broader spherulite size distribution in the blend. The PET crystallization rate was found to be depressed as evidenced by broadening of the crystallization peak during the DSC cooling scans; the degree of crystallinity was also reduced. The adverse effects on both crystallinity and crystallization rate of PET were attributed to the expenditure of energy in rejection or occlusion of polyolefin domains by the growing crystallite fronts of PET.

The blends of PET with polybutyleneterephthalate, PBT, were investigated by Escala and Stein [111] using various experimental techniques. The blends were found to be compatible exhibiting a single composition-dependent T_g. It was observed that the rate of crystallization and the degree of crystallinity of PET increased with increasing PBT content. The increase in the crystallization rate was attributed to the lowering of T_g, thereby increasing the crystallization temperature range.

In a number of investigations [112-118], blending of PET with amorphous polymers such as polycarbonate (PC), polyestercarbonate (PEC), and polyarylate (PAr), was found to result in reduction of the crystallization rate of PET and the degree of crystallinity (based on DSC studies). All these amorphous polymers exhibit T_gs higher than that of PET and they form partially or completely miscible blends with PET. Thus, a drop in the crystallization rate and in the degree of crystallinity were attributed to the increase of T_g leading to a narrower temperature range of crystallization for PET.

Thus, the crystallization of PET is retarded by the presence of molten polyolefins or miscible/amorphous polymers, although the reasons for the retardation could be different. In the case of immiscible PET/polyolefin blends the retardation of crystallization has been attributed to the energy expended in the rejection or occlusion of polyolefin domains within PET spherulites, whereas in the miscible PET blends with amorphous polymers, the retardation of PET crystallization has been attributed to the narrowing down of the temperature range of crystallization. It is interesting to note that PET crystallizes in the presence of superheated melt of polyolefins (40-80°C higher than T_m) whereas in the case of blends of PET with PC, PEC, and PAr (see NOMENCLATURE), PET crystallizes in presence of supercooled melt (30-50 °C higher than T_g) of the second component. Hence, as far as the crystal growth is concerned, it may be facilitated in presence of polyolefins whereas it may be restricted in presence of PC, PEC, and PAr. The first hypothesis is based on the larger spherulite size of PET observed in polyolefin blends whereas the second hypothesis is based on a significant increase in the crystallization halftime from 40 sec to about 5400 sec at 120 °C in the case of PET/PEC blends.

The blends of PET with two other amorphous polymers, namely, polymethylmethacrylate, PMMA, and aromatic polyamide have been investigated by Nadkarni et al. [119,120]. It was observed that a low concentration of PMMA (around 15 wt%) accelerated the crystallization of PET to such an extent that PET could be molded at low temperatures (19 °C) without mold heating. The blends also exhibited a higher degree of crystallinity than PET molded under identical conditions, indicating that PET crystallization in the presence of PMMA melt cannot be suppressed even at high quench rates [119]. The authors reported that blends of PET with an aromatic amorphous polyamide (85/15 vol%) exhibited a higher degree of crystallinity and increased PET crystallization rate [120]. They observed that the presence of molten polyamide reduced the induction time, indicating enhanced nucleation. The total isothermal crystallization time was also reduced indicating accelerated crystallization. These results, however, are at variance with the earlier results on miscible blends of PET with other amorphous thermoplastics. The discrepancy may be attributed to the differences in the extent of interactions between the component polymers governed by their chemical structures, and/or to the relative viscosities. As discussed earlier, the crystallization of PET takes place in the presence of supercooled melt of the amorphous polyamide. Hence the viscosity of the second component would be expected to influence the crystal growth rate.

8.5.4 Polybutyleneterephthalate Blends

Polybutyleneterephthalate is the other major thermoplastic polyester extensively used as an engineering polymer. Unlike PET, the PBT crystallizes inherently faster and therefore manifests better processability than PET. However, it displays a lower heat-distortion temperature and a lower modulus than PET. These properties can be improved by blending it with amorphous thermoplastics like PC with high glass transition temperatures, liquid crystal polymers (LCP), polyethyleneterephthalate (PET), copolyester, or amorphous polyamides.

The blends of PBT with amorphous thermoplastics having higher glass transition temperatures, such as PC, PAr, and Phenoxy, have been investigated by Wahrmund et al. [121], Kimura and Porter [122], and Robeson and Furtek [123]. The degree of crystallinity of PBT did not change by blending with PC, whereas in blends with PAr the crystallinity plotted as a function of composition showed a maximum. In PBT/Phenoxy blends, the degree of crystallinity decreased with increasing Phenoxy content. Although the crystallization of PBT takes place in presence of the supercooled melts of the amorphous polymers, the differences in the modification of the crystallization behavior in the different blends may be due to differences in miscibility of PBT with these amorphous polymers, and in their melt viscosities.

The blending of PBT with amorphous polyamide, Trogmid-T, has been investigated by Nadkarni et al. [120]. It was observed that incorporation of 15 vol% of the aromatic polyamide enhanced PBT nucleation as indicated by a lower degree of supercooling and enhanced overall crystallization rate. However, no significant change was observed in the degree of crystallinity of PBT.

Slogowski et al. [124] investigated miscible blends of PBT with an amorphous copolyester (based on Bisphenol-A, neopentyl glycol, and terephthalic acid). The authors reported that the degree of crystallinity of PBT decreased from 33% to 8% as the copolyester content increased up to 25 wt%. The spherulites of PBT in the blends were found to be coarser than those in PBT homopolymer indicating hindered crystal growth of PBT. These observations were attributed to an increase of T_g and concurrent depression of PBT melting point, thereby significantly reducing the range of PBT crystallization temperatures.

Pracella et al. [125] studied PBT blends with LCP (HTH 10) having $T_m = 216°C$. Although the melting points of PBT and the LCP were comparable, the crystallization of PBT took place in presence of supercooled LCP melt. In DSC cooling scans the crystallization-onset temperatures were found to be about 199°C for PBT and 156°C for the LCP [125]. It was reported that the onset temperature of crystallization of PBT decreased with increasing amount of the LCP, indicating an adverse effect on nucleation. The degree of crystallinity was also found to decrease with increasing LCP content.

Paci et al. [126] studied the thermal behavior of PBT blends with another LCP, the poly(biphenyl-4-4'sebacate), PBS. In the DSC scans the crystal-to-smectic transition and smectic-to-isotropic transition for PBS were observed at 170–230°C and 280°C, respectively. As a result, the crystallization of PBT took place in the supercooled liquid crystalline melt of PBS in its smectic state. The temperature of the onset of PBT crystallization in the DSC cooling scan was found to decrease with increasing PBS content, indicating retardation of the crystallization process. The isothermal crystallization studies suggested a decrease in the rate of PBT crystallization with increasing PBS content as indicated by longer crystallization halftimes in the blend. It was also observed that the crystallinity of PBT varied nonlinearly with PBS content, increasing up to 35 wt% of PBS and then decreasing. The variation in the crystallinity may be due to the increased hindrance of PBS phase to the crystallization of PBT, confirmed by the isothermal crystallization studies.

In summary, the crystallization behavior of PBT in the presence of molten phase of the second component is differently modified by different polymers. No discernible general trends appear to be present, although in most cases blending seems to retard PBT crystallization leading to a lower degree of crystallinity with respect to that of homopolymer.

8.5.5 Copolyester Blends

The blends of thermoplastic copolyester (Kodar 150) with PC have been investigated by Mohn et al. [127] and Barnum et al. [128]. It was observed that blending resulted in a lower degree of crystallinity, although the rate of crystallization from melt as well as from the glassy state was enhanced. In another study [120] the crystallization of the copolyester was found to be accelerated by addition of an amorphous aromatic polyamide at 15 vol%. The blends showed an enhanced nucleation and acceleration of the overall crystallization rate although no significant change was observed in the degree of crystallinity.

8.5.6 Polyphenylenesulfide Blends

Polyphenylenesulfide, PPS, is a high-performance engineering polymer. Its applications potential can be extended by blending it with other thermoplastics for cost dilution and improved toughness.

Nadkarni et al. [129] investigated blends of unfilled-grade of PPS with three grades of HDPE having different melt-flow indices (from MFI = 0.4 to 52 gm/10 min). Due to the large difference ($\sim 150\,^{\circ}$C) in the melting points of HDPE and PPS, the crystallization of PPS took place in presence of HDPE melt. The nonisothermal crystallization studies of blends with all three grades of HDPE indicated that the presence of HDPE melt adversely affected the nucleation process of PPS, as evidenced by the drop in the onset of crystallization temperature. However, the increase of the onset of melting temperature indicated that the crystal growth was facilitated, resulting in a larger crystallite size of PPS in the blends. The isothermal crystallization studies also suggested that the overall rate of crystallization of PPS was retarded and its crystallinity was reduced with increasing HDPE content. The retarding influence of HDPE melt on PPS crystallization decreased with increasing melt viscosity, that is, with decreasing MFI of HDPE. However, in the case of blends of glass-reinforced PPS with HDPE [130], it was observed that the isothermal crystallization of PPS was accelerated at low concentrations of HDPE w ≤ 25 wt%. The degree of crystallinity was found to decrease with increasing HDPE content.

The discrepancy in the results of isothermal crystallization studies of unfilled and glass-reinforced grades of PPS may be attributed to the differences in the inherent crystallizabilities of PPS samples and to the presence of glass fibers in the filled PPS leading to heterogeneous nucleation. Thus, in general, blending PPS with HDPE was found to affect the crystallization rate and degree of crystallinity of PPS.

In another study [131] of blends of unfilled PPS with PET, it was observed that, although nucleation of PPS was not enhanced, the overall crystallization process of PPS was accelerated as evidenced by narrowing of the crystallization-peak width for PPS-rich compositions. The isothermal crystallization studies, however, indicated enhanced nucleation process as evidenced by a drop in the induction time. The presence of PET melt also led to a decrease in the isothermal crystallization halftime of PPS. However, the degree of crystallinity of the blends was found to decrease with increasing PET content. In the case of blends of glass-reinforced PPS with PET, similar results were obtained [132].

It is interesting to note that although the crystallization of PPS takes place in the presence of the melt of the second component in PPS/HDPE and PPS/PET blends the modification of the PPS crystallization behavior is different. In PPS/HDPE blends the crystallization of PPS takes place in the presence of superheated melt of HDPE, which is about 100°C above its melting point, whereas in PPS/PET blends, it occurs in the presence of supercooled melt of PET. The homogeneous nucleation of PPS was found to be adversely affected in the presence of HDPE whereas it was found to be facilitated by the presence of PET melt. Also, the overall crystallization rate of PPS was found to be retarded in presence of HDPE, and accelerated in presence of PET. The degree of crystallinity was, however, found to decrease with increasing amount of the second component in both the blends. The observed effects may be partly ascribed to the differences in the melt viscosities of HDPE and PET at the PPS crystallization temperature, as well as to the molecular interactions between PPS and, on one hand, aromatic polymer like PET and, on the other hand, a nonpolar one like HDPE. The other molecular parameter that could affect the transport of a diffusing PPS molecule to the crystal-growth front is the chain length of the hindering second component. It may be noted that the average polymer-chain length of HDPE molecule is significantly (3-4 times) longer than that of a PET polymer chain.

8.5.7   Polyvinylidenefluoride Blends

The majority of investigations on PVDF concern its miscible blends with polymethylmethacrylate, PMMA, and polyethylmethacrylate, PEMA [133-147]. The blends of PVDF with these acrylates exhibit single-composition-dependent T_g [133]. Thus, the crystallization of PVDF takes place in the presence of supercooled melt of the acrylic polymers. It was observed that both the spherulitic growth rate and the PVDF crystallinity decreased as a result of blending it with PMMA [134-136]. The decrease in the growth rate was attributed to the presence of depletion layers at the growth fronts.

Kwei et al. [137] studied blends of PVDF with PEMA. They reported that the degree of crystallinity of PVDF decreased with increasing PEMA content up to 80 wt%. However, it was observed that for blends containing 40 wt% or less PEMA, the crystallization of PVDF prevailed even at high quench rates, whereas in the case of PVDF/PMMA blends PVDF crystallization could be suppressed at high quench rates to give a single amorphous phase. The differences in the crystallization behavior of PVDF in the presence of PMMA and PEMA may be attributed to the large differences of viscosities of the two methacrylates (PMMA was more viscous than PEMA) at the PVDF crystallization temperature.

A number of other studies on melting and crystallization behavior of PVDF/PMMA blends [138-141] confirm the retarding influence of PMMA on PVDF crystallization. The melting point of PVDF decreased significantly with increasing amounts of PMMA, and in the PMMA-rich blend compositions, no crystallization or melting peaks were observed [138]. The melting-point depression was from 173°C for homopolymer PVDF to 160°C for the 50 wt% blend. The heat of fusion for PVDF was significantly reduced even with about 25 wt% PMMA [138-140]. At blend compositions containing greater than 60 wt% PVDF, two melting peaks were observed [138] indicating two types of crystallites.

Roerdink and Challa investigated the influence of PMMA tacticity on melting and crystallization of PVDF in the blends [141]. All the PVDF blends with atactic, syndiotactic, or isotactic PMMA exhibited a single composition-dependent T_g and comparable melting-point depression. The crystallization temperature, T_c, of PVDF, in nonisothermal cooling scans, dropped with increasing PMMA concentration, the drop being the most dramatic for blends containing syndiotactic PMMA.

The retardation and suppression of PVDF crystallization in its miscible blends with PMMA are attributed to the significant reduction in the crystallization-temperature range as a result of simultaneous T_m depression and T_g elevation.

The blending of PVDF with other alkyl acrylates, PCL, polyvinylmethylether, PVME, and polyvinylmethylketone, PVMK, have also been reported to lead to a lower degree of crystallinity and retardation of PVDF crystallization [142-144]. The PVDF/PVMK blends were found to be miscible with a single composition-dependent T_g, whereas the PVDF/PCL and PVDF/PVME blends were immiscible. It was therefore concluded that the presence of the carbonyl group in the polymer chain is essential for interaction with PVDF leading to miscibility [144].

Recently, Briber and Khoury [145] studied the morphology of PVDF/polyethylacrylate (PEA) blends. They reported that the texture of spherulites of PVDF became more open as the PEA content increased, indicating reduced twisting of the lamellae and thus a less dense spherulitic structure with loose packing. Similar results were reported earlier by Morra and Stein [146-148]. Thus, blending of PVDF with acrylates results in a drop in the spherulitic growth rate and the degree of crystallinity of PVDF, accompanied by significant changes in the spherulitic texture.

Shingankuli [131] studied the crystallization behavior of PP/PVDF blends. Although the melting points of the component polymers were comparable, PVDF exhibited faster crystallizability than PP. Thus, the crystallization of PVDF took place in presence of PP melt. It was observed that the presence of PP melt resulted in enhanced nucleation of PVDF as evidenced by a drop in the degree of supercooling and narrowing of the crystallization peak in the nonisothermal DSC cooling scans. The isothermal crystallization of PVDF was also found to be accelerated as a result of blending. The overall degree of crystallinity was found to be more than that calculated using the additivity rule. Thus, the presence of PP melt was found to lead to enhanced nucleation, faster crystallization rate, and increased degree of crystallinity of PVDF in the PP/PVDF blends.

8.6   CRYSTALLIZATION IN THE PRESENCE OF SOLIDIFIED POLYMER

The crystallization of polymer in the presence of solidified domains of the other one takes place through a heterogeneous nucleation process. Since the rate of heterogeneous nucleation is generally higher than homogeneous nucleation, and since nucleation is the rate-controlling step in polymer crystallization, the crystallization rate would be expected to be higher in such blends when compared to the homopolymers.

8.6.1   Crystallization of Polyethylene in Blends

The crystallization of LDPE or HDPE in its blends with PP takes place in the presence of solidified PP. The crystallinity of LDPE was unaffected by blending [94]. In the case of blends of HDPE with PP, however, a drop in the crystallinity of HDPE was observed [96-99]. The crystallization rate of HDPE was found to be retarded as a result of blending with PP. During the isothermal crystallization after addition of 10 wt% PP, the halftime of HDPE was found to be three times longer than that in homopolymer. The retardation in the crystallization rate was attributed to the increase in the melt viscosity of HDPE due to the presence of solidified PP. Similar results were obtained by Shingankuli [131] for PPS/HDPE blends in which crystallization of HDPE took place in the presence of solidified PPS. The isothermal crystallization halftimes for HDPE in the unfilled PPS/HDPE blends containing w ≥ 10 wt% PPS were longer than those in the homopolymer. However, in the case of blends of glass-reinforced PPS/HDPE it was observed that the crystallization of HDPE in the blend was accelerated, but the degree of crystallinity of HDPE decreased as indicated by the drop in the heat of fusion from 134 to 92 (J/g) with increasing content of filled PPS [130]. Also, the melting peak width was found to become narrower as PPS content increased. The narrowing of melting-peak width indicated a narrow crystallite size distribution resulting from the heterogeneous nucleation.

In blends of polyethylene, PE, with polystyrene, PS, Aref-Azar et al. [149] reported that the presence of solidified PS did not affect PE crystallization kinetics in the PE-rich blends, whereas in the PS-rich blends, the crystallization rate of PE constituting the dispersed phase was markedly affected by the number and size of the dispersed phase domains.

8.6.2 Polypropylene Crystallization

The majority of published works on crystallization in PP blends concern the blends of PP with lower T_m polymers such as polyethylenes and olefinic elastomers. As a result, the crystallization of PP in presence of a solidified polymer has seldom been reported. The reported investigations on PP/polyamide-6, PA-6, blends [150], in which the crystallization of PP would take place in the presence of solidified PA-6, dealt with the effect of blending on the rheological properties; the crystallization behavior of PP in the blends was not studied.

Shingankuli [131] studied the crystallization behavior of PP in the presence of solidified PVDF. It was observed that crystallization of PP took place at a higher temperature relative to homopolymer PP, suggesting enhanced nucleation in the blends. The heat of crystallization was found to increase by about 30 to 40% with increasing PVDF content indicating higher crystallinity. The isothermal crystallization studies also confirmed the acceleration of the overall crystallization rate in terms of shorter crystallization halftimes for PP at all blend compositions.

8.6.3 Polyethyleneterephthalate Crystallization

The blending of PET with polyamides has received considerable attention because of their compatibility and improved mechanical properties [151-155]. Kamal et al. [151] studied the crystallinity of PET/PA-66 blends. In these blends PET crystallizes in the presence of solidified PA-66. The crystallinity of PET was found to increase with increase in PA-66 content up to 45 mole% of PA-66. For the polyamide-rich compositions no such enhancement of crystallinity was observed. The observed results may be explained as follows. At low concentrations of PA-66, PET is a continuous phase and hence the crystallized polyamide acts as a nucleating agent, thereby increasing the degree of crystallinity. However, at high concentrations of solidified polyamide, the mobility of PET chains is restricted, thereby leading to a lower degree of crystallinity.

Similar results were obtained by Shingankuli et al. [131,132] for PPS/PET blends where the crystallization of PET took place in the presence of solidified PPS. It was observed that the heat of crystallization of PET increased up to 50 wt% and then decreased with increasing PPS content.

8.6.4 Polybutyleneterephthalate Crystallization

Escala and Stein [111] reported that the crystallization of PBT was retarded as a result of blending with PET. The degree of crystallinity was also found to decrease with increasing PET content. They attributed the retardation to the increase in the T_g of PBT as a result of blending. It should be noted that the crystallization of PBT takes place in presence of solidified PET. Thus, the drop in the crystallinity may be ascribed to the hindrance to PBT crystal because of the presence of solidified PET.

8.6.5 Miscellaneous Blends

As discussed earlier, although the presence of the solidified second component may enhance nucleation through the heterogeneous mode, other factors such as hindrance to crystal growth and narrowing down of the crystallization temperature range due to T_g elevation in miscible blends can result in lower crystallinity and reduced crystallization rate. Similar observations were reported for PE/PB [83], PCL/PP [109], and PA-6/PA-11 [156] blends.

In the blends of LDPE with polybutene-1, PB crystallizes in the presence of solidified LDPE. Here the nonisothermal crystallization temperature of PB decreased with increasing LDPE content from 75°C for the homopolymer to 60°C in the blends. The degree of crystallinity was reduced and the spherulitic growth rate dropped, resulting in coarser spherulites for PB [83].

Kalfoglou reported a decrease in PCL crystallinity in PP-rich blends [109]. A similar drop in crystallinity of polyamide-11, PA-11, was reported by Inoue in PA-6/PA-11 blends [156].

8.7   GENERAL CONCLUSIONS

Although polymer blends represent a relatively new class of thermoplastic materials, their considerable application potential and commercial relevance have led to a number of scientific investigations aimed at understanding the structure-property relations. In semicrystalline polymer blends, the major physical process governing structure development is polymer crystallization. The published literature on crystallization in polymer blends clearly indicates that the crystallization behavior and the crystalline morphology of a polymer are significantly modified by the presence of the second component.

The critical factors governing the extent and direction of change in the rate of crystallization and morphology of a polymer in a blend include miscibility, the glass transition, and melting temperatures of the constituent polymers, their relative melt viscosities, chemical compatibility, inherent crystallizability and phase morphology in the as-compounded pellets. It is interesting to note that owing to the interplay of these factors, the morphology of a molded or extruded article from the blend is more sensitive to changes in the processing conditions. However, it is this sensitivity of the blend morphology to processing conditions that can be utilized effectively to obtain a broader range of property combinations from a single blend composition via selective manipulation of the process parameters.

The crystallization of polymer in a blend takes place under conditions different from those involved in its crystallization from the virgin melt. The physical state of the second component and the temperature range of crystallization are determined by considerations such as the difference in the melting points of the component polymers, their glass transition temperatures, and miscibility. Depending on the differences in the melting points and the inherent crystallizability, a polymer may crystallize either concurrently with the other component, in the presence of the solidified second component, or in the presence of the melt of the second component. Thus, in LLDPE/HDPE blends, concurrent crystallization of the two components is feasible since their melting points are close and the temperature ranges of crystallization are overlapping. However, in PP/PVDF blends, although the melting points of the two polymers are comparable, PVDF crystallizes first in the presence of supercooled PP melt owing to its higher crystallizability; thus, PP would crystallize in the presence of solidified PVDF.

The presence of a second component either in the molten or solid state affects both nucleation and crystal growth of the crystallizing polymer. The effect of blending on the overall crystallization rate is the net combined effect on nucleation and growth. In general, the presence of the second component physically hinders the transport of the crystallizing polymer molecule, thereby adversely affecting crystal growth. The extent of the adverse effect is governed by the blend composition, the relative melt viscosities of the component polymers, and miscibility. The effect of blending on nucleation is more subtle and complex, particularly in the presence of the melt of the second component. It may be presumed that the mode of nucleation of a polymer in the presence of the solidified domains of the second component is heterogeneous nucleation, and therefore the rate of nucleation should be higher than that in homopolymer. However, in presence of the molten second component, factors such as miscibility, relative melt viscosity, and inherent crystallizability all influence the formation of critical size nuclei.

The subtleties of the nucleation process in polymer blends may be highlighted with reference to PPS/PET and PPS/HDPE blends. It was reported [129-132] that the rate of nucleation of PPS (as indicated by the induction time for initiating isothermal crystallization in DSC scans) is enhanced by the presence of molten PET but depressed by the presence of molten HDPE. Over the temperature range of crystallization of PPS, the PET melt is in a "supercooled" state, whereas HDPE melt is at a temperature significantly above its melting point. This difference in the "nature" of the molten phase coupled with chemical-compatibility considerations might explain the observed differences. This conjecture may be supported by the observation that, in PP/HDPE blends, the nucleation of PP is enhanced by HDPE melt at temperatures below 127°C but retarded at higher temperatures [90].

In miscible polymer blends, exemplified by the blends of thermoplastic polyesters with amorphous polymers, such as polycarbonate, polyarylates, and aromatic polyamide, the effect of blending on the overall crystallization rate seems to be governed to a large extent by the elevation or suppression of the glass transition temperature of the base polymer. Thus, in blends of PET or PBT with PC, polyarylates, and Phenoxy, the overall crystallization rate is reduced because of narrowing down of the temperature range of crystallization resulting from elevation of the T_g, whereas in PET/PBT miscible blends, exhibiting lower T_g than that of PET, the overall crystallization rate of PET is enhanced. However, besides the change in the glass transition temperature, other factors such as chemical compatibility and relative melt viscosities of the components also have to be considered, in particular when considering the reported accelerating influence of the molten aromatic polyamide on the crystallization of thermoplastic polyesters, in spite of the higher T_gs exhibited by the miscible blends [120].

In a majority of blends, the degree of crystallinity of the crystallizing-polymer component is reduced because of the presence of the second component. There are a few exceptions such as blends of PET with an amorphous aromatic polyamide and also with PBT exhibiting increased crystallinity levels.

The spherulitic morphology in polymer blends is coarser as a result of the obstructing effect of the second phase on lamellar packing. The concurrent crystallization of the two chemically compatible component polymers may lead to spherulites containing both types of lamellae. Cocrystals may be formed only if the polymers are isomorphic and miscible.

In summary, although a few general trends may be discernible in the effect of blending on the rate of crystallization and crystalline morphology, considerable further research with systematically selected blend systems is required to elucidate the relative impact of the various factors such as miscibility, chemical compatibility, inherent crystallizability, and melt viscosity on the crystallization in polymer blends. The crystallization in ternary polymer blends also represents an unexplored area, relevant to understanding the effect of compatibilization in controlling phase morphology for property manipulation.

NOMENCLATURE

ABS	Acrylonitrile-butadiene-styrene copolymer
CAB	Cellulose-acetate-butyrate
EPDM	Ethylene-propylene-diene terpolymer
EPR	Ethylene-propylene rubber
HDPE	High density polyethylene
i-PS	Isotactic polystyrene
LCP	Liquid crystalline polymer
LDPE	Low density polyethylene
LLDPE	Linear low density polyethylene
PA-6	Polyamide-6
PA-11	Polyamide-11
PA-66	Polyamide-66
PAr	Polyarylate
PB	Poly-1-butene
PBT	Polybutyleneterephthalate
PBT-THF	Polybutyleneterephthalate-tetrahydrofuran copolymer
PC	Polycarbonate
PCL	Poly-ε-caprolactone
PEA	Polyethylacrylate
PEC	Polyestercarbonate
PEEK	Polyetheretherketone
PEK	Polyetherketone
PEMA	Polyethylmethacrylate
PET	Polyethyleneterephthalate
PETG	Copolyester of terephthalic acid, isophthalic acid, ethylene glycol, and cyclohexane dimethanol
PIB	Polyisobutylene
PMMA	Polymethylmethacrylate
PP	Polypropylene
PPE	Polyphenyleneether
PPS	Polyphenylenesulfide
PVAc	Polyvinylacetate
PVC	Polyvinylchloride
PVDF	Polyvinylidenefluoride
PVME	Polyvinylmethylether
PVMK	Polyvinylmethylketone
SBR	Styrene-butadiene rubber
SEBS	Styrene-ethylene-butylene-styrene triblock copolymer
SIBS	Styrene-isoprene-styrene-butadiene block copolymer
UHMWPE	Ultrahigh-molecular-weight polyethylene
DSC	Differential scanning calorimetry
MFI	Melt flow index
NMR	Nuclear magnetic resonance
SALS	Small-angle light scattering
SAXS	Small-angle X-ray scattering
vol%	Volume percent
WAXD	Wide angle X-ray diffraction

wt%	Weight percent
a	Unit cell dimension
$C(t)$	Cooling function
K	Crystallization rate constant
n	Avrami exponent
n'	Ozawa exponent
q	Cooling rate
t	Time
T_c	Crystallization temperature
T_g	Glass transition temperature
T_m	Melting temperature
w	Polymer concentration in wt%
x_t	Extent of crystallization

REFERENCES

1. Utracki, L. A., *Polym. Eng. Sci.*, **22**, 1166 (1982).
2. Utracki, L. A., *Polymer Alloys and Blends*, Hanser Publishers, Munich (1989).
3. Shen, M., Kawai, H., *AIChE J.*, **24**, 1 (1978).
4. Rudin, A., *J. Macromol. Sci., Rev. Macromol. Chem.*, **C19**, 267 (1980).
5. Shaw, M. T., *Polym. Eng. Sci.*, **22**, 115 (1982).
6. Robeson, L. M., *Polym. Eng. Sci.*, **24**, 587 (1984).
7. Barlow, J. W., Paul, D. R., *Polym. Eng. Sci.*, **21**, 985 (1981).
8. Noolandi, J., *Polym. Eng. Sci.*, **24**, 70 (1984).
9. Fayt, R., Jerome, R., Teyssie, Ph., *Polym. Eng. Sci.*, **27**, 328 (1987).
10. Wu, S., *Polym. Eng. Sci.*, **27**, 335 (1987).
11. Barlow, J. W., Paul, D. R., *Polym. Eng. Sci.*, **24**, 525 (1984).
12. Plochocki, A. P., *Polym. Eng. Sci.*, **26**, 82 (1986).
13. Platzer, N., in *Applied Polymer Science*, Tess, R. W., Poehlein, G. W., Eds., American Chemical Society, Washington, D.C. (1985).
14. Paul, D. R., Barlow, J. W., *J. Macromol. Sci., Rev. Macromol Chem.*, **C18**, 109 (1980).
15. Paul, D. R., in *Multicomponent Polymer Materials*, Paul, D. R., Sperling, L. H., Eds., Adv. Chem. Series, 211, American Chemical Society, Washington, D.C. (1986).
16. Krause, S., *J. Macromol. Sci., Rev. Macromol. Chem.*, **C7**, 251 (1972).
17. Min, K., White, J. L., Fellers, J. F., *Polym. Eng. Sci.*, **24**, 1327 (1984).
18. van Krevelen, D. W., Hoftyzer, P. J., *Properties of Polymers: Their Estimation and Correlation with Chemical Structure*, Elsevier Science Publishers, Amsterdam (1976).
19. Mandelkern, L., *Crystallization of Polymers*, McGraw-Hill, New York (1964).
20. Wunderlich, B., *Macromolecular Physics*, Vol. 1-3, Academic Press, New York (1973, 1976, 1980).
21. Schultz, J. M., *Polymer Materials Science*, Prentice-Hall, Englewood Cliffs, N.J. (1974).
22. Magill, J. H., *J. Polym. Sci.*, **A3**, 1195 (1965).
23. Devoy, C., Mandelkern, L., *J. Polym. Sci.*, **A2**, 7, 1883 (1969).
24. Harsey, E. D., Hybart, F. J., *J. Appl. Polym. Sci.*, **14**, 2133 (1970).
25. Katz, T., *J. Polym. Sci.*, **B8**, 789 (1970).
26. Pelzbauer, Z., Galeski, A., *J. Polym. Sci.*, Part C, **38**, 23 (1972).
27. Archambault, P., Prudhomme, R. E., *J. Polm. Sci., Polym. Phys. Ed.*, **18**, 35 (1980).

28. Miller, R. L., Seely, E. G., *J. Polym. Sci., Polym. Phys. Ed.*, **20**, 2097 (1982).
29. Kumar, S., Anderson, D. P., Adams, W. W., *Polymer*, **27**, 329 (1986).
30. Booth, A., Hay, J. N., *Polymer*, **10**, 95 (1969).
31. Borri, C., Bruckner, S., Crescenzi, V., Della Fortuna, G., Mariano, A., Scarazzato, P., *Europ. Polym. J.*, **7**, 1515 (1971).
32. Maneschalchi, F., Rosi, R., Mattiussi, A., *Europ. Polym. J.*, **9**, 601 (1973).
33. Gilbert, M., Hybart, F. J., *Polymer*, **15**, 407 (1974).
34. Hay, J. N., Fitzgerald, P. A., Wiles, M., *Polymer*, **17**, 1015 (1976).
35. Atanassov, A. M., *Polym. Bull.*, **17**, 445 (1987).
36. Jog, J. P., Nadkarni, V. M., *J. Appl. Polym. Sci.*, **30**, 997 (1985).
37. Jog, J. P., Nadkarni, V. M., *J. Appl. Polym. Sci.*, **32**, 3317 (1986).
38. Cebe, P., Hong, S. D., *Polymer*, **27**, 1183 (1986).
39. Chan, C. M., Venkatraman, S., *J. Polym. Sci., Polym. Phys. Ed.*, **25**, 1655 (1987).
40. Griffith, J. H., Ranby, B. G., *J. Polym. Sci.*, **44**, 369 (1960).
41. Sharples, A., Swinton, F. L., *Polymer*, **4**, 119 (1963).
42. Gordon, M., Hiller, J., *Polymer*, **6**, 213 (1965).
43. Hoshino, S., Meinecke, E., Powers, J., Stein, R. S., *J. Polym. Sci.*, Part A, **3**, 3064 (1965).
44. Hybart, F. J., Pepper, B., *J. Appl. Polym. Sci.*, **13**, 2643 (1969).
45. Booth, C., Dodgeson, D. V., Hiller, I. H., *Polymer*, **11**, 11 (1970).
46. Turska, E., Gogolewski, S., *Polymer*, **12**, 616, 629 (1971).
47. Ergoz, E., Fatou, J. G., Mandelkern, L., *Macromolecules*, **5**, 147 (1972).
48. Capizzi, A., Gianotti, G., *Makromol. Chem.*, **157**, 123 (1972).
49. Gallez, F., Legras, R., Mercier, J. P., *J. Polym. Sci., Polym. Phys. Ed.*, **14**, 1367 (1976).
50. Allen, R. C., Mandelkern, L., *J. Polym. Sci., Polym. Phys. Ed.*, **20**, 146 (1982).
51. Avrami, M., *J. Chem. Phys.*, **7**, 1103 (1939); **8**, 212 (1940); **9**, 177 (1941).
52. Morgan, L. B., *Phila. Trans. Roy. Soc.* (London), **247**, 13 (1954).
53. Mandelkern, L., Quinn, F. A., Flory, P. J., *J. Appl. Phys.*, **25**, 830 (1954).
54. Ziabicki, A., *J. Chem. Phys.*, **48**, 4368 (1968).
55. Ziabicki, A., *Fundamentals of Fiber Formation*, John Wiley & Sons, New York (1976).
56. Ozawa, T., *Polymer*, **12**, 150 (1971).
57. Nakamura, K., Katayama, K., Amano, T., *J. Appl. Polym. Sci.*, **17**, 1031 (1973).
58. Jeziorny, A., *Polymer*, **19**, 1143 (1978).
59. Eder, M., Wlochowicz, A., *Polymer*, **24**, 1593 (1983).
60. Kamal, M. R., Edwards, C., *Polym. Eng. Sci.*, **23**, 27 (1983).
61. Malkin, A., Begishev, V. P., Keapin, I. A., Andrianova, Z. S., *Polym. Eng. Sci.*, **24**, 1402 (1984).
62. Kenig, S., Kamal, M. R., *SPE J.*, **26**, 50 (1970).
63. Kamal, M. R., Lafleur, P. G., *Polym. Eng. Sci.*, **24**, 692 (1984).
64. Lafleur, P. G., Kamal, M. R., *Polym. Eng. Sci.*, **26**, 92 (1986).
65. Addonizio, M. L., Martuscelli, E., Silvestre, C., *Polymer*, **28**, 183 (1987).
66. Keller, A., in *Growth and Perfection of Crystals*, Doremus, R. H., Roberts, B. W., Turnbull, D., Eds., John Wiley & Sons, New York (1958).
67. Geil, P. H., *Polymer Single Crystals*, Interscience Publishers, New York (1963).
68. Mandelkern, L., in *Physical Properties of Polymers*, Mark, J. E., Eisenberg, A., Graessley, W. W., Mandelkern, L., Koenig, J. L., Eds., American Chemical Society, Washington, D.C. (1984).
69. Stein, R. S., Khambatta, F. B., Warner, F. P., Russell, T., Escala, A., Balizer, E., *J. Polym. Sci., Polym. Symp.*, **63**, 313 (1978).

70. Martuscelli, E., *Polym. Eng. Sci.*, **24**, 563 (1984).
71. Paul, D. R., Barlow, J. W., in *Polymer Alloys III*, Klempner, D., Frisch, K. C., Eds., Plenum Press, New York (1983).
72. Runt, J. P., Martynowicz, in *Multicomponent Polymer Materials*, Paul, D. R., Sperling, L. H., Eds., Adv. Chem. Series, 211, American Chemical Society, Washington, D.C. (1986).
73. Olabisi, O., Robeson, L. M., Shaw, M. T., *Polymer–Polymer Miscibility*, Academic Press, New York (1979).
74. Starkweather, H. W., Jr., *J. Appl. Polym. Sci.*, **25**, 139 (1980).
75. Kyu, T., Vadhar, P., *J. Appl. Polym. Sci.*, **32**, 5575 (1986).
76. Vadhar, P., Kyu, T., *Polym. Eng. Sci.*, **27**, 202 (1987).
77. Hu, S. R., Kyu, T., Stein, R. S., *J. Polym.Sci., Polym. Phys. Ed.*, **25**, 71, 89 (1987).
78. Siegmann, A., Nir, Y., *Polym. Eng. Sci.*, **27**, 1182 (1987).
79. Rego-Lopez, J. M., Gedde, U. W., *Polymer*, **29**, 1037 (1988).
80. Sham, C. K., Guerra, G., Karasz, F. E., MacKnight, W. J., *Polymer*, **29**, 1016 (1988).
81. Clampitt, B. H., *J. Polym. Sci.*, **A3**, 671 (1965).
82. Bhateja, S. K., Andrews, E. H., *Polym. Eng. Sci.*, **23**, 888 (1983).
83. Kishore, K., Vasanthakumari, R., *Polymer*, **27**, 337 (1986).
84. D'Orazio, L., Greco, R., Martuscelli, E., Ragosta, G., *Polym. Eng. Sci.*, **23**, 489 (1982).
85. Greco, R., Mancarella, C, Martuscelli, E., Ragosta, G., Jinghua, Y., *Polymer*, **28**, 1922 (1987).
86. Kalfoglou, N. K., *J. Macromol. Sci.*, **B22**, 343, 363 (1983).
87. Teh, J. W., *J. Appl. Polym. Sci.*, **28**, 605 (1983).
88. Jacoby, P., Bersted, B. H., Kissel, W. J., Smith, C. E., *J. Polym. Sci., Polym. Phys. Ed.*, **24**, 461 (1986).
89. Galeski, A., Pracella, M., Martuscelli, E., *Polymer*, **25**, 1323 (1984).
90. Bartczak, Z., Galeski, A., Pracella, M., *Polymer*, **27**, 537 (1986).
91. Lovinger, A. J., Williams, M. L., *J. Appl. Polym. Sci.*, **25**, 1703 (1980).
92. Gupta, A. K., Gupta, V. B., Peters, R. H., Harland, W. G., Berri, J. P., *J. Appl. Polym. Sci.*, **27**, 4669 (1982).
93. Noel, O. F., III, Carley, J. F., *Polym. Eng. Sci.*, **24**, 488 (1984).
94. Bartczak, Z. Galeski, A., *Polymer*, **27**, 544 (1986).
95. Siegmann, A., *J. Appl. Polym. Sci.*, **24**, 2333 (1979).
96. Siegmann, A., *J. Appl. Polym. Sci.*, **27**, 1053 (1982).
97. Gohil, R. M., Peterman, J., *J. Macromol. Sci.–Phys.*, **B18**, 217 (1980).
98. Karger-Kocsis, J., Kallo, A., Szafner, A., Bodor, G., Senyei, Z. S., *Polymer*, **20**, 37 (1979).
99. Martuscelli, E., Silvestre, C., Abate, G., *Polymer*, **23**, 229 (1982).
100. Kalfoglou, N. K., *Angew. Makromol. Chem.*, **129**, 103 (1985).
101. Greco, R., Mancarella, M., Martuscelli, E., Ragosta, G., Jinghua, Y., *Polymer*, **28**, 1929 (1987).
102. Coppola, F., Greco, R., Martuscelli, E., Kammer, H. W., Kummerlowe, C., *Polymer*, **28**, 47 (1987).
103. Martuscelli, E., Silvestre, C., Bianchi, L., *Polymer*, **24**, 1458 (1983).
104. Bartczak, Z., Galeski, A., Martuscelli, E., *Polym. Eng. Sci.*, **24**, 115 (1984).
105. Karger-Kocsis, J., Kiss, L., *Polym.Eng. Sci.*, **27**, 254 (1987).
106. Gupta, A. K., Purwar, S. N., *J. Appl. Polym. Sci.*, **29**, 1595 (1984).
107. Danesi, S., Porter, R. S., *Polymer*, **19**, 448 (1978).

108. Bianchi, L., Cimmino, S., Forte, A., Greco, R., Martuscelli, E., Rira, F., Silvestre, C., *J. Mater. Sci.*, **20**, 895 (1985).
109. Kalfoglou, N. K., *J. Appl. Polym. Sci.*, **30**, 1989 (1985).
110. Wilfong, D. L., Hiltner, A., Baer, E., *J. Mater. Sci.*, **21**, 2014 (1986).
111. Escala, A., Stein, R. S., in *Multiphase Polymers*, Cooper, S. L., Ester, G. M., Eds., Adv. Chem. Series, 176, American Chemical Society, Washington, D.C. (1979).
112. Nassar, T. R., Paul, D. R., Barlow, J. W., *J. Appl. Polym. Sci.*, **23**, 85 (1979).
113. Hanrahan, B. D., Angeli, S. R., Runt, J., *Polym. Bull.*, **15**, 455 (1986).
114. Murff, S. R., Barlow, J. W., Paul, D. R., *Makromol. Chem.*, **185**, 1041 (1984).
115. Ahroni, S. M., *J. Macromol. Sci. (Phys.)*, **B22**, 813 (1983).
116. Seymour, R. W., Zehner, B. E., *J. Polym. Sci., Polym. Phys. Ed.*, **18**, 2299 (1980).
117. Robeson, L. M., *J. Appl. Polym. Sci.*, **30**, 4081 (1985).
118. Murff, S. R., Barlow, J. W., Paul, D. R., *J. Appl. Polym. Sci.*, **29**, 3231 (1984).
119. Nadkarni, V. M., Jog, J. P., *Polym. Eng. Sci.*, **27**, 451 (1987).
120. Nadkarni, V. M., Shingankuli, V. L., Jog, J. P., *Polym. Eng. Sci.*, **28**, 1326 (1988).
121. Wahrmund, D. C., Paul, D. R., Barlow, J. W., *J. Appl. Polym. Sci.*, **22**, 2155 (1978).
122. Kimura, M., Porter, R. S., *J. Polym. Sci., Polym. Phys. Ed.*, **22**, 1697 (1984).
123. Robeson, L. M., Furtek, A. B., *J. Appl. Polym. Sci.*, **23**, 645 (1979).
124. Slogowski, E. L., Chang, E. P., Tkacik, J. J., *Polym. Eng. Sci.*, **21**, 513 (1981).
125. Pracella, M., Dainelli, D., Galli, G., Chiellini, E., *Makromol. Chem.*, **187**, 2387 (1986).
126. Paci, M., Barone, C., Magagnini, P. L., *J. Polym. Sci., Polym. Phys. Ed.*, **25**, 1595 (1987).
127. Mohn, R. N., Paul, D. R., Barlow, J. W., Cruz, C. A., *J. Appl. Polym. Sci.*, **23**, 575 (1979).
128. Barnum, R. S., Barlow, J. W., Paul, D. R., *J. Appl. Polym. Sci.*, **27**, 4065 (1982).
129. Nadkarni, V. M., Shingankuli, V. L., Jog, J. P., *Int. Polym. Proc.*, **2**, 53 (1987).
130. Nadkarni, V. M., Jog, J. P., *J. Appl. Polym. Sci.*, **32**, 5828 (1986).
131. Shingankuli, V. L., *Ph.D. thesis*, Bombay Univ., India (1990).
132. Shingankuli, V. L., Jog, J. P., Nadkarni, V. M., *J. Appl. Polym. Sci.*, **36**, 335 (1988).
133. Noland, J. S., Hsu, N. N. C., Saxon, R., Schmitt, J. M., in *Multicomponent Polymer Systems*, Platzer, N. A. J., Ed., Adv. Chem. Series 99, American Chemical Society, Washington, D.C. (1971).
134. Wang, T. T., Nishi, T., *Macromolecules*, **10**, 421 (1977).
135. Nishi, T., Wang, T. T., *Macromolecules*, **8**, 909 (1975).
136. Patterson, G. D., Nishi, T., Wang, T. T., *Macromolecules*, **9**, 603 (1976).
137. Kwei, T. K., Patterson, G. D., Wang, T. T., *Macromolecules*, **9**, 780 (1976).
138. Hirata, Y., Kotak, T., *Polym. J.*, **13**, 273 (1981).
139. Mijovic, J., Luo, H. L., Han, C. D., *Polym. Eng. Sci.*, **22**, 234 (1982).
140. Hourston, D. J., Hughes, J. D., *Polymer*, **18**, 1175 (1977).
141. Roerdink, E., Challa, G., *Polymer*, **19**, 173 (1978).
142. Bernstein, R. E., Paul, D. R., Barlow, J. W., *Polym. Eng. Sci.*, **18**, 677 (1978).
143. Paul, D. R., Barlow, J. W., Bernstein, R. E., Wahrmund, D. C., *Polym. Eng. Sci.*, **18**, 1225 (1978).
144. Paul, D. R., Barlow, J. W., Bernstein, R. E., Wahrmund, D. C., *Polym. Eng. Sci.*, **18**, 1221 (1978).
145. Briber, R. M., Khoury, F., *Polymer*, **28**, 38 (1987).
146. Morra, B. S., Stein, R. S., *J. Polym. Sci., Polym. Phys. Ed.*, **20**, 2243 (1982).
147. Morra, B. S., Stein, R. S., *J. Polym. Sci., Polym. Phys. Ed.*, **20**, 2261 (1982).
148. Morra, B. S., Stein, R. S., *Polym. Eng. Sci.*, **24**, 311 (1984).

149. Aref-Azar, A., Hay, J. N., Marsden, B. J., Walker, N., *J. Polym. Sci., Polym. Phys. Ed.*, **18**, 637 (1980).
150. Liang, B. R., White, J. L., Spruiell, J. E., Goswami, B. C., *J. Appl. Polym. Sci.*, **28**, 2011 (1983).
151. Kamal, M. R., Sahto, M. A., Utracki, L. A., *Polym. Eng. Sci.*, **22**, 1127 (1982).
152. Kamal, M. R., Sahto, M. A., Utracki, L. A., *Polym. Eng. Sci.*, **23**, 637 (1983).
153. Utracki, L. A., Bata, G. L., in *Polymer Alloys III*, Klempner, D., Frisch, K. C., Eds., Plenum Press, New York (1983).
154. Pillon, L. Z., Lara, J., Pillon, D. W., *Polym. Eng. Sci.*, **27**, 984 (1987).
155. Pillon, L. Z., Utracki, L. A., *Polym. Eng. Sci.*, **24**, 1300 (1984).
156. Inoue, M., *J. Polym. Sci.*, Part A, **1**, 3427 (1963).

CHAPTER 9

BIREFRINGENCE IN INJECTION-MOLDED POLYMER BLENDS

by Yasuhiko Kijima, Takao Kawaki

Tokyo Central Research Laboratories
Mitsubishi Gas Chemical Co., Inc.
Niijuku, Katsushika-ku, Tokyo 125
JAPAN

and

by Takashi Inoue

Department of Organic and Polymeric Materials
Tokyo Institute of Technology
Ookayama, Meguro-ku, Tokyo 152
JAPAN

The birefringence in melt-processed plastics can be reduced to zero by polymer blending without sacrificing transparency. Such a reduction of birefringence (or zero birefringence) can be attained as a result of compensation of positive and negative birefringences induced by chain orientation of both constituent polymers. More than ten pairs of the dissimilar polymers have been found to render birefringence-free blends. Both single-phase and two-phase systems, such as polyphenyleneether/ polystyrene blend, polycarbonate-polystyrene graft copolymer and related materials, were injection molded to form a substrate for optical disk and the optical anisotropy in the substrate was investigated for vertical and oblique incident beams. A graft copolymer of polycarbonate with poly(styrene- co-maleic anhydride) can be molded as low-birefringence disk substrate and, at present, it seems to be the best candidate for the moldable disk substrate.

9.1   INTRODUCTION

More than twenty years ago, the polymeric lens was mounted on the camera. Modern applications of plastics for optical devices are primarily in compact disks and optical fibers. These applications will continue to expand in the future.

In general, plastics have several advantages over inorganic glasses, e.g., 1) good processability, especially moldability into complicated shapes such as non spherical lens; 2) low density, rendering weight reduction; 3) excellent impact resistance; and 4) low labor cost in mass production. On the other hand, the weak points are: low refractive index, large temperature dependence of refractive index, large birefringence, insufficient heat resistance, and large thermal expansion coefficient, as demonstrated in Table 9.1. Some weak points originate from the intrinsic properties of polymers and these are inevitable. Others may be improved; the birefringence is an example.

In this chapter, we deal with birefringence-free plastics. First, we describe the background of the problem; how undesirable is the birefringence for optical application, especially in the case of the optical disk. Then we explain a concept of the birefringence reduction by blending dissimilar polymers. Finally, we show the birefringence behavior in the injection-molded blends of both single-phase and two-phase systems.

9.2   BACKGROUND: Noise Introduced by Birefringence

Polymer processing operations, such as injection molding and extrusion, usually induce birefringence owing to frozen-in molecular orientation. The birefringence, Δn, is often undesirable for optical applications, such as optical disks and lenses.

Magnetro-optical disk consists of a memory layer and a substrate [1,2]. The disk substrate is made of a plastic and it supports and protects the memory layer. The information is "written" or stored in terms of the local magnetization direction in the memory layer. The written information is "read" or reproduced by the Kerr effect, i.e., the polarization direction of light reflected from the memory layer varies with the magnetization direction and the variation is detected through the plastic substrate, as shown in Fig. 9.1. Hence the birefringence in the substrate strongly affects the signal-to-noise ratio of the reading system.

Table 9.1 *Physical Properties of Glassy Polymers and Inorganic Glasses*

Property[a]	Polymer				Glass		
	PMMA	SAN	PC	PS	BK7	SK16	F2
n^{20}	1.492	1.567	1.586	1.592	1.516	1.620	1.621
ν_d	58	35	31	31	64	60.3	36.1
$10^4 \cdot dn/dT \, (^\circ C)^{-1}$	-1.2	-1.4	-1.4	-1.5	0.03	0.02	0.03
Transmittance (%)	92	90	89	88	92	91	91
HDT ($^\circ C$)	65-100	80-95	120-140	70-100	565	648	432
$\kappa \, [10^{-4} cal/s \, ^\circ C \, cm]$	4-6	---	4.5	2.4-3.3			
$\alpha \cdot 10^6 \, (^\circ C)^{-1}$	70	70	70	80	7.4	6.6	8.7
$\rho \, (kg/m^3)$	1190	1070	1200	1060	2530	3650	3610

[a] *For explanation of the symbols see* NOTATION *on page 256.*

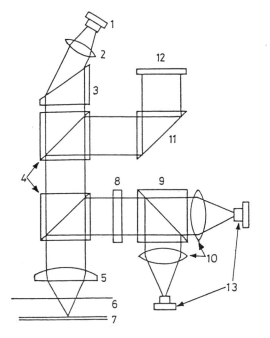

1. Laser diode
2. Collimator lens
3. Shaping prism
4. Half mirror
5. Object lens
6. Substrate
7. Memory layer
8. Half-wave plate
9. Beam splitter
10. Collimator
11. Totally reflecting prism
12. Detector
13. PIN photodiode

Figure 9.1 *Differential optical head for magneto-optical disk [1].*

During the last five years, the birefringence in disk substrates has been reduced to a fairly low level by improving the molding technique and the materials (e.g., by reduction of molecular weight or by addition of a plasticizer). However, the noise has remained even in a "zero birefringence" substrate. It should be noted that even in the "zero birefringence" materials, $\Delta n = 0$ only for the vertical incidence of the reading beam while a definite birefringence may persist for an oblique incidence. Thus the residual noise is caused by the birefringence induced by oblique incident beams. As shown in Fig. 9.1, the reading beam is focused on the memory layer through an objective lens with large aperture number in order to read the densely written information.

For proper characterization of optical properties in the injection-molded disk substrate one has to investigate the anisotropy in various directions, schematically indicated in Fig. 9.2. Figure 9.3 presents a typical example of the directional dependencies of birefringence, as a function of the angle of incidence Θ_t in a tangential plane (see Fig. 9.2). When the injection temperature of polycarbonate, PC, is increased, Δn for near vertical incidence ($\Theta_t \to 0$) is small, within the desired limit: $\Delta n \leq 1.7 \times 10^{-5}$, i.e., less than 20 nm in retardation for substrate of 1.2 mm thick (retardation $= \Delta n \times$ specimen thickness). However, at larger incidence angles, Θ_t, the birefringence exceeds the limit.

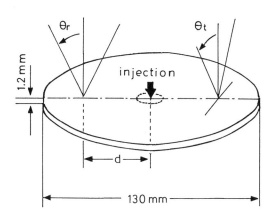

Figure 9.2 *Injection-molded disk substrate.*

Takeshima et al. [3] investigated the depth profile of chain orientation in an injection-molded disk of PC by the polarized laser Raman spectroscopy. In light of their results (reproduced in Fig. 9.4) the "zero birefringence" substrate, mentioned above, seems to be molded under a complicated flow. In other words, the overall birefringence for vertical incidence may be accidentally reduced to zero. Thus, the best solution to the reading noise problem is development of a birefringence-free polymer.

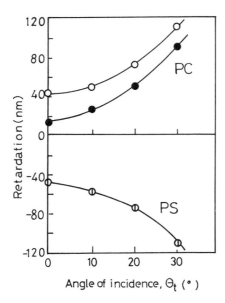

Figure 9.3 *Birefringence in the injection-molded disks of PC and PS, as a function of the angle of incidence,* θ_t: *(○) PC molded at 310° C, (●) PC at 320° C, (⊘) PS at 300° C (birefringence: retardation/1.2 mm thick, at d = 43 mm, viz. Fig. 9.2) [2].*

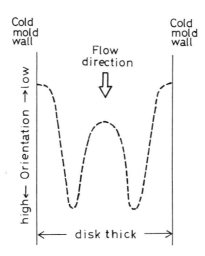

Figure 9.4 *Schematic representation of molecular orientation in the molded disk substrate of PC, on the basis of the laser Raman spectroscopy [3].*

9.3   CONCEPT OF BIREFRINGENCE-FREE POLYMER BLEND

9.3.1   Combination of Positive and Negative Birefringence Polymers

Hahn and Wendorf [4] as well as Saito and Inoue [5] found independently that the birefringence can be reduced to zero by blending dissimilar polymers without sacrificing transparency. The reduction of birefringence or zero birefringence comes from compensation of the positive and negative birefringences induced by the chain orientation of both constituent polymers.

In Fig. 9.5, the birefringence of polyvinylchloride/polymethylmethacrylate blends, PVC/PMMA, is shown as a function of the draw ratio λ [6]. The specimens here is a drawn-quenched film. The solution-cast film was uniaxially stretched rapidly (ca. 500% elongation/min) to a desired draw ratio at 90°C and then the drawn film quenched in an ice-water bath immediately after stretching. Then birefringence of the drawn-quenced film was measured at 20°C. The drawn PMMA had negative birefringence, whereas that of PVC was positive. PVC/PMMA blends, depending on composition, showed negative or positive birefringence. The PMMA/PVC blend having the composition 82/18 by weight had zero birefringence. That is, a birefringence-free polymer was obtained by blending appropriate amounts of negative and positive birefringence polymers.

Such behavior has been found in other miscible blends [6]. Birefringence-free polymer blends are listed in Table 9.2. The birefringence-free compositions were determined from the results of stretching experiments similar to those presented in Fig. 9.5. That is, from the almost linear relation between birefringence and λ, as in Fig. 9.5, the slope was obtained, then the slope was plotted against the blend composition; the zero-crossing point gave the birefringence-free composition.

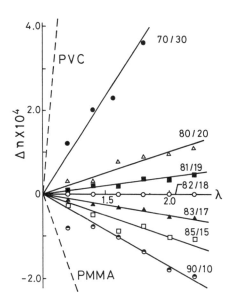

Figure 9.5 *Birefringence vs. draw ratio of PMMA/PVC blends at different compositions [6].*

Table 9.2 *Birefringence-Free Polymer Blends*

POLYMER PAIR[a] nega.Δn/posi.Δn	Composition[b] (wt% ratio)	Stretching temp. (°C)
PMMA/PVDF	80/20	90
PMMA/VDF·TrFE-58	90/10	90
PMMA/PEO	65/35	90
PS/PPO	71/29	180
S·LMI-19/PPE	73.5/26.5	180
S·PMI-16/PPE	74.5/25.5	180
S·CMI-17/PPE	75.5/24.5	180
S·VHC/PC	35/65	180
S·MA-8/PC	77/23	180
AS-25/NBR-40	40/60	90
PS/PVME	-------	---

[a] *Combination of negative-birefringence polymer and positive-birefringence polymer.*
[b] *Birefringence-free composition.*

In the miscible blends when the positive-birefringence polymer chain and the negative one undergo an equivalent deformation accompanied by an equivalent relaxation, there is a specific composition at which a complete compensation of birefringence can be achieved at any stage of stretching or melt flow, no matter what processing conditions. Thus, in relation to the design of birefringence-free plastics, it is of interest to characterize the orientation of individual chains in the deformed and relaxed blends. This will be discussed in the following section.

9.3.2 Molecular Orientation

One may characterize chain orientation by the second moment of the orientation function f defined by:

$$f = (3 < \cos^2 \Theta > -1)/2 \tag{9.1}$$

where Θ is the angle between the chain axis and the stretching direction. The orientation function, f, is evaluated by the infrared dichroic ratio $D = A_{\parallel}/A_{\perp}$ ($A_{\parallel}$ and $A_{\perp}$ being the absorbance parallel and perpendicular to the stretching direction, respectively) of any absorption band as:

$$f = \frac{D - 1}{D + 2} \cdot \frac{2 \cot^2 \psi + 2}{2 \cot^2 \psi - 1} \tag{9.2}$$

where ψ is the angle between the transition moment vector of the absorbing group and the chain axis.

Figure 9.6 shows a typical example of the polarized Fourier-transform infrared spectra of a drawn-quenched blend of PVDF/PMMA. Infrared dichroism is seen for component polymers at various absorbing bands. Assuming an appropriate value of ψ, one can calculate the orientation function by Eq (9.2). The calculated orientation functions of the two polymers are shown to be nearly equal. Similar results have been obtained for the specimens stretched to various draw ratios [6].

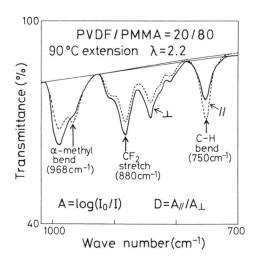

Figure 9.6 *Polarized Fourier-transform infrared spectra of a stretched 20/80 PVDF/PMMA blend [5].*

In order to discuss the chain-relaxation behavior from infrared dichroism, it is convenient to employ the normalized orientation function $F(t)$ defined by the ratio of f_t and f_0 (f_t and f_0 being the orientation function at relaxation time, t, and that immediately after the deformation at $t = 0$, respectively). From Eq (9.2):

$$F(t) = f_t/f_0 = (D_t - 1)(D_0 + 2)/(D_t + 2)(D_0 - 1) \qquad (9.3)$$

Since in many cases value of ψ is unknown, Eq (9.3) provides a simpler method of characterization of the relaxation process than Eq (9.2).

In Fig. 9.7 the normalized orientation function $F(t)$ of a 20/80 PVDF/PMMA blend is shown as a function of the relaxation time, t, at 110°C [7]. The dissimilar polymers showed identical time variation of $F(t)$. In the case when the molecular weight of the partner polymer was large (HM-PMMA; $M_w = 369$ kg/mol) the $F(t)$ of PVDF decreased slowly, whereas when the partner had a low molecular weight (LM-PMMA; $M_w = 110$) it decreased more rapidly. This implies that PVDF chains do not relax independently but cooperatively with PMMA.

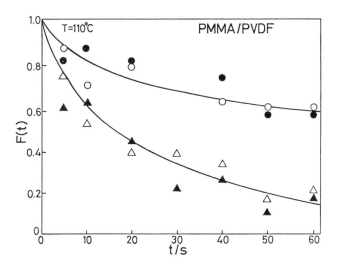

Figure 9.7 *Normalized orientation function, F(t), versus relaxation time, t, for 20/80 PVDF/PMMA blends at 110°C [7]. (●) PVDF and (○) PMMA in PVDF/HM-PMMA system; (▲) PVDF and (△) PMMA in PVDF/LM-PMMA system. Orientation functions were calculated from dichroism at 750 cm^{-1} (PMMA) and 880 cm^{-1} (PVDF).*

These results suggest that PVDF and PMMA chains undergo an equivalent deformation and they relax at same speed. Similar behavior has been observed for other birefringence-free blends, such as PVC/PMMA and PMMA/poly(VDF·TrFE-58) [7]. The cooperative deformation and relaxation are favorable phenomena for designing zero-birefringence blends.

9.3.3 Birefringence-free Composition

Birefringence of polymer blends may be formulated by

$$\Delta n = \Delta n_A{}^\circ f_A \, \phi_A + \Delta n_B{}^\circ f_B \, \phi_B + \Delta n_F \tag{9.4}$$

where $\Delta n_i{}^\circ$ is the intrinsic birefringence of the i-polymer, f_i is the orientation function, ϕ_i is the volume fraction, and Δn_F is the form birefringence. Δn_F is assumed to be zero in the miscible blend. So, when f_A is equal to f_B, as in Fig. 9.7, one is able to estimate birefringence-free composition. In order to do so, the intrinsic birefringence of A and B polymers must be known.

The intrinsic birefringence is approximately related to the polarizability difference, $\Delta \alpha$, as in the Lorentz–Lorenz equation:

$$\Delta n^\circ = \frac{2}{9} \cdot \frac{(\bar{n}^2 + 2)^2}{\bar{n}} \cdot \frac{\rho}{M} \cdot N_A \, \Delta \alpha \tag{9.5}$$

where $\bar{n}$ is the average refractive index, ρ is the density of polymer, M is the molecular weight of monomer, N_A is the Avogadro's number, and $\Delta \alpha$ is defined as:

$$\Delta \alpha = \alpha_X - (\alpha_Y + \alpha_Z)/2 \tag{9.6}$$

where α_i is the principal polarizability of monomer represented as a sum of contributions from mer constituent bonds. The values of principal polarizabilities of constituent bonds were listed by Denbigh [8]. We are able to estimate the contribution from the individual bond using the transformation of the polarizability tensor $(\alpha)_{X'Y'Z'}$ for the $(X'Y'Z')$ coordinate of a constituent bond into the (XYZ) coordinate of the chain axis by [9,10]:

$$(\alpha)_{XYZ} = R_Z(\theta) \, R_{X}{}'(\phi) \cdot \text{diag}(\alpha)_{X'Y'Z'} \cdot R_X{}'^{-1}(\phi) \, R_Z{}^{-1}(\theta) \tag{9.7}$$

with

$$R_X{}'(\phi) = \begin{bmatrix} 1 & 0 & 0 \\ 0 & \cos\phi & -\sin\phi \\ 0 & \sin\phi & \cos\phi \end{bmatrix}, \quad R_Z(\theta) = \begin{bmatrix} \cos\theta & -\sin\theta & 0 \\ \sin\theta & \cos\theta & 0 \\ 0 & 0 & 1 \end{bmatrix} \tag{9.8}$$

where $R_X{}'(\phi)$ is the transformation for a rotation, ϕ, of the bond about the X' axis and $R_Z(\theta)$ is that for a rotation, θ, about the Z axis. One can calculate $\Delta n°$ from Eqs (9.5) to (9.8) and then the birefringence-free composition $\phi_B(\Delta n = 0)$ using Eq (9.4) with the assumption: $\Delta n_F = 0$ and $f_A = f_B$. Typical results are shown in Table 9.3.

In Table 9.3, the observed values of $\phi_B(\Delta n = 0)$ are also listed. The value of $\phi_B(\Delta n = 0)$, estimated by assuming a trans zig-zag conformation, agrees well with the observed value for PMMA/PVC and PMMA/PVDF. On the other hand, for the PMMA/VDF·TrFE-58 system, better agreement was obtained assuming a trans-gauche conformation. When the exact information on the chain conformation is available, much better agreement would be expected.

Table 9.3 *Birefringence-free Composition: PMMA[a]/B–polymer Blends[d] [6]*

B-polymer[b]	$\phi_B(\Delta n = 0)$ (obs.)[c]	Trans-zigzag		Trans-gauche	
		$\Delta n°$	$\phi_B(\Delta n=0)$	$\Delta n°$	$\phi_B(\Delta n=0)$
PVDF	17	0.060	17.6	0.202	5.8
PVC	15	0.101	11.0	0.226	5.2
VDF·TrFE–58	8	0.039	24.3	0.195	6.0

[a] $\Delta n°$ *of PMMA is estimated using the results by Read [11]*, [b] *Positive-birefringence polymer*, [c] *From Table 9.2; in vol. fraction*, [d] *Since we have no information on the chain conformations in the miscible polymer blends, the calculation was carried out for two chain conformations: planar trans-zigzag and trans-gauche.*

9.4 BIREFRINGENCE IN INJECTION-MOLDED BLENDS

9.4.1 Single-Phase Systems

The concept of birefringence-free polymer blends discussed in the previous section, can be confirmed in the melt-processed materials. Figure 9.8 shows the photographs of injection-molded dumbbell specimens placed between cross-polarizers [6]. Pure PMMA and the 60/40 PMMA/PVDF blend are seen as the optically anisotropic materials. In contrast, the 80/20 molded blend is optically isotropic, as expected from the results of stretching experiments in Table 9.2.

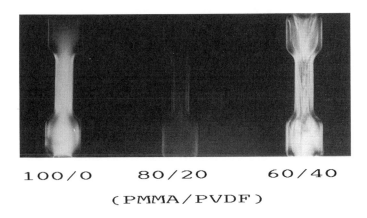

100/0   80/20   60/40

(PMMA / PVDF)

Figure 9.8 *Photograph of injection-molded specimen placed between crossed polarizers. Flat dumbbell specimen, ca. 1.5 mm thick, was prepared by a miniature molder (Mini Max Molder, CS-183MM) set at 200° C [6].*

However, from the practical point of view, the PMMA/PVDF blend is less interesting because PVDF has a low glass transition temperature ($\approx$ -40°C) and addition of PVDF reduces the heat resistance. Among the birefringence-free combinations of dissimilar polymers in Table 9.2, the practically interesting systems are limited: PPE/PS, PPE/styrene copolymer, and PC/styrene copolymer. However, the latter two systems exhibit the LCST (lower critical solution temperature)-type phase behavior [12] and they are immiscible at the injection temperature T_i, i.e., LCST < T_i, hence the injection molding would yield opaque materials. PPE/PS is the only promising combination.

Figure 9.9 shows the birefringence of the injection-molded PPE/PS blends as a function of blend composition [2]. Both birefringences by vertical incidence ($\theta_t = 0$) and oblique incidence ($\theta_t = 30°$) increased with PPE content and changed the sign from negative to positive, suggesting that low birefringence and noise-free disk substrate can be obtained by molding a PPE/PS blend with composition of ca. 40/60. However, there is the problem of thermal deterioration. It is hard to mold the PPE/PS blend containing no oxidized material. One has to find a stabilizer which would not damage the optical clarity.

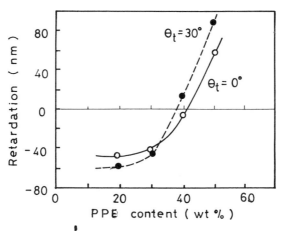

Figure 9.9 *Vertical ($\theta_t=0°$) and oblique ($\theta_t=30°$) retardation in injection-molded disk substrate of PPE/PS blend [2]. Δn: retardation/1.2 mm thick, at $d=43$ mm (see Fig. 9.2). Injection molding: temperature $300°C$, maximum pressure 1840 kg/cm^2, vol/shot$=68$ cm^3.*

9.4.2   Two-Phase Systems

PC is a highly anisotropic polymer with positive polarizability difference, $\Delta\alpha$, estimated from Eqs (9.6)-(9.8) to be ca. 65×10^{-25} cm^3. To compensate for this positive birefringence, a highly anisotropic polymer with negative $\Delta\alpha$ should be used. As indicated in Table 9.2, PS has a negative birefringence with an estimated $\Delta\alpha = -25 \times 10^{-25}$ to -72×10^{-25} cm^3 (depending on the rotational angle of the phenyl group [13]). Hence PS could be the partner for PC to attain zero birefringence.

Unfortunately PS is immiscible with PC. Simple melt mixing of PS with PC results in a two-phase material with large domain size, say, on the order of micrometers (see Fig. 9.10b). The two-phase material is opaque and undesirable for optical use. However, it is well known that, if dissimilar polymers are combined to form block or graft copolymer, the domain size can be reduced to the order of several tens of nm.

A PC-PS graft copolymer was produced by the copolymerization of styrene with a PC modified with a vinyl group at the chain end (PC macromer). The transmission electron micrograph of the PC-PS graft copolymer in Fig. 9.10a indicates, as expected, regular two-phase morphology with small domains [2].

Figure 9.11 shows birefringence in the injection-molded PC-PS graft copolymers as a function of PC content. It suggests that the graft copolymer with composition ranging from 60/40 to 70/30 wt% yields low birefringence material and hence noise-free substrate. The zero birefringence composition for vertical incidence deviates from that for an oblique one. Deviation is slightly larger than in the case of PPE/PS blends in Fig. 9.9. This may be due to the form birefringence, Δn_F, caused by the two-phase nature of the copolymer. In Fig. 9.12 the birefringence characteristics of the disk substrate of a PC-PS (60:40) graft copolymer is presented. Birefringence here is shown as a function of the distance from the center of disk (injection gate). Oblique birefringence in both tangential and radial planes (see Fig. 9.2) is shown. Low birefringence was attained throughout the molded substrate. It is interesting that, as shown in Fig. 9.13, the birefringence behavior was not affected by the injection temperature.

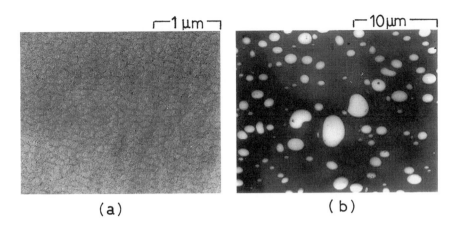

Figure 9.10 *Transmission electron micrograph (with RuO$_4$ strain); (a) PC-PS graft copolymer, (b) PC/PS blend.*

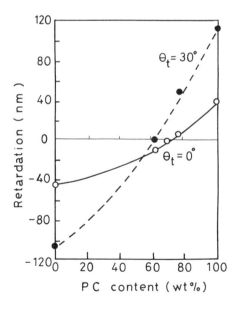

Figure 9.11 *Birefringence in injection-molded PC-PS graft copolymers with different compositions [2]. Conditions of molding and birefringence analysis were the same as Fig. 9.9.*

As described above, the graft copolymer forms the microdomain structure and it possesses good optical clarity. Reduction of difference in the refractive indices between the two microphases of the graft copolymer will improve transparency and reduce the form birefringence. Such matching of the refractive indices can be achieved by copolymerization of PC macromer with a binary monomer system, e.g., styrene (S) and maleic anhydride (MA). In this way, Mitsubishi Gas Chemical Co. Inc. developed a PC-SMA graft copolymer designed specifically for the disk substrate. In Table 9.4 the optical, thermal, and mechanical properties of the MGC resin are listed [14].

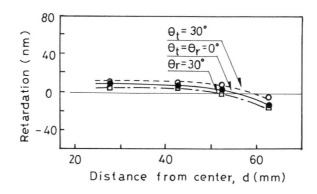

Figure 9.12 *Birefringence in injection-molded PC-PS graft copolymer as a function of distance from injection gate.*

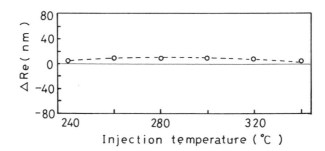

Figure 9.13 *Temperature dependence of the retardation difference between vertical and oblique incidences, in injection-molded PC-PS graft copolymer.*

Table 9.4 Physical Properties of MGC[a] Resin

Item	Method	MGC resin	PC	PMMA
Optical properties				
Total light transmission (%)	ASTM D-1746	91	91	92
Index of refraction	ASTM D-542	1.59	1.59	1.49
Birefringence (nm/1.2 mm)	ellipsometry			
Vertical incidence	$\theta_t = 0°$	< 15	< 20	< 15
Oblique incidence	$\theta_t = 30°$	< 15	< 80	< 15
Thermal properties				
Vicat softening point (°C)	130–145	145	110	
Mechanical properties				
Tensile strength (MPa)	ASTM D-638	59	61	43
Ultimate elongation (%)	ASTM D-638	6	31	2
Bending strength (MPa)	ASTM D-790	98	90	65
Flexural modulus (MPa)	ASTM D-790	2450	2160	3240
Rockwell hardness	ASTM D-785	75	65	82
Water absorption (%)		0.25	0.34	2.2

[a] *Mitsubishi Gas Chemical Co.*

9.5   CONCLUSIONS

We are able to design low birefringence materials by blending or combining the dissimilar polymers with different signs of optical anisotropy. One possibility has been explored for the production of injection-molded substrate for noise-free optical disks. From a practical point of view, at present, PC-SMA graft copolymer seems to be the best candidate for the disk material.

A PPE/PS blend might be another candidate when an excellent stabilizer is found to prevent thermal deterioration without sacrificing the optical clarity. The number of miscible pairs of high-temperature polymers is growing. Among them, one may discover new birefringence-free combinations in the future.

In this chapter, we have mostly dealt with the molecular aspects of birefringence-free blends in relation to injection molding. Recent reviews [1,15] are recommended for the readers who wish to know more details of other problems related to the birefringence-free plastics, such as relevant optics and the relationship between process conditions and the optical properties of molded materials.

NOTATION

AS-25	Poly(acrylonitrile-co-styrene) (25 wt%AN)
NBR-40	Poly(acrylonitrile-co-butadiene) (40 wt%AN)
PC	Polycarbonate
PEO	Polyethyleneoxide
PMMA	Polymethylmethacrylate
PPE	Polyphenyleneether
PS	Polystyrene
PVC	Polyvinylchloride
PVDF	Polyvinylidenefluoride
PVME	Polyvinylmethylether
S·CMI-17	Poly(styrene-co-cyclohexyl maleimide) (17 wt%CMI)
S·LMI-19	Poly(styrene-co-lauloyl maleimide) (19 wt%LMI)
S·MA-8	Poly(styrene-co-maleic anhydride) (8 wt%MA)
S·PMI-16	Poly(styrene-co-phenyl maleimide) (16 wt%PMI)
S VHC	Poly(styrene-co-vinylphenyl hexafluoro dimethylcarbinol) (10 wt%VHC)
VDF·TrFE-58	Poly(vinylidene fluoride-co-trifluoroethylene) (58 mol%VDF)
D	Infrared dichroic ratio
f	Orientation function
F	Normalized orientation function
M_w	Weight average molecular weight (kg/mol)
n^{20}	Index of refraction at 20°C
α	Linear expansion coefficient
α_i	Principal polarizability
$\Delta\alpha$	Polarizability difference
Δn	Birefringence
Δn_F	Form birefringence
$\Delta n°$	Intrinsic birefringence
Θ	Angle between chain axis and stretching direction
θ_r	Angle of incident beam in radial plane
θ_t	Angle of incident beam in tangential plane
κ	Thermal conductivity coefficient
ν_d	Dispersion coefficient
ρ	Density
ϕ	Volume fraction
ψ	Angle between the chain axis and the transition moment vector of IR absorbing group

REFERENCES

1. Kijima, Y., Kawaki, T., *Nikkei New Materials*, Sep. 26, 56 (1988).
2. Kijima, Y., Kawaki, T., Inoue, T., *Proc. 4th Annual Polymer Proces. Soc. Meeting*, Orlando, FL, May 8-11 (1988).
3. Takeshima, M., Funakoshi, N., *Kobunshi Ronbun.*, **42**, 317 (1985).
4. Hahn, B. R., Wendorf, J. H., *Polymer*, **26**, 1619 (1985).

5.   Saito, H., Inoue, T., Soc. Polym. Sci. Japan, Div. Adhes. Coat., Meeting, Dec. 12 (1985).

6.   Saito, H., Inoue, T., *J. Polym. Sci.*, B, *Polym. Phys.*, **25**, 1629 (1987).

7.   Saito, H., Takahashi, M., Inoue, T., *J. Polym. Sci.*, B, *Polym. Phys.*, **26**, 1768 (1988).

8.   Denbigh, K. G., *Trans. Faraday Soc.*, **36**, 936 (1940).

9.   Erman, B., Marvin, D. C., Irvine, P. A., Flory, P. J., *Macromolecules*, **15**, 664 (1982).

10.   Smith, R. P., Mortensen, E. M., *J. Chem. Phys.*, **32**, 502 (1960).

11.   Read, B. E., *J. Polym. Sci.*, Part C, **16**, 1887 (1967).

12.   Ishikawa, M., Saito, H., Inoue, T., *Polym. Prepr. Jpn.*, **36**, 2570 (1987).

13.   Gurnee, E. F., *J. Appl. Phys.*, **25**, 1232 (1954).

14.   Anonymus, *Mitsubishi Gas Chemical Co.*, *Technical Note* (1989).

15.   Greener, J., Kesel, R., Contestable, B. A., *AIChE J.*, **35**, 449 (1989).

CHAPTER 10

TOUGHNESS ENHANCEMENT IN POLYCARBONATE/ POLYMETHYLMETHACRYLATE BLEND VIA PHASE SEPARATION

by Thein Kyu, Jeanne M. Saldanha, and Mark J. Kiesel

Center for Polymer Engineering
University of Akron,
Akron, OHIO 44325
U.S.A.

Miscibility studies of polycarbonate/polymethylmethacrylate (PC/PMMA) blends were carried out by means of time-resolved light scattering. The PC/PMMA blend reveals a virtual lower critical-solution temperature (LCST). Temperature-jump experiments were conducted on the 70/30 PC/PMMA blend from a single-phase to two-phase region. The early stage of phase separation was dominated by the diffusion process and explained in the context of the linearized theory. The late stage of spinodal decomposition was found to be a nonlinear phenomenon, in which the time evolution of maximum scattering wavenumber and intensity obeyed the power-law scheme with exponents of -1 and 3, respectively. The process was identified to be a coarsening of domains driven by surface tension. Melt mixing of PC/PMMA blends at the two-phase temperature region invariably resulted in phase separation and exhibited two-phase morphology. The mechanical properties of the melt-blended 70/30 and 90/10 compositions, notably modulus, yield stress, tensile strength, and toughness, were found to be improved when compared to those of pure PC. The single-phase blends, prepared by compression molding of the solvent-cast films at temperatures below the LCST curve (approximately $40 \sim 50^{\circ}C$ higher than the respective T_g), showed inferior properties relative to the two-phase blends. The melt mixing of the solvent-cast films above the LCST temperature exhibited bicontinuous interconnected structure, leading to further improvement in toughness, strength, and modulus.

10.1   INTRODUCTION

Toughened plastics are of interest because of their practical significance [1-3]. Some commercially successful examples include high-impact polystyrene (HIPS) and copolymer of acrylonitrile-butadiene-styrene (ABS). In those polymers, soft rubbery segments are either copolymerized in block sequence or grafted to the backbone chains. The rubbery domains are generally microphase-separated from the glassy matrix. When the material is subjected to an external force, crazes develop at the peripheral of rubbery domains. It is this craze formation that promotes the toughness of the materials in a manner dependent on the type and direction of craze growth.

Recently a similar development was witnessed in brittle/ductile polymer alloys. The toughness of polycarbonate (PC) was improved by incorporating styrene-acrylonitrile (SAN) glassy copolymer [4]. Koo and co-workers [5] improved PC toughness by adding brittle polymethylmethacrylate (PMMA). The authors attributed this behavior to the brittle-ductile transition of the brittle dispersed phase associated with the compressive shear stress. Very recently, Ma et al. [6] and Gregory et al. [7] observed the massive craze development in the glassy brittle phase during tensile deformation of PC/SAN multilayer laminates. It seems the lateral craze propagation in glassy phase was prohibited by the neighboring ductile layers, thereby increasing the toughness of the laminates. The craze phenomenon undoubtedly plays an important role in the toughness enhancement of brittle/ductile polymer alloys, thus rendering a similar character to that of the rubber-modified plastics.

Meanwhile, miscibility studies on the PC/PMMA blends have been in progress in our laboratories [8,9]. The existence of a lower critical-solution temperature (LCST) in PC/PMMA blends was noticed simultaneously in three laboratories including ours [8-11].

Phase equilibria and the kinetics of phase separation in PC/PMMA mixtures have been thoroughly investigated. The early and late stages of spinodal decomposition have been evaluated using linear [12-14] and nonlinear theories [15-17] as well as the recent scaling theories [18,19]. These studies demonstrated a possibility of controlling the phase-separated domain structure by maneuvering the processing conditions below and above the LCST window. The controlled morphology of the blends is expected to exert profound influence on the mechanical performance, particularly on toughness.

It is well known that the criteria for toughness improvement are the particle size, uniformity, dispersion, and interfacial adhesion between the particles and the matrix. This field has received widespread attention because of its practical significance. There are numerous efforts in the literature dealing with the control of multiphase structure for improved properties [20,21]. Some typical examples include block copolymers in which domain structure and its size may be tailored via synthetic routes by grafting or copolymerizing in block sequence of varying length and composition. In the case of immiscible blends, surface functionalization has been widely practiced to improve interfacial adhesion between the disperse phase and the matrix through cross-linking. Polyblend compatilization has been sought as an alternative way, where block copolymers may be added to control interface miscibility and surface tension in order to reduce particle size.

A new approach to be described here is the control of blend morphology via spinodal decomposition (SD). SD is an unstable and spontaneous process that occurs owing to the instability at infinitesimal composition fluctuations. When phase separation occurs via SD, initially the individual domains are uniform and finely dispersed, thus favorable for toughness enhancement. The control of the domain size may be feasible, once phase-separation kinetics have been determined. Since the interdiffusion of chain molecules is dominant in the early stage of SD, the interconnected spinodal structure may be frozen in at appropriate early stages of spinodal decomposition. This interconnected structure is basically similar to that of an interpenetrating polymer network (IPN) through which the interfacial properties may be improved. In this chapter, we outline the phase-separation kinetics and discuss the effect of spinodal structure on the mechanical performance of PC/PMMA blends.

10.2   EXPERIMENTAL

10.2.1 Materials

The materials used in this study were commercial grades of bisphenol-A polycarbonate (Lexan 141, General Electric Co.) and polymethylmethacrylate (Plexiglass V811, Rohm and Haas Co.). Their properties are tabulated in Table 10.1. Single-phase blends of PC/PMMA were prepared by dissolving the pellets in a mutual solvent tetrahydrofuran (THF) and casting them at about 50°C in an oven. The polymer concentration ranged from 2 to 5 wt%. Compression molding of the miscible blends was undertaken at about 155°C. A rectangular mold (length, 20 mm; width, 4 mm; and thickness, 1 mm) was used. The specimens were melted for 5 min, followed by compression molding for an additional 5 min at an elevated pressure. The molded specimens were transparent and showed a single glass transition temperature, T_g. Two-phase blends were prepared by melt mixing the dry pellets in a Mini-Max molder at 265°C. The apparatus was described in detail by Koo et al. [5]. The melt-blended PC/PMMA invariably gave a two-phase morphology without interconnected spinodal structure. Visually, the blend shows a pearlescent appearance. For tensile testing, cylindrical dumbbell-shaped specimens were prepared by injection molding at the mold temperature of approximately 110 °C. The cylindrical specimens were kept for about 1 min at the mold temperature. Then, the mold was dismounted and quenched in ice water. The total length of the specimens was 20 mm with a diameter of 1.5 mm. Partially phase-separated blends with interconnected spinodal structure were obtained by melt mixing the solvent-cast single-phase blends using the Mini-Max molder at the two-phase region (265 °C). These blends exhibited dual T_gs analogous to those of the pure components.

Table 10.1 *Properties of Polymers*

Polymers	Supplier	M_w (kg/mol)	M_w/M_n
PC (Lexan 141)	General Electric	58	2.70
PMMA (Plexiglass V811)	Rohm & Haas	87	2.13

10.2.2 Equipment and Methods

The kinetic of phase separation was studied by means of time-resolved light scattering. The setup consisted of a He–Ne laser light source with a wavelength of 632.8 nm. The time evolution of scattering profiles was monitored by a two-dimensional Vidicon camera (1252B, EG & G Princeton Applied Research Co.) with the aid of an Optical Multichannel Analyzer (OMA III, 1460) and an off-line IBM-PC computer. A set of sample hot stages were used in the temperature (T)-jump experiments; one was controlled at the experimental temperatures and the other was used for preheating. A detailed description of this set-up was reported elsewhere [8].

Differential scanning calorimetric studies of the blends were conducted in a DuPont thermal analyzer (Model 9900) with a heating module (Model 910). A heating rate of 20 °C/min was selected arbitrarily. Tensile properties were determined in a Monsanto tensile tester (Model T-10) at a cross-head speed of 3 mm/min. Optical micrographs were photographed on a Leitz microscope (Model Laborlux-12). The cylindrical dumbbell specimens were fractured in liquid nitrogen. The morphology of the fractured surface of the blends was characterized using a scanning electron microscope (Model ISI SX-40).

10.3   RESULTS AND DISCUSSION

Figure 10.1 shows the DSC scans of the solvent-cast PC/PMMA blends of various compositions obtained at 20 °C/min. A single T_g was observed in all blends. The T_g varies systematically with composition as typical for miscible pairs. In a subsequent run after cooling from 300 °C, dual T_gs appear in the intermediate compositions. Obviously, thermally induced phase separation must have taken place during the first heating scan. This observation has been further supported by optical microscopic and light-scattering investigations. As can be seen in Fig. 10.2, the phase-separated domains with a high level of interconnectivity were observed under optical microscope. The average domain size increases with increasing temperature while the scattering halo, corresponding to the periodic concentration fluctuations of the blends, shifts toward the center. This phenomenon may be attributed to the coarsening process.

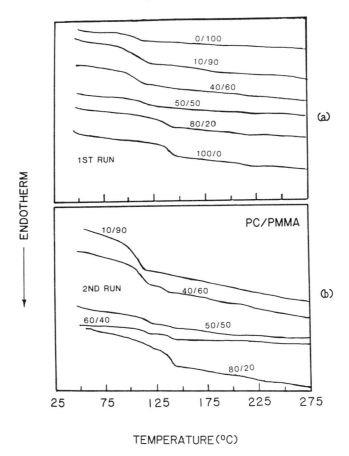

Figure 10.1 *DSC scans of (a) solvent-cast PC/PMMA blends and (b) the second runs after cooling from 300° C for various compositions. The heating rate was 20° C/min.*

To establish an equilibrium phase diagram, a cloud-point measurement was conducted at several heating rates. Figure 10.3 shows a typical plot of temperature versus composition-phase diagram obtained at 2°C/min. The cloud-point curve is similar in character to a binodal with the lower critical solution temperature (LCST) shown as a concave minimum around 240°C at the 40/60 ratio of PC/PMMA. As can be seen in Fig. 10.4, the cloud-point temperature of 40/60 PC/PMMA shows a strong heating-rate dependence, i.e., extrapolation to the zero heating rate is no longer practical. As an example, if one anneals the blend for 2 hrs at each temperature and visually examines the cloudy appearance of the films, the cloud point would appear around 160°C. Because of the strong dependence of cloud-point temperature on the heating rate, we were unable to establish the true phase diagram. Hence,

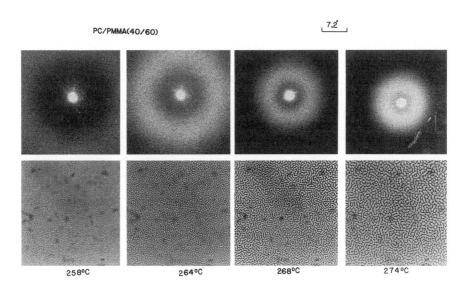

Figure 10.2 *Optical micrographs and the corresponding light scattering picture of the 40/60 PC/PMMA blends at various temperatures.*

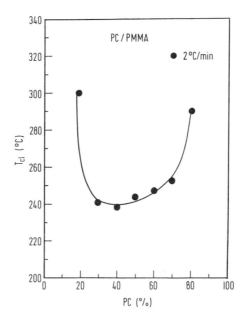

Figure 10.3 *A cloud-point phase diagram for PC/PMMA blends obtained at 2° C/min.*

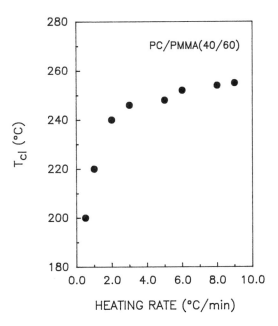

Figure 10.4 *Heating-rate dependence of cloud-point temperature of 40/60 PC/PMMA.*

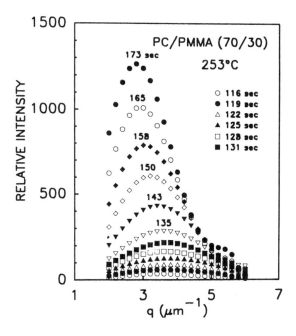

Figure 10.5 *Time evolution of scattering curves of 70/30 PC/PMMA blend after following a T-jump from 200 to 253° C.*

the processibility window may be limited to 160°C or below for the present PC/PMMA blends.

   We felt that it is of crucial importance to investigate the kinetics of phase separation in order to determine the optimum resident time for melt processing (e.g., compression molding) and gain control of the phase morphology of the blends. Figure 10.5 illustrates time evolution of the scattering for 70/30 PC/PMMA following a T-jump to 253°C. The scattering maximum first develops at a wavenumber (q) of approximately 4 μm^{-1} and appears stationary for a short period, then rapidly moves to lower scattering wavenumbers. The behavior of peak invariance has been predicted in the linearized Cahn-Hilliard theory [13] in which the increase in the scattered intensity is associated with the gradient of concentration fluctuations without accompanying changes in the fluctuation size. The linearized theory predicts that the scattered intensity at the initial period should increase exponentially; i.e.:

$$I(q,t) = I(q,t=0) \exp\{2R(q)t\} \qquad (10.1)$$

with

$$R(q) = -Mq^2[\partial^2 f/\partial c^2 - 2\kappa q^2] \qquad (10.2)$$

where R(q) represents the amplification factor characterizing the growth of concentration fluctuation, f is local free energy density, κ, the coefficient of gradient of concentration fluctuation, M, the mobility, c, the concentration, and t, the phase-separation time. The wavenumber q is defined as:

$$q = (4\pi/\lambda)\sin\theta/2 \qquad (10.3)$$

where λ and θ are the wavelength of incident light and the scattering angle measured in the medium, respectively. In the linearized theory, it is assumed that the fluctuation is extremely small, i.e., the order parameter $(c-c_0)$ is infinitesimal such that the contribution of higher-order nonlinear terms is insignificant. Hence, this approach may be applicable only at the early stage of phase separation. In the present case, the behavior of phase separation is predominantly nonlinear in character since the time for the peak invariance is very short. In an earlier publication [9], we showed that lowering the T-jump depth prolongs the early period of SD.

   The linear theory of Cahn-Hilliard was further modified by Cook [14] who pointed out the need for taking into consideration the contribution arising from thermal fluctuations. The modified equation reads:

$$S(q,t) - S_S(q) = [S(q,t=0) - S_S(q)]\exp\{2R(q)t\} \qquad (10.4)$$

where $S_S(q) \sim I_S(q)$ is the virtual structure factor, i.e., an extrapolated structure factor into the two-phase region assuming the continuity of the system. This aspect has been thoroughly discussed by Binder [22] and by Sato and Han [23]. For the large temperature jump as in the present case, the contribution from thermal fluctuations to the scattering is negligibly small, thus Eq (10.1) may be applicable. Moreover, the present set-up required the beam stopper to protect the detector from damage which resulted in parasitic scattering. The scattering measured before the onset of phase separation was subtracted from the experimental signal in order to correct for the parasitic scattering. This procedure should eliminate the scattering arising from thermal fluctuations, although this phenomenon may be insignificant for large T

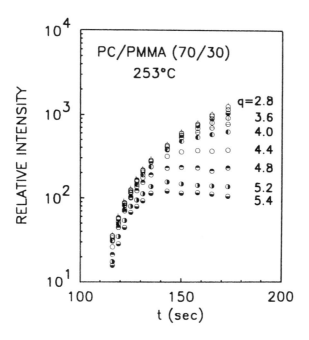

*Figure **10.6** Semilogarithmic plots of scattered intensities versus phase-separation time for 70/30 PC/PMMA blend.*

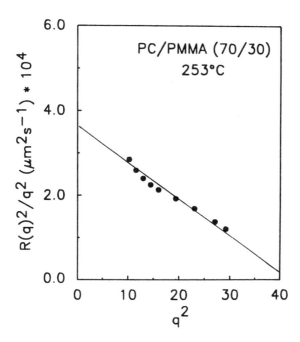

Figure 10.7 $R(q)/q^2$ versus q^2 plots of 70/30 PC/PMMA blend.

jumps. The corrected scattering curves were plotted in semilogarithmic scale in Fig. 10.6. The data at the early period may be approximated by a linear slope, however, deviations are seen at a later time. The values of R(q) as determined from the initial slopes are plotted in Fig. 10.7 in accordance with Eq (10.2). The plot of $R(q)/q^2$ versus q^2 appears fairly linear. It is therefore concluded that the early stage of SD is dominated by the diffusion process and may be explained in the context of the linearized theory.

The regime at which the scattering peak moves to a lower scattering angle corresponds to the late stage of SD. This phenomenon may be best explained in terms of the nonlinear statistical theories [15,16], which predict a change of the maximum wavenumber and the corresponding increase of the maximum intensity in a power-law form rather than an exponential fashion, i.e.:

$$q_m(t) \propto t^{-\alpha} \tag{10.5}$$

and

$$I_m(t) \propto t^\beta \tag{10.6}$$

where the subscript m stands for the maximum value. The nonlinear theory of Langer et al. [15] predicted the value of $\alpha = 0.21$. On the basis of cluster dynamics, Binder and Stauffer [16] obtained the relationship of $\beta = 3\alpha$ with $\alpha = 1/3$ and $\beta = 1$. Siggia [17] derived the same equation using the percolation approach taking into consideration the hydrodynamic effects. He predicted values of $\alpha = 1/3$ at the early stage of growth and $\alpha = 1$ at intermediate growth regime.

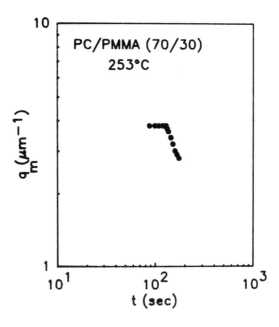

Figure 10.8 *Log-log plots of the maximum wavenumber versus phase-separation time for 70/30 PC/PMMA blend.*

Figures 10.8 and 10.9 exhibit the double logarithmic plots of q_m and I_m against t, respectively. The initial period corresponds to the early stage of SD, where q_m is constant. In a later period, the slopes change rapidly to -1 and 3, in accordance with the values suggested by Siggia [17] based on the percolation approach. This process has been identified to be the coarsening driven by surface tension. The same behavior has been observed in the 40/60 PC/PMMA composition [9]. According to the scaling tests of the late stages of SD, self-similarity was attained as the structure function became universal with time [24]. The analysis of the shape of the scattering function of both compositions showed the behavior typical for off-critical mixtures.

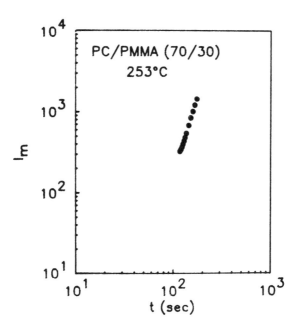

Figure 10.9 *Log-log plots of the maximum scattered intensity versus phase-separation time for 70/30 PC/PMMA blend.*

The diffusion process is dominant in the early stage of SD, whereas the surface tension plays an important role in the coarsening process of the late stages of SD. It is natural to examine the influence of the phase-separated structure on the mechanical performance of the blends. As mentioned previously, the compression molding of the solvent-cast films below the LCST at about 155°C yields a single-phase blend with outstanding transparency. Figure 10.10 shows the stress-strain curves of the compression-molded miscible blends in comparison with those of the pure PC and PMMA. The pure PC exhibits a typical ductile behavior while the pure PMMA reveals a brittle character. An addition of 10 wt% PMMA to PC resulted in some improvements of the Young's modulus, yield strength and, tensile strength, but the elongation at break decreased. This implies that the toughness (the area under the stress–strain curve) is inferior when compared to that of the pure PC. When PMMA content increases further, the toughness deteriorated even more.

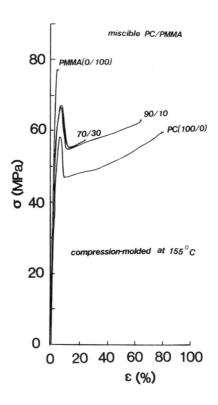

Figure 10.10 *The stress–strain diagram of the compression–molded single–phase blends of PC/PMMA obtained at room temperature at a cross–head speed of 3 mm/min.*

When the dried PC and PMMA pellets were melt mixed in a Mini-Max molder at 265°C, the blends invariably resulted in two-phase structures without spinodal structure. Figure 10.11 shows the stress-strain diagram for various PC/PMMA blends. Both 90/10 and 70/30 PC/PMMA compositions exhibit improved Young's modulus, tensile strength, yield strength, and plateau modulus without sacrificing the elongation at break. Hence, the area under the stress–strain curve, which represents the energy to break the materials, is greater for the above two blends as compared with that of the pure PC, indicating the improved toughness of the blends. Figure 10.12 shows the comparison of the mechanical properties for all compositions. This kind of behavior was first reported by Koo and co-workers [5] who attributed the toughness enhancement to the brittle–ductile transition of the brittle phase. The authors further pointed out that the diameter of the dispersed phase must be smaller than 1 μm in order to see such an effect.

As the phase behavior of PC/PMMA is now fairly well established, attention can be focused on the role of interconnected spinodal structure. This structure may be achieved through controlled spinodal decomposition by bringing the single phase PC/PMMA to the two-phase temperature regime. Therefore, the solvent–cast blends films were melt mixed at 265°C in a Mini-Max molder for 2 min and subsequently injection molded into cylindrical

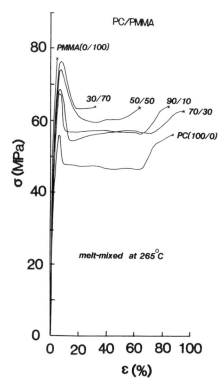

Figure 10.11 *The stress–strain diagram of the two–phase PC/PMMA blends without spinodal structure obtained at room temperature at a cross–head speed of 3 mm/min.*

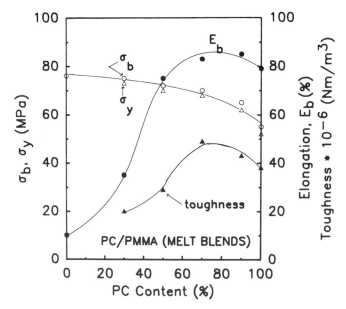

Figure 10.12 *Tensile properties as a function of blend composition of PC/PMMA.*

dumbbell-shaped specimens. In this case, the blend crosses over the LCST-type phase boundary from a single- to a two-phase region. Previous kinetic studies demonstrated that phase separation occurs through spinodal decomposition. The coarsening process is rapid, thereby resulting in the phase domains as large as 3 to 5 μm within 10 min. Hence, we restricted the mixing time to 2 min and the molten polymers were injection molded into cylindrical shaped in order to freeze the spinodal, i.e., the interconnected structure. Figure 10.13 shows the fractured surface of the above specimen (70/30 PC/PMMA) in which the dispersed phase is somewhat diffused and interconnected, suggestive of the spinodal structure. This morphology is consistent with the optical micrograph shown in Fig. 10.2. The average size of the dispersed domains at the fractured surface is in the range of 0.2 μm and appears uniform throughout the surface.

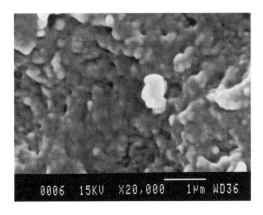

Figure 10.13 *A scanning electron micrograph of the fractured surface of the 70/30 PC/PMMA prepared through spinodal decomposition.*

The stress-strain diagram of the 70/30 PC/PMMA prepared through spinodal process is shown in Fig. 10.14 in comparison with those of the single-phase and two-phase melt blends. As expected, the toughness of the single-phase blend turns out to be inferior relative to those of the phase-separated mixtures. This has led to the conclusion that some level of phase segregation is needed to improve the toughness of the blend materials. However, it should be borne in mind that this improved toughness via controlled morphology was demonstrated by the tensile stress-strain measurements at low strain rates, in which the effect of orientation was not considered. In an early study by Koo et al. [5] the improved toughness was seen in stress-strain testing of the melt-blended PC/PMMA, which is consistent with our observation. The authors subsequently confirmed the toughness enhancement in the impact testing. Hence, it is fair to conclude that the blend has improved toughness at impact speeds. The significance of orientation of multilayer laminates consisting of polystyrene (PS) and polypropylene (PP) was pointed out by Schrenk and Alfrey [25]. It seems that at lower elongations localized failure occurred in brittle PS layers; however, these flaws were randomized among the numerous layers. Moreover, the transverse craze growth was prohibited by the ductile PP layers provided that the ductile layers were strong enough to

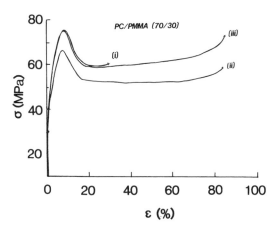

Figure 10.14 *A comparison of stress-strain curves of (i) the compression-molded single-phase blend (70/30 PC/PMMA), (ii) the melt-mixed two-phase blend, and (iii) the solution-cast/melt-mixed blend prepared via spinodal decomposition.*

resist the stresses caused by a crack propagating in the brittle phase. The authors added that the mutual reinforcement depended on molecular orientation of the brittle phase. In the studies of Koo et al. [5] as well as in the one discussed in this chapter, the stress-strain and impact tests were carried out in the tensile mode, thus the orientation effects, if any, should be considered in interpreting the toughness.

It is well known that the molecular orientation in the melt extrudate of amorphous materials is generally small and even smaller in their blends (due to the redistribution of stresses between dissimilar molecules) [26]. In the case of injection molding, the mold temperature was about 110°C, which is in close proximity to the blend's T_g. Thus, the relaxation rate of molecules being fast, no significant orientation is expected. This argument has been supported by the fact that the fracture surface in Fig. 10.13 shows no oriented fibrils or preferred orientation of the domains. Because of the geometry of the cylindrical dumbbell specimens, we were unable to carry out the impact test in the cross-direction.

It should be emphasized that the blend with the interconnected SD structure shows a better toughness relative to the melt blend without it. Conceptually, the frozen-in interconnected SD structure is reminiscent of the interpenetrating network [27]. The molecular penetration at the peripheral of the domain boundaries might enhance the adhesion between the domains and the matrix, thereby providing added improvement of toughness. In polymer phase separation by spinodal decomposition, the uniformity and the dispersion of the domains are generally outstanding in these PC/PMMA blends. The average domain size, in principle, is controllable via phase-separation kinetics; in the present study the average size was approximately 0.2 μm in diameter. It may be concluded that interfacial strength can be improved further through the creation of interconnected structure.

10.4   CONCLUSIONS

We have demonstrated that the biphasic structure is essential for improvement of toughness of polymer alloys. The important criteria for toughness enhancement, such as uniform size and fine dispersion, can be met by controlling the kinetics of spinodal decomposition. The controlled SD structure provides improved interfacial adhesion between the disperse phase and the matrix polymers due to the interconnected network structure. This approach may be employed as an alternate means to other widely used techniques, such as chemical modification at the interface, crosslinking, or addition of compatibilizers for enhanced interfacial adhesion.

ACKNOWLEDGMENT

Support of this work by the Edison Polymer Innovation Corporation (EPIC) is gratefully acknowledged.

NOTATION

HIPS	High impact polystyrene
ABS	Acrylonitrile-butadiene-styrene
PS	Polystyrene
PP	Polypropylene
PC	Polycarbonate
SAN	Styrene-acrylonitrile
PMMA	Polymethylmethacrylate
LCST	Lower critical solution temperature
SD	Spinodal decomposition
f	Local free energy density
I	Scattered intensity
I_m	Maximum intensity
q	Scattering wavenumber
q_m	Maximum wavenumber
t	Phase separation time
$R(q)$	Amplification factor
M	Onsager type mobility
c	Concentration
κ	Coefficient of concentration gradient
λ	Wavelength
θ	Scattering angle
S	Structure factor
S_s	Virtual structure factor
α, β	Kinetic exponents of growth
T_g	Glass transition temperature

REFERENCES

1.   Ward, I. M., *Mechanical Properties of Solid Polymers*, John Wiley & Sons, New York (1971).
2.   Deanin, R. D., Crugnola, A. M., Eds., *Toughness and Brittleness of Plastics*, Am. Chem. Soc., Adv. Chem. Series, 154, Washington, D.C. (1976).
3.   Bucknall, C. B., *Toughened Plastics*, Applied Science Publishers, London (1977).
4.   Kurauchi, T., Ohta, T., *J. Mater. Sci.*, **19**, 1669 (1984).
5.   Koo, K. K., Inoue, T., Miyasaka, K., *Polym. Eng. Sci.*, **27**, 741 (1985).
6.   Ma, M., Hiltner, A., Baer, E., Im, J., *SPE Techn. Pap.*, **34**, 1525 (1988).
7.   Gregory, B., Hiltner, A., Baer, E., *J. Mater. Sci.*, **22**, 532 (1987).
8.   Kyu, T., Saldanha, J. M., *J. Polym. Sci., Polym. Lett. Ed.*, **26**, 33 (1988).
9.   Kyu, T., Saldanha, J. M., *Macromolecules*, **21**, 1021 (1988).
10.   Chiou, J. S., Barlow, J. W., Paul, D. R., *J. Polym. Sci., Polym. Phys. Ed.*, **25**, 1459 (1987).
11.   Kambour, R. P., Gundlach, P. E., Wang, I. C. W., White, D. M., Yeager, G. W., *Am. Chem. Soc., Div. Polym. Sci., Polym. Prepr.*, **28**(2), 140 (1987).
12.   Cahn, J. W., *J. Chem. Phys.*, **42**, 93 (1965); *Trans Metall. Soc.*, **242**, 166 (1968).
13.   Hilliard, J. E., in *Phase Transformations*, ASME Seminar, p. 497, Oct. (1968).
14.   Cook, H. E., *Acta Metall.*, **18**, 297 (1970).
15.   Langer, J. S., Bar-on, M., Miller, H. S., *Phys. Rev.*, **A11**, 1417 (1975).
16.   Binder, K., Stauffer, D., *Phys. Rev. Lett.*, **33**, 1006 (1973).
17.   Siggia, E. D., *Phys. Rev.*, **A20**, 595 (1979).
18.   Furukawa, H., *Phys. Rev. Lett.*, **43**, 136 (1979).
19.   Furukawa, H., *Physica*, **123A**, 497 (1984).
20.   Gould, R. F., Ed., *Multicomponent Polymer Systems*, Am. Chem. Soc., Adv. Chem. Series 99, Washington, D.C. (1971).
21.   Utracki, L. A., Weiss, R. A., Eds., *Multiphase Polymers: Blends and Ionomers*, Am. Chem. Soc., Symp. Series 395, Washington, D.C. (1989).
22.   Binder, K., *J. Chem. Phys.*, **79**, 6387 (1983).
23.   Sato, T., Han, C. C., *J. Chem. Phys.*, **88**, 2057 (1988).
24.   Kyu, T., Saldanha, J. M., *Am. Chem. Soc., Div. Polym. Chem., Polym. Prepr.*, **29**(1), 454 (1988).
25.   Schrenk, W. J., Alfrey, T., *Polym. Eng. Sci.*, **9**, 393 (1969).
26.   Min, K., White, J. L., Fellers, J. F., *J. Appl. Polym. Sci.*, **29**, 2117 (1984).
27.   Sperling, L. H., Huelck, V., Thomas, D. A., in *Polymer Networks: Structure and Mechanical Properties*, Newman, S., Chompff, A. J., Eds., Plenum Press, New York (1971).

CHAPTER 11

PREDICTING FIBER ORIENTATION IN INJECTION MOLDING

by M. Vincent and J. F. Agassant

Centre de Mise en Forme des Materiaux
Unite associee au CNRS no. D13740
Ecole Nationale Superieure des Mines de Paris
Sophia Antipolis, VALBONNE 06560
FRANCE

In the first part, a short review of observed fiber orientation in short-fiber-reinforced polymer parts is presented. A complex orientation pattern was observed throughout the thickness, with usually and roughly a core region oriented perpendicular to the flow direction, and a skin region oriented parallel. In the second part, the bulk flow was assumed to be independent of the microstructure formed by the fibers. Calculation of a single ellipsoid motion in a Newtonian fluid is presented. It gives valuable results concerning the specific influence of shear and elongational flows, despite strong assumptions, which are discussed. In the third part, the calculation was applied to injection molding. In the last part, different approaches, which couple the flow and the microstructure related to fiber orientation, are presented. They are based on anisotropic behavior laws. Presently they are tested on more simple flows than those occurring in injection molding.

11.1   INTRODUCTION

The great variety of composites with polymeric matrices can be roughly divided into two groups: the first concerns low-tonnage and high-cost composites with expensive materials (carbon or aramid fibers for instance) and processing techniques involving long manual operations. The polymeric matrix is either a thermoset (epoxy for instance) or an expensive engineering resin (e.g., polyetheretherketone). These composites are used in aeronautics, aerospace, etc. The second group concerns high-tonnage composites. This means that the basic materials are relatively inexpensive (unsaturated polyester resin, commodity thermoplastics, and glass fiber) and the processing is automated, for example injection molding or extrusion. This subdivision is obviously an oversimplification since not only low-cost reinforced thermoplastic are injection molded, but also specialty thermoplastics, such as polyethersulphones or polyetheretherketones, reinforced with carbon or aramid fibers.

This chapter focuses on the second group of reinforced polymers processed by injection molding. It concerns mainly short-glass-fiber reinforced thermoplastics, though there exist glass fiber reinforced thermosets (Bulk Molding Compound or BMC). In the first case, the final fiber dimensions are about 500 μm long and 10 μm in diameter; in the second case fibers are made into bundles of about 500 to 1000 filaments, 10 to 20 mm long. The fiber loading is usually 20 to 40 wt%. Significant improvement in strength and stiffness can be obtained relative to that of the matrix.

During the flow in the mold, each fiber moves and orients according to the local stresses, depending on the matrix, other fibers, and the mold walls. After cooling, one observes a complex distribution of orientation, changing from one spot to another. Usually, in the core region, the fibers orient perpendicularly to the flow direction, and more or less parallel in the skin. The properties of the molded part, such as the elastic modulus, the notch sensitivity, or the thermal coefficient of expansion, are strongly dependent on the orientation distribution, as well as on the fiber concentration and fiber length distribution, which can be inhomogeneous. Experimental measurements of the elastic modulus of highly oriented samples indicate differences of approximatively three orders of magnitude depending on whether the traction direction is parallel or perpendicular to the fibers [1-3]. The value corresponding to a random orientation lies between these two limits. It is thus obvious that an understanding of the orientation phenomena should lead to a better use of the potential properties of the reinforced materials. Furthermore, knowledge of precise fiber orientation is essential for predicting local elastic properties [1-3] and the thermal coefficient of expansion, which is necessary for calculating the mechanical properties and thermal stresses in complex molded parts.

Predicting fiber orientation in injection molding is complex. Injection molding is an unsteady nonisothermal process, with a moving boundary, in complex geometries. A precise flow calculation is still a research subject for unfilled thermoplastics. Furthermore, the reinforced polymers are non-Newtonian and nonhomogeneous materials. The velocity and stress fields are functions of orientation, migration, and segregation phenomena (see the numerous studies about the rheology of reinforced polymers reviewed in [4]), induced by the velocity and stress fields.

This chapter reviews the orientation of short fibers in injection molding in thermoplastics where most of the studies have been conducted. The next part will briefly describe the experimental observations. Then a decoupled fiber-motion calculation (the flow field and the fiber motion are computed separately) will be presented. Next, these models will be applied to the injection molding. The last part will provide information on models that simultaneously calculate the flow field and the fiber orientation.

11.2 EXPERIMENTAL OBSERVATION OF FIBER ORIENTATION IN MOLDED PARTS

Numerous observations have been reported for different geometries, various polymers, fibers, and injection molding conditions [5]. The important features of fiber orientation will be summarized.

11.2.1 Experimental Techniques

A great number of authors used optical reflection microscopy, on cross sections prepared by metallographic polishing techniques [5-13]. Owen and Whybrew [14] and later Blanc et al. [15] improved the quality of the observation by removing the glass fibers located on the surface with hydrofluoridric acid and then coating the specimen with ink. Fisher and Eyerer [16] used interference contrast light microscopy. Kamal et al. [17] used transmission microscopy to observe microtomed layers heated above the melt temperature.

Darlington and co-workers [18-23] as well as Folkes et al. [24,25] prepared the same kind of thin slices for contact microradiography (CMR). This technique has also been used by Akay [26] as well as by Hegler and Menning [27]. The use of a small amount of glass fibers with a high lead content permitted McGee et al. [28,29] and Jackson et al. [30] to observe directly the whole thickness of a SMC sample. This is probably of less interest in injection molding where great differences are observed throughout the thickness.

These techniques need different lengths of time for the preparation of the specimen. They all have advantages and drawbacks. Even if the fibers do not belong exactly to the cross section - in fact, they are commonly observed to lie more or less in a plane parallel to the plane of the part - the CMR technique allows the visualization, but it can hide different orientations occurring in the thickness of the microslice. On the contrary, the classical metallographic technique permits a precise investigation in the thickness. Several authors [8-10,16] determined a three-dimensional fiber orientation by studying the shape of the ellipses representing the cross section of the cylindrical fibers with the observed plane.

11.2.2 Quantification of the Orientation

It is often difficult to distinguish the orientation distribution in different photographs without quantification. Most studies present only photographs or schematic representations. Some authors present histograms [4,16,22] obtained either manually, semi-automatically with a computer digitizing tablet [31], or automatically using an image-analysis technique [16]. For the latter method, a good quality photograph is required, and in practice, operator intervention is often needed in the binarization step by thresholding. Jackson et al. [32] used a probability distribution function for generalization of the histogram.

Another method of quantification uses the orientation function. Vincent and Agassant [6] employed the following definition of planar orientation:

$$f_{or} = (2 < \cos^2\theta > - 1)/2; \quad \text{where} \quad < \cos^2\theta > = (\Sigma N_i \cos^2\theta_i)/\Sigma N_i \quad (11.1)$$

with N_i the number of fibers with an angle θ_i relative to a reference direction (usually the local flow direction). A similar description was introduced through the Krenchel factors, used by Hegler and Menning [27], or the Hermans descriptor used by Kau [33], Fakirov and Farikova [9], or Pipes et al. [34].

The orientation function is a very concise representation of the distribution of orientation, but obviously it is restrictive, since different histograms or probability orientation functions can lead to the same value of the orientation factor. This is why Advani and Tucker [35] introduced a tensor description. For planar orientation, a second rank tensor is defined by:

$$a_{11} = < \cos^2\theta >; \quad a_{22} = < \sin^2\theta >; \quad a_{12} = a_{21} = < \sin\theta\cos\theta > \quad (11.2)$$

A non-zero value of a_{12} means that there is no symmetry relative to the reference direction.

11.2.3 Review of the Literature Data

Owing to the numerous parameters investigated in references [2,5-35], it is difficult to present a synthesis of all the observations. The following results should be stressed: 1) short fibers usually belong to planes parallel to the surface of thin parts, except at the front; and 2) orientation is usually very different throughout the thickness. This is apparent in a center-gated disk [5,6,24-26]. The distribution of the orientation function in the thickness of a disk is presented in Fig. 11.1. In the core, the value is very near -0.5, meaning that the orientation is markedly transverse to the flow direction. In the skin, fibers are slightly oriented in the flow direction, with an orientation factor value of about 0.1 to 0.2. Under the skin, the orientation is quite marked in the flow direction, with a value of 0.3. This orientation is clearly related to the rate of strain field. Figure 11.2 represents the computed distribution of the shear rate, $\dot{\gamma}$, and of the elongational rate, $\dot{\alpha}$, in the θ direction during mold filling throughout the disk thickness. The elongational rate is at a maximum in the core, and this leads to a general result: elongational flows nearly perfectly orient fibers in the direction of elongation. The shear rate reaches a maximum just under the skin. This leads to another general result: shearing flows align fibers in the flow direction, but with less efficiency, though the shear rate is about 100 times greater than the elongational one.

These different orientations between the skin and the core can also be observed in a rectangular plaque. This can appear surprising, especially when there is a feeding reservoir at the entrance to the plaque: the flow front remains nearly perpendicular to the walls during the

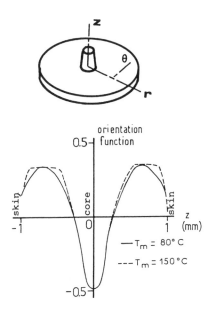

Figure 11.1 *Measured evolution of the orientation function in the thickness of a 2 mm thick disk, for a polyamide-6,6 filled with 30 wt% of glass fiber, at two mold temperatures $T_M = 80^\circ C$ and $150^\circ C$.*

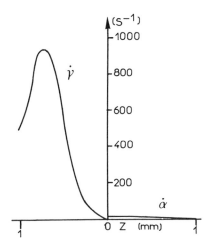

Figure 11.2 *Shear rate (left) and elongational rate (right) evolution in half-thickness of a center-gated disk.*

filling of the plaque, which means that, except for the effects of the lateral walls and of the "fountain flow" at the front, shear deformations dominate. Figure 11.3 shows that the core thickness (with an orientation perpendicular to the flow direction) decreases progressively along the plaque, without vanishing. The existence of the core region is due to the difference of cross section between the feeding reservoir and the plaque, which generates high elongational rates. Its thickness decreases progressively along the plaque but never vanishes. Without a feeding reservoir, the flow front is at first semicircular, and then becomes perpendicular to the walls of the plaque. The core thickness is the same as in the disk near the entry, but it decreases less rapidly than in the previous plaque geometry. This points out the influence of the thermomechanical history on the fiber-orientation phenomena.

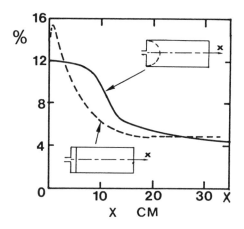

Figure 11.3 *Evolution on the x axis of the observed core thickness with orientation perpendicular to the flow, in a rectangular plaque mold, with two different gate geometries. Polyamide-6,6 filled with 30 wt% glass fiber.*

The influence of molding parameters may also be important. Figure 11.1 showed the influence of the mold temperature on the orientation function. This can be explained by the particular flow field near the front ("fountain flow" effect) which carries the fiber from the core towards the surface where a frozen skin forms. Its thickness depends on the mold temperature. Darlington [23] pointed out the influence of the flow rate and related it to this frozen layer and to the velocity profile induced by thermal effects. All these mechanisms can lead to a much more complex layered structure than that described above (Kamal et al. [17], Kenig [36]). On the contrary, Vincent and Agassant [6] did not observe any significant effect of the flow rate in a molded disk.

Akay [26] pointed out the influence of the polymer matrix viscoelasticity on orientation. Darlington and Smith [23] reported fiber migration across the streamlines.

In conclusion, it appears that a precise predicting model of fiber orientation should take into account a thermomechanical description of the flow in the mold without neglecting three-dimensional flows in gates or near the moving boundaries. The model must take into account a rheological behavior which in turn is a function of microstructure and in particular of the fiber orientation. This is an ambitious task, and some hypotheses have to be made. In the next part, the fiber orientation, apart from the flow calculation, is presented.

The influence of fiber orientation on the flow behavior of fiber-filled Newtonian and viscoelastic liquids is discussed by Mutel and Kamal in Chapter 12.

11.3   DECOUPLED FIBER-MOTION CALCULATION

Several mold-filling calculations are now available. These solve the mechanical and thermal equations in complex geometries considering the homogeneous fluid. This kind of calculation provides the pressure, stress, and velocity fields at each time step during the filling stage. The results are introduced as data in a fiber-motion calculation. The most frequently used theory has been developed by Jeffery [37]. It provides good qualitative results despite some strong assumptions.

11.3.1 Description of Jeffery's Theory

The following assumptions were made: 1) the particle shape is ellipsoidal; 2) the particle assumes the translational velocity of the adjacent fluid which displaces it; 3) away from the particle, the fluid is in a steady-state motion, and the components of the rate of strain and rotation tensors are taken to be constant in space through a volume which is large compared to the particle dimensions; 4) interactions between particles are neglected; 5) mass and inertia are neglected for both the particle and the fluid; and 6) the fluid is Newtonian. A discussion of these assumptions will be presented in Section 11.3.4.

Jeffery solved the Stokes equations with the following boundary conditions: (a) there is a nonslip contact on the surface of the particle, so that the fluid velocity is equal to the velocity of the rigid body, which is a function of the three components of its rotation velocity and (b) at an infinite distance relative to the particle size, the velocity is equal to the undisturbed fluid velocity.

The velocity components and pressure were expressed as linear combinations of particular solutions of the Laplace's equation. Its coefficients were calculated using the boundary conditions (the particle-rotation velocity is then an unknown parameter). Then, the Newtonian law yields the stress field in the perturbed fluid, and in particular on the surface of the particle. If the particle is subjected to no force except those exerted by the fluid, then the resultant couple must vanish. This yields the following equations:

$$\omega_1 = \nu + f(b^2 - c^2)/(b^2 + c^2)$$

$$\omega_2 = \mu + g(c^2 - a^2)/(c^2 + a^2) \qquad (11.3)$$

$$\omega_3 = \lambda + h(a^2 - b^2)/(a^2 + b^2)$$

where a, b, c are the axes of the ellipsoid, ω_1, ω_2, ω_3 are the particle-rotation velocity components in a frame linked to the particle; f, g, h and λ, μ, ν are components of, respectively, the rate of strain tensor $[\dot{\epsilon}]$ and the rotation tensor $[\Omega]$ of the undisturbed fluid in a coordinate system affixed to the particle, as indicated below:

$$[\dot{\varepsilon}] = \begin{bmatrix} d & h & g \\ h & e & f \\ g & f & k \end{bmatrix} \qquad [\Omega] = \begin{bmatrix} 0 & \lambda & -\mu \\ -\lambda & 0 & \nu \\ \mu & -\nu & 0 \end{bmatrix}$$

A cylindrical fiber will be considered an ellipsoid of revolution, with a the half length, and b = c the radius, so that Eq (11.3) becomes:

$$\omega_1 = \nu; \quad \omega_2 = \mu - \text{rg}; \quad \omega_3 = \lambda + \text{rh} \tag{11.4}$$

with $r = (p^2 - 1)/(p^2 + 1)$ and $p = a/b$ the fiber aspect ratio (notice that if p is about 30 to 50, r is close to 1).

11.3.2 Application of Jeffery's Theory to Simple Flow

Simple shear flow. In the laboratory frame (Fig. 11.4), the undisturbed velocity field is u(0,0, $\dot{\gamma}$y'). The corresponding rate of strain tensor and rotation tensor are:

$$[\dot{\varepsilon}]_R{}' = \begin{bmatrix} 0 & 0 & 0 \\ 0 & 0 & \dot{\gamma}/2 \\ 0 & \dot{\gamma}/2 & 0 \end{bmatrix} \qquad [\Omega]_R{}' = \begin{bmatrix} 0 & 0 & 0 \\ 0 & 0 & -\dot{\gamma}/2 \\ 0 & -\dot{\gamma}/2 & 0 \end{bmatrix}$$

[$\dot{\varepsilon}$] and [Ω] are then expressed in the axis affixed to the particle using a transformation matrix whose terms are functions of the Euler angles (see Fig. 11.4), where θ, σ, and Ψ are respectively, the angle between x (main axis of the fiber) and x', the angle between the planes (x', y') and (x', x), and the angle between the planes (x', x) and (x, y). Note that for $\theta = 0$, σ and Ψ are not defined. A specific resolution is required but is not described here.

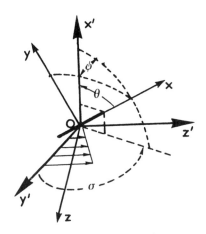

Figure 11.4 The Euler angles.

One obtains λ, μ, ν, h, and g as functions of p and $\dot{\gamma}$. w_1, w_2, w_3 can also be written as functions of θ, σ, and Ψ, and of their time derivatives $\dot{\theta}$, $\dot{\sigma}$, $\dot{\Psi}$. Substituting into Eq (11.4) yields:

$$\dot{\sigma}\cos\theta + \dot{\Psi} = (\dot{\gamma}/2)\cos\theta$$

$$\dot{\theta} = \dot{\gamma}\sin 2\,\theta\sin 2\,\sigma\,[(p^2 - 1)/4(p^2 + 1)] \qquad\qquad\qquad (11.5)$$

$$\dot{\sigma} = \dot{\gamma}\,(p^2\cos^2\sigma + \sin^2\sigma)/(p^2 + 1)$$

In the following, the angle Ψ (which represents the angular position of the fiber relative to its axis) will be neglected. Integrating the second and third dependence in Eq (11.5) yields:

$$\tan\sigma = p\tan\,[p\,\dot{\gamma}t/(1 + p^2) + C_\sigma] \qquad\qquad (11.6)$$

$$\tan\theta = C_\theta\,[(p^2 - 1)/(\sin^2\sigma + p^2\cos^2\sigma)]^{1/2} \qquad\qquad (11.7)$$

where C_σ and C_θ are integration constants dependent on the initial fiber position.

Eq (11.6) indicates that the projection of the fiber on the shearing plane (y', z') follows a periodic rotation of period:

$$t_p = 2\,\pi\,(p + 1/p)/\dot{\gamma} \qquad\qquad (11.8)$$

Eq (11.5) shows that $\dot{\sigma}$ can never be zero; under Jeffery's assumptions, a fiber has no equilibrium position in a shearing flow. The evolution of σ with time is presented in Fig. 11.5. The fiber rotates rapidly when it is perpendicular to the flow ($\sigma = 0$, π, ...), and slowly when it is parallel to the flow ($\sigma = \pi/2$, $3\pi/2$, ...). So the fiber spends most of its time aligned with the flow. This is consistent with the near perfect orientation observed in the injected part. Eqs (11.6) and (11.7) give the three-dimensional fiber motion. Cohen et al. [38] represented this motion with the ellipse traced out by the intersection of the particle main axis with the surface of a sphere centered at the center of gravity of the particle for different values of C_θ (named orbit constant). If C_θ is very small, then the fiber is nearly parallel to the velocity axis (z axis in Fig. 11.6), whereas when C_θ tends to infinity, the fiber is very nearly in the shearing plane.

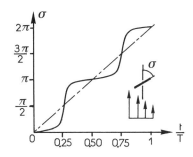

Figure 11.5 *Evolution of orientation in the shear plane as a function of the reduced time, t/t_p.*

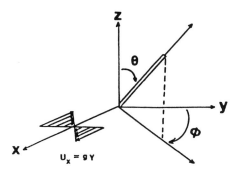

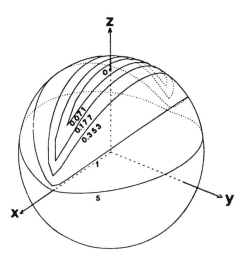

Figure 11.6 *Representation of the three-dimensional fiber motion in the shear flow (after Cohen et al. [38]).*

Purely elongational flow. In the laboratory frame, the undisturbed fluid velocity is u $(\dot{\alpha}x, -\dot{\alpha}y/2, -\dot{\alpha}z/2)$. The same calculations as for the shear flow transform Eq (11.4) with the Euler angles:

$$\dot{\sigma} = 0; \quad \dot{\Psi} = 0; \quad \dot{\theta} = -(3/4)\,\dot{\alpha}\,r\sin 2\theta \tag{11.9}$$

into (after integration):

$$\tan \theta = C\exp\left(-3\,\dot{\alpha}\,r\,t/2\right); \quad \sigma = \text{const.}; \quad \Psi = \text{const.} \tag{11.10}$$

As t tends to infinity, θ tends to zero when $\dot{\alpha}$ is positive (Fig. 11.7), or to $\pi/2$ when $\dot{\alpha}$ is negative, i.e., in purely elongational flow a fiber tends to a stable equilibrium position. The result explains the perfect transverse orientation observed in the core of molded disks.

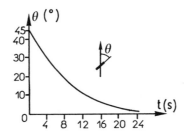

Figure 11.7 *Evolution of orientation with time in purely elongational flow (elongational rate = 0.1 sec⁻¹, aspect ratio = 10).*

11.3.3 Fiber Motion in Both Shear and Elongational Flow

In the general case of flows in complex geometries both shear and extension are present. First, the flow in two-dimensional (or axisymmetric) convergent or divergent channels will be considered. The study will provide interesting results on fiber behavior in important geometries such as those of diverging mold sprues or converging gates.

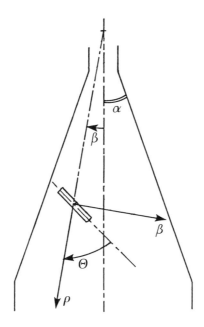

Figure 11.8 *Two-dimensional divergent channel geometry.*

If we restrict the calculation to the case of a fiber belonging to the (ρ, β) plane (Fig. 11.8), then the resolution remains analytical. For a two-dimensional divergent channel, Eq (11.4) gives:

$$\dot{\theta} = -r(\dot{\alpha}\sin2\theta + \dot{\gamma}\cos2\theta) + \dot{\gamma} \tag{11.11}$$

where θ is the Euler angle between the fiber and the ρ axis. It is assumed that the center of gravity of the fiber moves at a constant β value, which means that the perturbation at the entrance and the exit of the channel are neglected. This assumption is also used to calculate the undisturbed velocity field, by solving the Stokes equations, and to determine the elongational rate $\dot{\alpha}$ and the shear rate $\dot{\gamma}$:

$$\left. \begin{aligned} \dot{\alpha} &= -A/\rho^2\,(\cos2\beta - \cos2\alpha) \\[2mm] \dot{\gamma} &= -A/\rho^2\sin2\beta \\[2mm] A &= \dot{Q}/L\,(\sin2\alpha - 2\alpha\cos2\alpha) \end{aligned} \right\} \tag{11.12}$$

where $\dot{Q}$ is the volumetric flow rate and L the width. The solution of Eq (11.11) is different when the angle β (which fixes the fiber path) is greater or smaller than a value β_0 included between 0 and α defined by:

$$\cos2\beta_0 = r^2\cos2\alpha + [(1 - r^2)\,(1 - r^2\cos^2 2\alpha)]^{1/2} \tag{11.13}$$

From Eqs (11.11) and (11.13) one obtains:

$$\left. \begin{aligned} \beta = 0: &\quad \tan(\theta + \tau/2) = C\rho^{-2r}, \\[2mm] 0 < \beta < \beta_0: &\quad \tan(\theta + \tau/2) = \left[X + Y + (X - Y)\,C\rho^{-2Y/\dot{\alpha}}\right] \Big/ \left[\left[1 + C\rho^{-2Y/\dot{\alpha}}\right]\dot{\gamma}\right] \\[2mm] \beta_0 < \beta < \alpha: &\quad \tan(\theta + \tau/2) = [X - Y'\tan(Y'\,\text{Log}\,\rho/\dot{\alpha} + C)]/\dot{\gamma} \end{aligned} \right\} \tag{11.14}$$

where τ is defined by:

$$\cos\tau = \dot{\alpha}/(\dot{\alpha}^2 + \dot{\gamma}^2)^{1/2}$$

$$\sin\tau = \dot{\gamma}/(\dot{\alpha}^2 + \dot{\gamma}^2)^{1/2}$$

$$X = r(\dot{\alpha}^2 + \dot{\gamma}^2)^{1/2}$$

$$Y = [r^2\dot{\alpha}^2 + (r^2 - 1)\dot{\gamma}^2]^{1/2}$$

$$Y' = [(1 - r^2)\dot{\gamma}^2 - r^2\dot{\alpha}^2]^{1/2}$$

and C is a constant of integration depending on the initial orientation of the particle.

   Interpretation. For β greater than β_0, the fiber rotates without a stable equilibrium position; it is primarily oriented in the flow direction but it periodically rotates through 180°. The interval of time between two rotations is greater than the rotation time. This is a shear-type behavior.

   For β smaller than β_0, the fiber tends toward a stable equilibrium position when ρ becomes infinite. When the fiber moves on the divergent channel axis ($\beta = 0$), it becomes oriented perpendicular to the flow. This is consistent with the classical result obtained in purely elongational flow. When β is not zero, θ tends to a value θ_∞ which depends on α, β and a/b. Figures 11.9a,b show that fiber–equilibrium position weakly depends on the shape factor and on the angle of the divergent channel, but strongly on the angle β. Near the axis, the fiber becomes perpendicularly oriented to it, whereas when β is close to β_0, the equilibrium position is almost parallel to the flow lines.

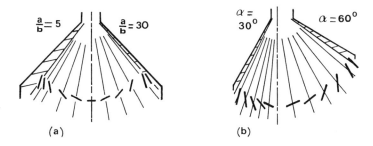

(a)   (b)

Figure 11.9 *Fiber equilibrium position in a divergent channel for different aspect ratio (a) and divergence angle (b).*

In a region where the shear rate, $\dot\gamma$, is greater than the elongational rate $\dot\alpha$, the fiber has an elongationlike behavior (it reaches a stable equilibrium position) even for β close to β_0 (Fig. 11.10). In fact, in Eq (11.14) $\dot\alpha^2$ is weighted by a coefficient, r^2, which is close to 1, and $\dot\gamma^2$ by a coefficient, $r^2 - 1$, which is very close to 0. This explains the great importance of an elongational flow, which is frequently reported in the literature.

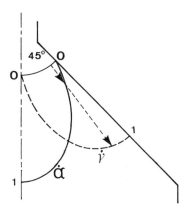

Figure 11.10 *Calculated for a Newtonian fluid normalized shear $\dot\gamma$ and elongational rate $\dot\alpha$ in polar representation.*

Another remark concerns the influence of the flow rate. Since in Eq (11.14) it vanishes, the fiber rotation does not depend on it. Obviously, the stronger the flow rate, the more rapidly the fiber will reach the end of the diverging area, but its final orientation will be the same regardless of the flow rate.

Figure 11.11 is a representation of the angle β_0. For fibers of large shape factor, β_0 is very close to α. The calculation gives similar results for an axisymmetric channel. In Figs. 11.12 and 11.13 the results for a mold sprue, with an angle $\alpha = 2.86°$, for a fiber with a shape factor of 20 (so that $\beta_0 = 1.53°$) are presented. As the mold sprue does not have an infinite length, depending on the initial orientation, the fiber is more or less close to its equilibrium position. These results are consistent with experimental observations (Fig. 11.14); near the axis, fibers are perpendicular to it, whereas they are nearly parallel to the flow direction in an important peripheral layer.

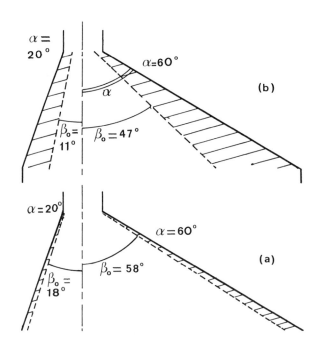

Figure 11.11 *Angle β_0 for two values of the divergent channel angle, α, and fiber aspect ratio 30 (a) and 5 (b).*

Solution of Eq (11.4) in the case of convergent channels is similar. The same Eq (11.11) has to be solved, leading to the same definition of the angle β_0. For β less than β_0, the fiber tends to a stable equilibrium position. It is strictly aligned with the flow on the axis, which is once again consistent with the results obtained in purely elongational flow. Outside the axis, the equilibrium position is very nearly aligned with the flow; 1 or 2 degrees off, depending on α and a/b. For a finite convergent length, these equilibrium positions may not be reached.

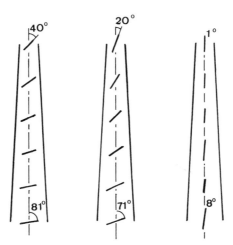

Figure 11.12 *Calculated fiber orientation along a mold sprue axis for three initial orientations 40°, 20°, and 1°.*

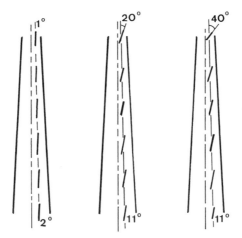

Figure 11.13 *Calculated orientation in a mold sprue for a fiber moving on a flow line β = 1°, for three initial orientations 1°, 20°, and 40°*

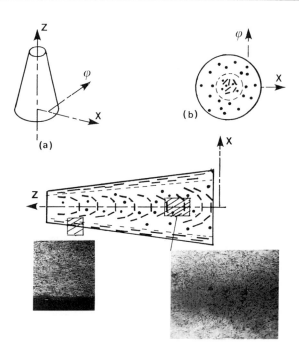

Figure 11.14 *Observed orientation in a disk–mold sprue, injected with polyamide filled with 30 wt% glass fibers (divergent flow).*

Figures 11.15 and 11.16 present evolution of the orientation in an axisymmetric channel ($\alpha = 45°$; $r = 0.99$; then $\beta_0 = 40°$) for fibers of different initial orientations moving on the axis ($\beta = 0°$) or for $\beta = 20°$, respectively. Whatever the initial orientation, the fibers align in the flow direction. This result is consistent with observations made for the entrance region of a capillary rheometer.

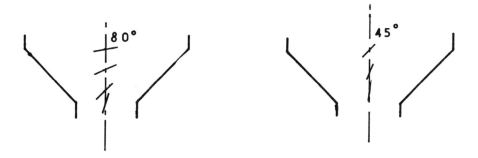

Figure 11.15 *Calculated fiber orientation along the axis of an axisymmetric convergent channel, for two initial orientations.*

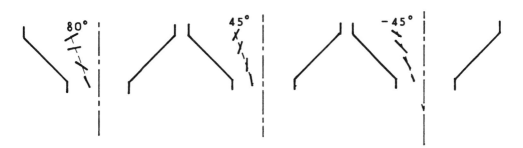

Figure 11.16 *Calculated orientation in a convergent channel for a fiber moving on a line* $\beta = 20°$, *for three initial orientations.*

The main conclusions from the study of these particular geometries are: 1) the major influence of the elongational flow even when the elongational rate is small (in comparison to the shear rate), and 2) the influence of the initial orientation of the fiber.

11.3.4 Discussion of Jeffery's Theory

Jeffery's theory has been experimentally verified by Anczurowski and Mason [39] in simple shear flow of a Newtonian fluid: a prolate spheroid, made by polymerizing a liquid droplet deformed in an electric field, had the period of rotation predicted by the theory. Only qualitative agreement was found during the flow in a divergent channel [7], since rods, not spheroids, were used. It is obvious that glass-fiber-filled thermoplastics are much more complex systems than those cited above. Thus Jeffery's assumptions seem far too much restrictive: 1) fibers have a cylindrical shape, and not a spheroidal one; 2) usually, their dimensions are not small in comparison with the flow dimensions; 3) the fiber-length distribution is nonuniform; 4) the fiber content is important, and it may be nonuniform; and 5) melted thermoplastics are non-Newtonian fluids.

Numerous experimental and theoretical studies have been carried out to understand the complex flow behavior of multiphase systems. The main results can be summarized as follows.

Influence of the particle shape. In the shear flow of a Newtonian fluid, Goldsmith and Mason [40] experimentally showed that the rotation of rigid rods is in agreement with the theory, provided that the equivalent aspect ratio (calculated from the measured period of rotation via Eq (11.8)) is used instead of the actual aspect ratio. The equivalent aspect ratio was smaller than the actual one; up to 50% for very long fibers. A phenomenological relation between the two shape factors was given by Harris and Pittman [41].

The theoretical considerations by Bretherton [42] proved that rotation of a rigid body of revolution in a uniform simple shear is identical to that of an equivalent ellipsoid of revolution.

Another theoretical approach was developed by Vincent et al.[43]. A two-dimensional finite element method, FEM, was used to calculate the velocity field in the neighborhood of a single particle (Fig. 11.17), the stresses acting on it, and its motion. FEM permitted calculation of the motion of a particle of any shape, in flows with nonuniform rate of strain or with a shear-thinning behavior. For instance, in a simple shear flow, the computed rotation period of a fiber of shape factor 6 is 29% smaller than that of an ellipsoid with the same shape factor, obtained from Jeffery's theory. This result is in good agreement with experiments [40].

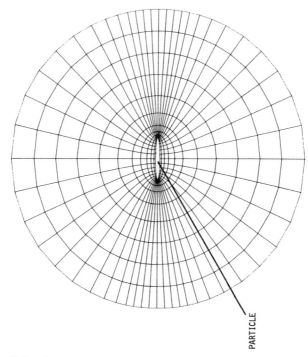

Figure 11.17 *The finite element mesh.*

Influence of a nonuniform rate of strain and of the presence of the flow walls. Goldsmith and Mason [40] observed that particle motion in flow through a tube is the same as in a simple shear flow, providing the particle size is small compared to that of the tube diameter. When the particle size is greater, two phenomena are observed: the translational velocity of the particle is smaller than that of the fluid, and there is an increase in the period of rotation. These phenomena are brought about by interaction between particles and the tube wall.

Goldman et al. [44] obtained an exact solution of Stokes equations for the translational and rotational velocities of a sphere moving near a plane wall in a simple shear flow. Their results are in agreement with the experimental data [40]. The numerical calculation of Vincent et al. [43] is also in agreement with these experiments: an ellipsoid of shape factor 10, in a Couette flow with nonuniform shear field is submitted to a rotation with period 17% greater than that calculated from Jeffery's theory (with the shear rate of the undisturbed flow at the center of gravity of the particle).

Influence of the suspending fluid rheology. Gauthier et al. [45,46] experimentally studied behavior of particles in pseudoplastic and viscoelastic fluids using Couette and Poiseuille flows. In all cases the authors observed a drift in the orbit constant C_θ (Eq (11.7)) towards a value corresponding to the minimum energy dissipation. In other words, fibers align in a Couette flow parallel to the axis, while in a tube perpendicular to its axis and tangent to the circular cross section. Another consequence of the non-Newtonian behavior of the suspending medium is a lateral migration of the fibers towards the region of higher shear strain in pseudoplastic liquids, and towards lower shear strain in viscoelastic liquids [45,46]. Molten polymers are simultaneously pseudoplastic and viscoelastic and it is difficult to decouple these opposite migration tendencies.

Theoretical considerations of Saffman [47] showed that in a non-Newtonian liquid, a single fiber aligns along the vorticity axis. Leal [48] calculated the rotation of an axisymmetric particle in a simple shearing motion of a Rivlin-Ericksen second-order fluid as well as the rate of change in the orbit constant C_θ. Surprisingly, for a shear rate smaller than a critical one, the particle aligns along the vorticity axis, and for higher rates it aligns in the flow direction. Brunn [49] calculated the motion of a slightly deformed sphere in a second-order fluid, in elongation and shear. In this latter case, the author found an alignment parallel to the vorticity axis. Vincent et al. [43] found a weak influence of pseudoplasticity on the rotation period in simple shear flow.

Influence of particle interactions. The above-mentioned studies are valid for a single particle. For concentrated suspensions in a Newtonian fluid, Goldsmith and Mason [40] observed that, in tubes, fibers align nearly parallel to the flow direction. Similar results were reported by Folgar and Tucker [50]. In a Couette flow the distribution of orientation presents a peak for an orientation parallel to the flow, though individual fibers irregularly make full rotation. The authors proposed a phenomenological model, which takes into account the randomizing effect of interactions. An additional term was introduced in Jeffery's equations, which has the form of a diffusion term, proportional to the second invariant of the strain-rate tensor, and to a material parameter which must be experimentally determined. When applied to a simple shear flow, the model predicts achievement of the constant orientation distribution in a good agreement with experiments.

Particle migration is also known to occur in concentrated suspensions. Kubat and Szalanczi [51] as well as Hegler and Menning [52] observed an increase in fiber concentration near the tip of a molded part, especially for large-size particles at high concentration. Darlington and Smith [23] observed a layer without fibers. In the literature devoted to suspensions, a lateral drift across the streamlines was reported, depending on the concentration and on the behavior of the suspending fluid [40,45,46]. Several theoretical investigations are available (e.g., [52]).

11.3.5 Concluding Remarks

All these investigations in shearing flows show that if any of the above effects are taken into account the particles will have preferred orientation, independent of the initial conditions, unlike the prediction of Jeffery's analysis. In elongational flows, Jeffery's theory seems to give good qualitative results, though studies on this subject are not numerous. These reasons added to the relative simplicity of use of Jeffery's equations, explain why this theory has been so frequently applied for calculating the flow of reinforced thermoplastics.

11.4   APPLICATION TO INJECTION MOLDING

11.4.1   Introduction

Most models for predicting fiber orientation in molded parts of general geometries are based on a modelization of the filling stage, taking into account the shear and elongational rates at any time and in any location within the filled area. Generally, they use Jeffery's equations. The limits of the calculation come from the four following simplifications: 1) the filling stage is calculated with an equivalent homogeneous rheology (this point will be discussed in Section 11.5); 2) most of the filling-stage programs give poor description of the flow in gates, in ribs, and near the flow front (fountain effects); 3) all the phenomena occurring after the filling stage, e.g. packing and cooling, are neglected, which means that the fiber orientation at the end of this stage is the fiber orientation in the final part; and 4) Jeffery's assumptions are restrictive as discussed in Section 11.3.

11.4.2 Application to a Center-Gated Disk

The center-gated geometry is general because it is found near the gate of numerous complex industrial molds. Both shearing and elongational flows are present, but the calculation remains reasonably simple.

Computation of a center-gated disk for nonstationary, nonisothermal flow has been developed by several authors (e.g., [6]). It leads to the radial velocity component $u(\rho, z)$ (Fig. 11.18), the rate of strain tensor $[\dot{\varepsilon}]$ and the rotation tensor $[\Omega]$ in the laboratory frame (ρ, β, z):

$$[\dot{\varepsilon}] = \begin{bmatrix} \dot{\alpha} & 0 & \dot{\gamma} \\ 0 & -\dot{\alpha} & 0 \\ \dot{\gamma} & 0 & 0 \end{bmatrix} \qquad [\Omega] = \begin{bmatrix} 0 & 0 & \dot{\gamma} \\ 0 & 0 & 0 \\ -\dot{\gamma} & 0 & 0 \end{bmatrix} \tag{11.15}$$

with $\dot{\alpha} = \partial u / \partial \rho$ and $\dot{\gamma} = (\partial u / \partial z)/2$. Introducing the Euler angles defined in Fig. 11.18, and using Eq (11.4) yields:

$$\left. \begin{aligned} \dot{\theta} &= -r \left[\dot{\gamma} \sin \sigma \cos 2\theta + \dot{\alpha} (1 + \cos^2 \sigma) \sin 2\theta / 2 \right] + \dot{\gamma} \sin \sigma \\[2em] \dot{\sigma} &= (\cos \sigma / \sin \theta) \left[2 \cos \theta / (p^2 + 1) + r \dot{\alpha} \sin \theta \sin \sigma \right] \end{aligned} \right\} \tag{11.16}$$

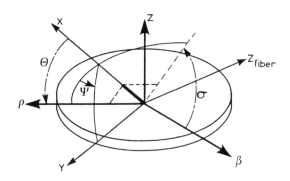

Figure 11.18 Disk geometry and Euler angles.

When the fiber enters the mold at a certain level, z (between core and skin), with a given initial orientation (Fig. 11.19), its displacement within the mold is calculated assuming that the center of gravity moves with the calculated fluid velocity at a constant z. At each time step, the values of $\dot\alpha$ and $\dot\gamma$ are calculated for the fiber center of gravity. Then, Eq (11.16) is solved using a Runge-Kutta method, giving the orientation during filling. The evolution of the angles θ and σ with ρ is presented in Fig. 11.20. It is worth noting that: (a) for z = 0: in a very short time the fibers in the core region reach an equilibrium position, transverse to the flow direction ($\theta = 90°$) and parallel to the plane of the disk ($\sigma = 0°$) (Fig. 11.20a); the final orientation does not depend on the initial one; (b) for z = 0.4 mm: (intermediate region), the fibers tend to a similar position (Fig. 11.20b); (c) for z = 0.5 mm the final orientation is quite different: σ tends to 0° which means that the fibers belong to the (ρ, β) plane; θ takes values between 0 and 180°, so that the orientation is nearly isotropic (Fig. 11.20c); (d) for z = 0.7 mm, in the skin region, the fibers follow a pseudoperiodic movement: σ varies from +90 to -90° meaning that the fibers do not belong to the (ρ, β) plane, but at the same time, θ is very near 0 and 180°, so that the fibers are mostly parallel to the flow direction (Fig. 11.20d). Because of the quick variation of σ and θ, the orientation is not perfect. For z = 0.9 mm, the movement is very similar. The model predicts different core and surface orientations, qualitatively in good agreement with observations, but with an overestimation for the core region (see Fig. 11.21).

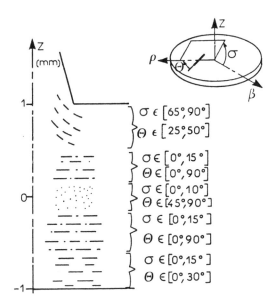

Figure 11.19 *Observed fiber orientation at a disk-mold entrance (polyamide-6,6 with 30 wt% glass fibers).*

In order to increase the accuracy of this model, Barbe et al. [53] coupled it with an expert system. The inference engine modifies the intermediate results of the calculation, according to a knowledge base which takes into account experimental observations. For instance, in the case of the disk, a coefficient is applied to the elongational flow rate, in order to diminish its efficiency. This means physically that, at the end of the sprue, fiber interactions in a small volume interfere with a fast rotation. Then, the thickness of the core region is predicted

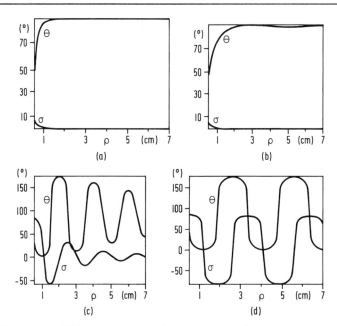

Figure 11.20 Evolution of fiber orientation (aspect ratio 10) during flow of a polyamide-6,6 containing 30 wt% of glass fibers in a center-gated disk (radius, 15 cm; thickness, 2 mm) for different positions between the core and the skin. (a) z = 0 mm; (b) z = 0.4 mm; (c) z = 0.5 mm; and (d) z = 0.7 mm.

(Fig. 11.21). The use of an expert system is also interesting in three-dimensional geometries (gates, ribs, strong thickness variations) when the flow- and fiber-orientation computations are not precise enough. In return, the knowledge base must be supplied with observed orientation for each material and for a large set of processing conditions.

11.4.3 Application to Other Mold Geometries

The case of a plaque has been studied by Barbe et al. [53]. They calculated the orientation function values (defined in [6]) from the motion of 36 fibers with different initial orientation values determined experimentally. The computed results were compared with the experimental values of the core-region thickness where orientation was transverse to the flow. For the film gate (Fig. 11.22), the agreement was good. For a point gate, large elongational stresses near the entrance perfectly oriented the fibers perpendicularly to the flow and this orientation remains in the whole plaque, contrary to observation. In fact, for interacting fibers this perfect orientation is impossible, so a disorientation was artificially imposed by the expert system. The final results were in better agreement with the experiments (Fig. 11.23).

Givler et al. [54,55] and Gillespie et al. [56,57] used a two-dimensional finite-element method to solve the mechanical equations for the flow. Jeffery's equations were solved with a Runge-Kutta scheme giving planar orientation. The calculation has been applied to the isothermal flow in an infinite expansion, around a hole in a plaque mold, and in the fountain flow near the front. In the latter case, the calculations were made in a reference frame linked to the front, so that they were time-independent; the fibers were almost perfectly aligned along the streamline near the free surface.

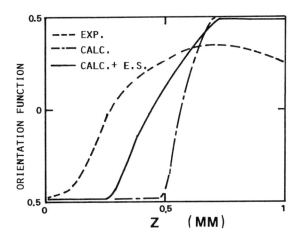

Figure 11.21 *Evolution of the orientation function in the thickness of a center-gated disk (measurement and computation without and with expert system (ES)).*

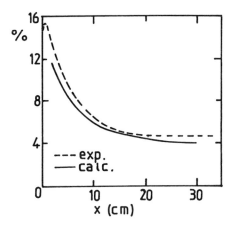

Figure 11.22 *Comparison of observed and calculated core-thickness evolution in a rectangular mold with a film gate. Polyamide-6,6 filled with 30 wt% glass fibers (after [53]).*

11.5 COUPLED FIBER-ORIENTATION AND FLOW CALCULATIONS

The aim of this kind of approach is to take into account the influence of the structure of the fluid on the flow. One needs to build a continuum-mechanics-constitutive equation which accounts for experimental rheological observations for anisotropic materials. This class of materials is very wide and depends on the scale of the study. Here it will be used for fiber suspension in polymer melts. The same theory can be employed, for example, for liquid crystal polymers.

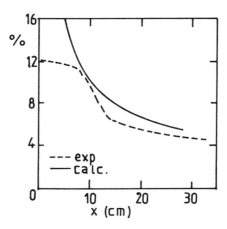

Figure 11.23 *Same as Fig. 11.20 for a point gate.*

The basic anisotropic fluid theory was formulated by Ericksen [58]. He considered an incompressible fluid having at each material point a single preferred direction given by a unit vector, n. He assumed that the stress at each point, as well as the material derivative of n, $\dot{n}$, are functions of the velocity gradients and of n. Using standard invariance procedures, and considering linearity in the rate tensor $\dot{\epsilon}_{ij}$, he obtained:

$$\left.\begin{array}{l} \sigma_{ij} = -p\,\delta_{ij} + 2\,\mu_0\,\dot{\epsilon}_{ij} + (\mu_1 + \mu_2\,\dot{\epsilon}_{km}\,n_k n_m)\,n_i n_j + 2\,\mu_3\,(\dot{\epsilon}_{jk}n_k n_i + \dot{\epsilon}_{ik}n_k n_j) \\[2mm] \dot{n}_i = \omega_{ij}n_j + \mu_4\,(\dot{\epsilon}_{ij}n_j - \dot{\epsilon}_{km}n_k n_m n_i) \end{array}\right\} \quad (11.17)$$

ω_{ij} are the rotation tensor components and μ_1, μ_2, μ_3, and μ_4 are rheological coefficients. If μ_1 is different from zero, then the constitutive equation describes a material of the Bingham type. Ericksen provided a solution for a simple shear velocity field. The evolution of the director n with time was identical to what Jeffery's theory predicted for the main axis of an ellipsoid, provided that $\mu_4 = (a^2 - b^2)/(a^2 + b^2)$. The same result can be obtained for a purely elongational flow. Leslie [59] found the steady solution for a flow in a convergent or divergent channel, for $\mu_4 > 1$. A comparison of Ericksen's and Jeffery's theories has been done for general flows by Lipscomb et al. [60]. The theories are consistent provided that μ_0, μ_2, μ_3, and μ_4 are expressed as functions of the particle aspect ratio. Akay and Leslie [61] extended Ericksen's theory to the case of a viscoelastic fluid, with an eight-coefficients constitutive equation.

The problem with such a theory is that the calculation of an orientation function must be carried out by repeating the calculation for different initial conditions. Lipscomb et al. [60] obtained a bulk stress constitutive equation by volume averaging the stresses described by Ericksen's theory. The theory incorporates the statistical orientation distribution function. The two-dimensional prediction of the recirculating vortex size in an abrupt contraction is in good agreement with experiments made at low volume fraction (less than 0.1 wt% of glass fiber in corn syrup).

Dinh and Armstrong [62] also developed a rheological equation of state for a semiconcentrated fiber suspension in a Newtonian fluid. Following Batchelor [63], the bulk stress was obtained with a cell model, which gives an estimate of the drag coefficient according to particle interactions. This theory predicts some observed transient rheological behavior. Bibbo et al. [64] extended this theory taking into account the walls.

Though all these studies incorporate the director n, or its statistical distribution function, few details are given on orientation. Moreover, these studies were restricted to steady, isothermal flows, in relatively simple geometry. Lastly, several rheological coefficients must be experimentally determined.

11.6   CONCLUSIONS

Even if Jeffery's theory is based on a set of restrictive assumptions, which seem to be violated during the flow of reinforced thermoplastics, the decoupled computation presented in Section 11.3 can obtain a reasonable prediction of fiber orientation in injected parts. This kind of approach has been successfully applied to the injection of reinforced thermoplastics as well as to Bulk Molding Compound (BMC) with long fibers [15].

More sophisticated approaches, based on anisotropic fluid theories, are useful for understanding complex phenomena resulting from the coupling of fiber orientation and flow, such as transient rheological effects. These developing theories pose a problem of determination of several rheological coefficients.

NOTATION

a, b, c	Ellipsoid axis (a, length; b, diameter)
$a_{11}, a_{22}, a_{12}, a_{21}$	Orientation tensor components
$C, C_\sigma; C_\theta$	Integration constants; C_θ, orbit constant
f, g, h	Rate of strain tensor components in a coordinate system affixed to the particle
f_{or}	Orientation function
L	Width of the two-dimensional channel
n	Director
n_i	Components of n
$\dot{n}$	Time derivative of n
$\dot{n}_i$	Components of $\dot{n}$
$p = a/b$	Fiber aspect ratio
$\dot{Q}$	Volumetric flow rate
t	Time
t_p	Rotation period
T_m	Mold temperature
u	Velocity
x, y, z, x', y', z':	Coordinates
X, Y, Y'	Intermediate quantities, Eq (11.14)
α	Half angle of the divergent (convergent) channels
$\dot{\alpha}$	Elongational rate
β	Determines the fiber path in divergent (convergent channels)
$\dot{\gamma}$	Shear rate
$\dot{\varepsilon}_{ij}$	Rate of strain tensor components in the laboratory frame
λ, μ, ν	Rotation tensor components in a coordinate system affixed to the particle
$\mu_0, \mu_1, \mu_2, \mu_3, \mu_4$	Rheological coefficients

ρ	Radius in a cylindrical coordinate system
σ_{ij}	Stress tensor components in the laboratory frame
θ, σ, Ψ	Euler angles
$\dot{\theta}, \dot{\sigma}, \dot{\Psi}$	Time derivatives of θ, σ, and Ψ
$\omega_1, \omega_2, \omega_3$	Particle rotation velocity (in a coordinate system affixed to it)
ω_{ij}	Rotation tensor components in the laboratory frame

REFERENCES

1.   Vigneron, D., *PhD thesis*, Universite de Technologie de Compiegne, France (1983).
2.   Folkes, M. J.,*Short Fibre Reinforced Thermoplastics*, Research Studies Press, New York (1982).
3.   MacNally, D., *Polym. Plast. Technol. Eng.*, **8**, 101 (1977).
4.   Kamal, M. R., Mutel, T., *J. Polym. Eng.*, **5**, 293 (1985).
5.   Vincent, M., *PhD thesis*, Ecole des Mines de Paris, France (1984).
6.   Vincent, M., Agassant, J. F., *Polym. Compos.*, **7**, 76 (1986).
7.   Vincent, M., Agassant, J. F., *Rheol. Acta*, **24**, 603 (1985).
8.   Vaxman, A., Narkis, M., Siegman, A., Kenig, S., *J. Mat. Sci., Lett.*, **7**, 25 (1988).
9.   Fakirov, S., Farikova, C., *Polym. Compos.*, **6**, 41 (1985).
10.   Sanou, M., Chung, B., Cohen, C., *Polym. Eng. Sci.*, **25**, 1008 (1985).
11.   Leuchter, H., *Eng. thesis*, IKV, Rhein-Westf., Technische Hochschule Aachen, (1981).
12.   Hegler, R. P., *Kunststoffe*, **74**(5), 271 (1984).
13.   Haskell, W. E., Petrie, S. P., Lewis, R. W., *SPE Techn. Pap.*, **28**, 292 (1982).
14.   Owen, M. J., Whybrew, K., *Plast. Rubber*, **1**, 6 (1976).
15.   Blanc, R., Philipon, S., Vincent, M., Agassant, J. F., Alglave, H., Muller, H., Froelich, D., *Int. Polym. Proc.*, **2**, 21 (1987).
16.   Fischer, G., Eyerer, P.,*SPE Techn. Pap.*, **32**, 532 (1986).
17.   Kamal, M. R., Song, L., Singh, P., *SPE Techn. Pap.*, **32**, 133 (1986).
18.   Darlington, M. W., McGinley, P. L., *J. Mat. Sci.*, **10**, 906 (1975).
19.   Darlington, M. W., McGinley, P. L., Smith, G. R., *J. Mat. Sci.*, **11**, 877 (1976).
20.   Darlington, M. W., Gladwell, B. K., Smith, G. R., *Polymer*, **18**, 1269 (1977).
21.   Christie, M. A., Darlington, M. W., McCammond, D., Smith, G. R., *Fibre Sci. Technol.*, **12**, 167 (1979).
22.   Bright, P. F., Darlington, M. W., *Plast. Rub. Proc. Appl.*, **1**, 139 (1981).
23.   Darlington, M. W., Smith, A. C., *Polym. Compos.*, **8**, 16 (1987).
24.   Folkes, M. J., Russell, D. A. M., *Polymer*, **21**, 1252 (1980).
25.   Bright, P. F., Crowson, R. J., Folkes, M. J., *J. Mat. Sci.*, **13**, 2497 (1978).
26.   Akay, G., in *Interrelations Between Processing Structure and Properties of Polymeric Materials*, Seferis, J. C., Theocaris, P. S., Eds., Elsevier Science Publishers, Amsterdam (1984).
27.   Hegler, R. P., Menning, G., *SPE Techn. Pap.*, **31**, 781 (1985).
28.   McGee, S. H., McCullough, R. L., *J. Appl. Phys.*, **55**, 1394 (1984).
29.   Denton, D. L., Munson, S. H., McGee, S. H., in *High Modulus Fiber Composites in Ground Transportation and High Volume Applications*, Wilson, D. W., Ed., ASTM STP 873, Philadelphia (1985).
30.   Jackson, W. C., Folgar, F., Tucker, III, C. L., in *Polymer Blends and Composites in Multiphase Systems*, Han, C. D., Ed., ACS, Washington (1984).
31.   Folgar, F., *PhD thesis*, Univ. Illinois at Urbana - Champaign, Urbana, IL (1983).

32. Jackson, W. C., Advani, S. G., Tucker, III, C. L., *J. Comp. Mat.*, **20**, 539 (1986).
33. Kau, H. T., *SPE Techn. Pap.*, **31**, 1224 (1985).
34. Pipes, R. B., McCullough, R. L., Taggart, D. G., *Polym. Compos.*, **3**, 34 (1982).
35. Advani, S. G., Tucker, III, C. L., *J. Rheol.*, **31**, 751 (1987).
36. Kenig, S., *Polym. Compos.*, **7**, 50 (1986).
37. Jeffery, G. B., *Proc. Roy. Soc.* (London), **A102**, 161 (1922).
38. Cohen, C., Chung, B., Stasiak, W., *Rheol. Acta*, **26**, 217 (1987).
39. Anczurowski, E., Mason, S. G., *Trans. Soc. Rheol.*, **12**, 209 (1968).
40. Goldsmith, H. L., Mason, S. G., in *Rheology: Theory and Applications*, Vol. 4, Eirich, F.R., Ed., Academic Press, New York (1967).
41. Harris, J. B., Pittman, J. F. T., *J. Colloid Interface Sci.*, **50**, 280 (1975).
42. Bretherton, F. P., *J. Fluid Mech.*, **14**, 284 (1962).
43. Vincent, M., Germain, Y., Agassant, J. F., in *Progress and Trends in Rheology II*, Giesekus, H., Hibberd, M. F., Eds., *Suppl. Rheol. Acta*, **26**, 144 (1988).
44. Goldman, A. J., Cox, R. G., Brenner, H., *Chem. Eng. Sci.*, **22**, 653 (1967).
45. Gauthier, F., Goldsmith, H. L., Mason, S. G., *Trans. Soc. Rheol.*, **15**, 297 (1971).
46. Gauthier, F., Goldsmith, H. L., Mason, S. G., *Rheol. Acta*, **10**, 344 (1971).
47. Saffman, P. G., *J. Fluid Mech.*, **1**, 540 (1956).
48. Leal, L. G., *J. Fluid Mech.*, **69**, 305 (1975).
49. Brunn, P., *Rheol. Acta*, **18**, 229 (1979).
50. Folgar, F. P., Tucker, III, C. L., *J. Reinf. Plast. Compos.*, **3**, 98 (1984).
51. Kubat, J., Szalanczi, A., *Polym. Eng. Sci.*, **14**, 873 (1974).
52. Hegler, R. P., Menning, G., *Polym. Eng. Sci.*, **25**, 395 (1985).
53. Barbe, A., Agassant, J. F., Vincent, M., *4th Annual Meeting of the Polymer Processing Society*, Orlando, FL, USA, May 8-11 (1988).
54. Givler, R. C., Crochet, M. J., Pipes, R. B., *J. Compos. Mat.*, **17**, 330 (1983).
55. Givler, R. C., in *Transport Phenomena in Materials Processing*, Chen, M. M., Mazumder, J., Tucker, C. L., Eds., ASME, New York (1983).
56. Gillespie, J. W., Vanderschuren, J. A., Pipes, R. B., *SPE Techn. Pap.*, **30**, 648 (1984).
57. Gillespie, J. W., Vanderschuren, J. A., Pipes, R. B., *Polym. Compos.*, **6**, 82 (1985).
58. Ericksen, J. L., *Kolloid−Z.*, **173**, 117 (1960).
59. Leslie, F. M., *J. Fluid Mech.*, **18**, 595 (1964).
60. Lipscomb, G. G., Denn, M. M., Hur, D. U., Boger, D. V., *J. Non−Newtonian Fluid Mech.*, **26**, 297 (1988).
61. Akay, G., Leslie, F. M., in *Advances in Rheology*, Vol. 3, Mena, B., Garcia-Rejon, A., Rangel-Nafaille, C., Eds., UNAM, Mexico (1984).
62. Dinh, S. M., Armstrong, R. C., *J. Rheol.*, **28**, 207 (1984).
63. Batchelor, G. K., *J. Fluid Mech.*, **41**, 545 (1970).
64. Bibbo, M. A., Dinh, S. M., Armstrong, R. C., *J. Rheol.*, **29**, 905 (1985).

CHAPTER 12

RHEOLOGICAL PROPERTIES OF FIBER-REINFORCED POLYMER MELTS

by A. T. Mutel and M. R. Kamal

Department of Chemical Engineering
McGill University
Montreal, QC, H3A 2A7
CANADA

*Suspension theories are reviewed with emphasis on the relationships between
the bulk rheological properties and the instantaneous state of microstructure.
These theories generally involve two relations: one to describe microstructure
development in terms of flow kinematics and structural properties, while the
other gives the macroscopic stress in terms of bulk-flow kinematics and the
instantaneous state of microstructure. Theories for suspensions in Newtonian
fluids assume that the flow at the microscopic level is governed by Stokes
equations, which are quasi-time independent. The bulk stress consists of two
parts: due to the suspending fluid and the other due to the particles. Jeffery's
development for dilute suspensions in Newtonian fluids was followed by
various proposed relationships between the specific viscosity of dilute fiber
suspensions and the aspect ratio. The predictions of these theories are in good
agreement with experimental data for volume fractions below 0.001 and aspect
ratios below 200. Various extensions of the above theory have been proposed to
describe the behavior of nondilute fiber suspensions. Recent developments in
this area treat particle interactions as a randomizing process over Jeffery's
orbits for isolated particles. These relations, which have been extended to treat
nondilute fiber suspensions in non-Newtonian fluids, have been successful in
predicting qualitatively many of the rheological phenomena observed during
the flow of these suspensions.*

12.1   INTRODUCTION

The utilization of reinforced plastics in commercial products has experienced substantial
growth in the recent years. The major factor behind this growth is due to the superior
mechanical and thermal performance of plastic composites. However, the final mechanical
properties of composite products are anisotropic due to the anisotropic nature of reinforcing
agents. The ultimate mechanical properties are achieved with continuous fibers and a
controlled fiber orientation. Although high-performance products having balanced properties
may be obtained for continuous fibers with controlled fiber orientation, the production of
continuous fiber composites is usually labor intensive and involves batch or semicontinuous
processes. Short- and long-fiber composites may be produced utilizing conventional
processing techniques, such as extrusion and injection molding. Fiber orientation distribution
in a short-fiber composite product is determined by the flow and thermal history experienced
by the composite melt during processing. Hence, the rheological and thermal properties of the
composite melt, as well as the geometry of the part and the processing conditions, have strong
effect on the final fiber-orientation distribution in the part. The rheological and thermal
properties of the melt are affected by the internal structure, as determined by the size
distribution and the spatial and orientational distributions of the fibers. Hence, the internal
structure, processing conditions, and the rheological behavior are all interrelated. An
understanding of these interactions is needed in order to control and optimize process
conditions and product performance.

Fiber-reinforced polymer melts are suspensions of anisometric particles in a
viscoelastic fluid. Glass, carbon, aramid, and polyamide fibers are the most commonly used
reinforcing agents. Among them, glass and carbon fibers can be considered rigid. The
rheological behavior of suspensions is governed by flow-induced changes in the internal
structure associated with the particles, as well as the rheological properties of the suspending

fluid. The structural changes are specific to the flow type and small deviations from Newtonian behavior of the suspending fluid may have a strong effect on the structural dynamics.

12.2   SUSPENSION THEORIES: A REVIEW

Suspension theories developed since the beginning of the century have dealt with the development of microstructure under flow and the relationship between the bulk rheological properties and the instantaneous state of the microstructure. This approach leads to two functional relations, which constitute a model for the bulk rheological behavior of a suspension.

The first relation describes microstructure development in terms of flow kinematics and structural parameters:

$$K_{ij} = \dot{K} \{K_{ij}, \nabla U_{ij}; \tau^i, \tau'\} \tag{12.1}$$

Here, K is an nth order tensorial quantity describing the microstructure, $\dot{K}$ is the time derivative of K, ∇U is the velocity gradient, τ^i is the relaxation time(s) associated with the suspending fluid and τ' is the characteristic relaxation time associated with a restoring force, such as Brownian forces. A unit vector ℓ is necessary to describe the microstructure of fibers with equal dimensions.

The second relation gives the macroscopic stress tensor in terms of bulk-flow kinematics and the instantaneous state of microstructure:

$$\sigma_{ij} = \sigma \{K_{ij}, \nabla U_{ij}; \tau^i, \tau'\} \tag{12.2}$$

The above two functional relationships together comprise a constitutive equation for a general suspension. However, Eqs (12.1) and (12.2) are too general and certain assumptions are needed for applications to specific systems.

Suspension theories have so far been developed for either very dilute suspensions, for which the concentration limit is too low for any practical purposes, or for concentrated systems for certain flow types. Another difficulty with existing suspension theories is in the matching of microscopic and macroscopic flow fields when the suspending fluid exhibits non-Newtonian properties. The relations for the bulk stress tensor, in terms of the structural state, are not available for non-Newtonian suspending fluids in a general flow field. Recent work on suspension theories has been focused on extending the concentration limit to suspensions of practical interest [1-4] and incorporating the elastic properties of the fluid at the slow-flow limit [3-7]. Although, the suspension theories developed so far cannot yield quantitative predictions of rheological properties of suspensions in viscoelastic fluids, the study of suspension theories is useful for the analysis and interpretation of experimental observations in terms of interactions between the microstructure and the observed rheological behavior.

A brief review of suspension theories for Newtonian fluids will be given along with the experimental observations. The rheological behavior of fiber-filled polymer melts will be considered in the second part along with structural observations.

12.2.1 Suspension of Fibers in Newtonian Fluids

Two important simplifications are possible when the suspending fluid is Newtonian: 1) The flow at the microscopic level is governed by Stokes equations. Hence, the microscale velocity field is linear in applied forces and any quantity linearly derived from flow will also be linear. This applies in particular to the bulk stress and structural dynamics. 2) Stokes equations are quasi-time independent. Only the instantaneous state of structure is relevant. However, although Eqs (12.1) and (12.2) are both linear and instantaneous, the bulk stress has a nonlinear dependence on the bulk deformation rate, showing history effects.

Bulk stress can be divided into two parts: one contribution due to the suspending fluid itself, $\sigma_0 = \eta_0 E$, and a contribution due to the particles, $\sigma_p = (1/V) \Sigma S$. Here, V is a representative volume element of the suspension, S is the contribution of each particle to the stress tensor, η_0 is the viscosity of the suspending fluid, and E is the strain rate tensor. The stress contribution from a particle depends only on the particle orientation, ℓ, and on the relative positions and orientations of other particles, ℓ', assuming that all the particles have the same shape, size, and constitution. The mean value of the structural contribution is then effectively an average of the value of S for the reference particle over all the values of ℓ and ℓ' weighted according to the joint probability distribution of ℓ and ℓ':

$$S_{m,ij} = \int S_{ij}(\ell, \ell') \, p(\ell, \ell' \mid 0) \, d\ell \, d\ell' \tag{12.3}$$

The state of microstructure is represented by the joint probability distribution $p(\ell, \ell' \mid 0)$. Two methods have been used to evaluate this integral, assuming that the state of microstructure is known. The first is to evaluate the energy dissipation due to particles, which is reflected by an increase in the effective viscosity. The disadvantage of this method is that energy is a scalar property, hence yielding only the increased viscous dissipation in the system. The other method employs ensemble averages. The bulk stress and velocity gradient in the suspension are defined as averages over an ensemble of realizations. This method allows for calculation of the full stress tensor, unlike the energy-dissipation method. Bulk stress tensor for the contribution of fiber-like particles was given by Batchelor [8,9]:

$$\sigma_{ij} = 2\eta_0 \, \phi_v \, [A < \ell_i \ell_j \ell_k \ell_l > : E_{ij} + \\ B(< \ell_i \ell_j > \cdot E_{ij} + E_{ij} \cdot < \ell_i \ell_j >) + CE_{ij} + D < \ell_i \ell_j >] \tag{12.4}$$

where ℓ is the unit vector along a fiber, $< \ell \ell >$ and $< \ell \ell \ell \ell >$ are the second and fourth moments of orientation distribution, A, B, C, and D are constants. The moments of orientation distribution are calculated from the distribution function using the following relation:

$$<P(\theta, \phi)> = \int_0^\pi d\theta \int_0^{2\pi} (\sin\theta) \, P(\theta, \phi) \, p(\theta, \phi) \, d\phi \tag{12.5}$$

where $p(\theta, \phi)$ is the probability distribution function and θ and ϕ are the spherical angles describing the orientation of a fiber, as shown in Fig. 12.1.

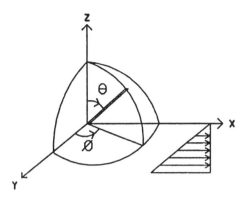

Figure 12.1 *Spherical polar coordinates describing the orientation of a fiber.*

12.2.2　Dilute Suspensions

If the concentration of particles in the suspension is sufficiently small, the distribution of field properties is approximately the same as if the particle was alone in an infinite matrix. The dilute–suspension assumption is valid for $L^3 n \ll 1$. Here, L is the length of fibers and n is the number of particles per unit volume. In the dilute–suspension limit, the value of S for the reference particle no longer depends on the configuration of other particles ℓ' and:

$$S_{ij} = \int S_{ij}(\ell)\, p(\ell)\, d\ell \tag{12.6}$$

If the suspension is subjected to a bulk flow with a linear mean velocity gradient, the rate of change of ℓ is given by [5]:

$$d\ell_i/dt = \dot{\ell}_i = \Omega_{ij} \cdot \ell_j + (p-1)/(p+1)\,[\ell_j \cdot E_{ji} - \ell_i(\ell_j \cdot E_{ji} \cdot \ell_i)] \tag{12.7}$$

where Ω is the asymmetric part of the bulk flow, while E is the symmetric part and p is the aspect ratio. The evaluation equation for $p(\ell)$ is derived from the conservation of particle number [9]:

$$\delta p/\delta t = -\nabla \cdot (\dot{\ell} p) \tag{12.8}$$

　　For pure straining motion (extensional flows), $\Omega = 0$, the particle will rotate towards a stable equilibrium position in the direction of the principal axis of extension. The stress contribution from the particles will be Newtonian. The stress contribution of particles calculated using Eq (12.6) and either the energy-dissipation or ensemble-averages method leads to the same reduced specific viscosity given by:

$$(\eta - \eta_0)/\eta_0\, \phi_v = [\eta] = (p^2/3 \ln 2p)\, \{1 - 1.5(\ln 2p)^{-1}\}^{-1} \tag{12.9}$$

It is worth noting that the reduced specific viscosity, $(\eta - \eta_0)/\eta_0\, \phi_v = \eta_{sp}/\phi_v$, in Eq (12.9) is equated with the intrinsic viscosity, $[\eta]$. Such identification is strictly valid at infite dilution.

As the particle concentration increases, η_{sp}/ϕ_v becomes concentration dependent and the symbol $[\eta]$ looses its customary identification.

Theoretical predictions suggest that the stress contribution from particles with large aspect ratios can be relatively large. However, the overall stress contribution at this level, $\phi_v[\eta]$, will be small because the theory is only valid for $\phi_v \ll 1$. On the other hand, if the bulk motion is a simple shear flow, each fiber undergoes a periodic rotation with nonuniform angular velocity, and spends most of the orbit period in directions close to the stream lines [10,11]. There is an infinite family of such orbits for a fiber, each labelled with an orbit constant C. The average value of S(ℓ) over an orbit and the distribution of particles among orbits need to be known to evaluate the integral in Eq (12.6). The average shear-stress contribution of a particle rotating in an orbit defined by C is given by:

$$S_{12}(C) = [p_e/2 \, (\ln 2p_e - 1.5)] \, [C/(C^2 + 1)^{1/2}] \tag{12.10}$$

Here, p_e is the equivalent aspect ratio for cylindrical particles. The total stress contribution from the particles can be calculated using Eq (12.6) rearranged to give:

$$S_{12} = \int S_{12}(C) \, p(C) \, dC \tag{12.11}$$

if the probability distribution of orbits, $p(C)$, is known.

There are no preferred orbits if particle interactions, Brownian forces, inertial effects, or external couples are negligible. The probability distribution of orbits at any time is defined by the initial distribution. The suspension properties are periodic in time, average values being Newtonian in form.

Two relations were suggested for the shear stress contribution from particles initially oriented over all possible directions with equal probability [12,13]:

$$[\eta] = 1.15 \, p/[\pi \ln (2p)] \tag{12.12}$$

$$[\eta] = p/[\pi \ln (2p - 1.8)] \tag{12.13}$$

The difference between Eqs (12.12) and (12.13) arises from different approximations used for Stokes coefficients. The maximum stress contribution from particles is given by Guth [14]:

$$[\eta] = p/(2 \ln 2p - 3) + 2 \tag{12.14}$$

if all the particles are in the plane normal to the vorticity axis, $C = \infty$. The intrinsic viscosity becomes equal to 2, when the particles are oriented near the vorticity axis, $C = 0$, corresponding to minimum viscous dissipation. Intrinsic viscosities under extensional and shear flows predicted by Eqs (12.9), (12.12), (12.13), and (12.14) are shown in Fig. 12.2 as a function of the aspect ratio of fibers. The stress contribution from the particles in extensional flows is one to two orders of magnitude larger than the stress contribution in shear flows. The effect of aspect ratio is also much more pronounced in extensional flows. Dilute suspension viscosities of fibers ($\phi_v < 0.001$) are in good agreement with theoretical predictions for $p < 200$ [15,16].

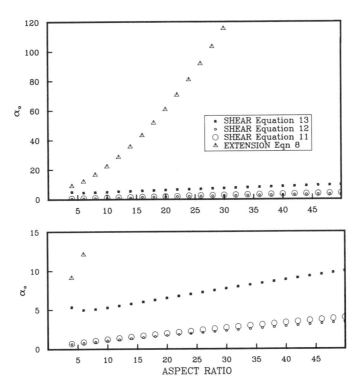

Figure 12.2 *Intrinsic viscosities for fiber suspensions under shear and extensional flows predicted by dilute suspension theories.*

12.2.3 Nondilute Suspensions

Fiber suspensions in Newtonian fluids generally exhibit non-Newtonian rheological behavior at concentrations above the limit for the dilute suspension assumption. Suspensions of fibers with small aspect ratios (approximately < 40) still exhibit shear stresses, σ, linear in shear rate, $\dot{\gamma}$, at small concentrations where only pair interactions are dominant [17-19]. Nevertheless, nonzero first normal stress differences have been observed for the same suspensions [18,19]. Interactions among the fibers increase the total viscous dissipation, leading to higher viscosities compared to dilute suspensions.

Many relative-viscosity-concentration relations represent semi-empirical extensions of dilute suspension theories through inclusion of $O(\phi^2_v)$ terms for the first-order interactions among fibers. Blakeney [16] extended Burger's [13] relationship for random initial orientation distribution by adding an $O(\phi^2_v)$ interaction term:

$$\eta_r = 1 + [\eta]\,\phi_v + k\,([\eta]\,\phi_v)^2 \tag{12.15}$$

where $[\eta]$ is the same as in Eq (12.13) and k was calculated as 0.73 by Simha [20] for randomly oriented systems. Guth [14] suggested a similar interaction term:

$$\eta_r = 1 + [\eta] \, \phi_v + k[\eta] \, \phi_v^2 \tag{12.16}$$

with $$k[\eta] = 0.04 \, p/(2 \ln 2p - 3)$$

The interaction term extends the applicable concentration range only up to ≈ 0.01.

Brodnyan [21] proposed the following relation for nondilute suspensions of fibers based on Simha's [22] analysis for dilute systems:

$$\ln \eta_r = [2.5 \, \phi_v + 0.399 \, (p-1)^{1.48} \, \phi_v/(1 - k \, \phi_v)] \tag{12.17}$$

where k is inversely proportional to the maximum packing volume fraction and, for slender particles, takes on the value of 1.91. When compared with Yang's [23] experimental results the correlation was successful only up to $\phi_v = 0.005$.

The σ - $\dot{\gamma}$ relationship becomes nonlinear as the concentration or aspect ratio is increased. Fiber suspensions, in general, exhibit shear-thinning viscosities with asymptotic plateaux at high and low shear rates [17-19,24-26]. The extent of shear thinning and the existence of a low shear rate plateau depend on the aspect ratio, rigidity of the fibers, and the viscosity of the suspending fluid, as well as on the fiber concentration. The low shear rate plateau is usually not observed for suspensions of fibers with small aspect ratios at high concentrations in low viscosity fluids. For these suspensions, shear thinning or yielding is observed at low shear rates followed by a high shear rate plateau. Suspensions of fibers with large aspect ratios exhibit shear thinning viscosities even at very small concentrations and usually a low shear rate asymptote (yield) is observed at high concentrations, especially with nonrigid fibers. Both η and its negative slope, $-d\eta/d\dot{\gamma}$, are enhanced by an increase in aspect ratio, concentration, or flexibility of the fibers and a decrease in the viscosity of the suspending fluid [17,26]. The effect of a $\Delta \phi_v$ increase in concentration on the magnitude of viscosity is higher at larger concentrations for low-aspect-ratio fibers, $d^2\eta/d\phi_v^2 > 0$, while crowding has a diminishing effect for large aspect-ratio fibers, $d^2\eta/d\phi_v^2 < 0$. The effect of concentration decreases with increasing shear rate.

Ziegel [26] developed a semi-empirical relation for the shear dependence of viscosity using the energy dissipation method. Total energy dissipation was taken to be the work required to rotate free particles in the medium and to disrupt agglomerates. The shear rate dependence of viscosity arises from the change in equilibrium between formation and disruption of agglomerates with shear. A second-order kinetic model was used to describe the agglomerate formation and disruption processes:

$$[\eta] = p^{5/7}(3 \, \xi z \beta/\dot{\gamma} + 1 - z)/9.668 \tag{12.18}$$

where z is a kinetic parameter, β is related to the average life time of a link, and ξ is a friction coefficient. The predictions of the equation are in good agreement with experimental data for glass and asbestos fibers and glass flakes for $\phi_v < 0.1$. Kitano and Kataoka [19] reported good agreement between the predictions of Eq (12.18) and experimental results for vinylon fibers in silicon oil.

Quemada [27,28] proposed the following relation based on a shear rate dependent maximum packing fraction:

$$\eta_r = [1 - (\phi_v [\eta]_0 + [\eta]_\infty \, \dot{\gamma}_r^m)/2(1 + \dot{\gamma}_r^m)]^{-2} \tag{12.19}$$

where $\dot{\gamma}_r$ $[\equiv \dot{\gamma}/\dot{\gamma}_c]$ is a relative shear rate, with $\dot{\gamma}_c$ as a critical shear rate which represents an effective diffusion coefficient associated with the relaxation of the structure formed by the particles. $[\eta]_0$ and $[\eta]_\infty$ are intrinsic viscosities at $\dot{\gamma}_r = 0$ and ∞, respectively. The parameter m takes the value of 0.5 for suspensions of fibers.

The Cross equation has also been successful in describing the shear dependence of viscosity for fiber suspensions [29]:

$$(\eta - \eta^\infty)/(\eta^0 - \eta^\infty) = [1 + (k\dot{\gamma})^n]^{-1} \tag{12.20}$$

where η^0 and η^∞ are shear viscosities at $\dot{\gamma} = 0$ and ∞, respectively.

Dinh and Armstrong [30,31] developed a constitutive equation for a semiconcentrated suspension of rigid fibers in a Newtonian fluid when the condition $(1/p)^2 < \phi_v < (1/p)$ is satisfied. Jeffrey's analysis was used to describe the fiber motion. Fiber interactions were modeled by using an effective medium viscosity in the Stokes resistance coefficient, which was obtained by matching the predictions of the theory with Batchelor's [32] model for a concentrated fiber suspension in extensional flows. The stress contribution of the particles was calculated from the orientation distribution function using Batchelor's [8] equation [Eq (12.4)]. The constitutive equation is given by:

$$\underline{\sigma}_{ij} = -\eta_0 \dot{\gamma} \left\{ 1 + \frac{1}{48} \frac{n\,L^3}{\ln\,(2h/D)} : \int \frac{\ell_i\,\ell_j\,\ell_i\,\ell_j\,d\ell}{[1 + \gamma_{ij}^0 : \ell_j\,\ell_i]^{3/2}} \right\} \tag{12.21}$$

with
$$\ell_i = \Gamma_{ij} \cdot \ell_{0,j}\, (\Gamma_{ij}^t \cdot \Gamma_{ij} : \ell_{0,i}\,\ell_{0,i})^{-1/2}$$

and
$$\Gamma_{ij} = \delta x_i / \delta x_j^0$$

where h is the lateral distance between particles, ℓ and ℓ_0 are the instantaneous and initial orientation vectors for a particle, and γ^0 is the Cauchy strain tensor. Γ is a deformation gradient tensor for which x_i and x^0_j are the coordinates of a fluid particle at times t and t_0, respectively. Fiber thickness is not included, and hence, when the fibers are in the shear planes they stop rotating and do not contribute to the shear stresses. Model predictions are strain-dependent functions with a steady-state orientation distribution of all fibers in the shear planes for shear flow or in the direction of strain rate for uniaxial extension. In simple shear flow with a random initial orientation distribution, the contribution of particles to the shear stresses remains constant up to a strain of approximately one, followed by a small overshoot and a power-law decrease towards the equilibrium value of zero. A sharp increase at small strains is predicted in normal stresses, followed by an almost exponential decrease to zero. In extensional flows, stresses develop much faster and finite steady-state particle stresses are predicted due to the alignment in the principal axis.

Folgar [2] proposed modeling the interactions among particles as a randomizing process over Jeffery's orbits for isolated particles. The resulting equation is a convection-diffusion equation in orientation space, which can be written in the form of a Fokker-Planck equation:

$$\partial\psi/\partial t + \nabla \cdot (\dot{\ell}\psi - D\nabla\psi) = 0 \tag{12.22}$$

Here, ψ is the orientation distribution function, ℓ is the unit vector along the axis of a fiber, and D is the diffusion coefficient. The convection in orientation space is caused by the rotation of fibers with shear couples according to Eq (12.7). The diffusion is caused by the interactions among particles, with D representing the intensity of interactions. Folgar assumed that the diffusion coefficient, D, is a linear function of shear rate. This assumption is valid only at low concentrations and it leads to steady-state properties independent of the shear rate. The Fokker–Planck equation is subject to two boundary conditions:

1)   there are no fibers in the direction of the vorticity axis, $\psi\,|_{\theta=0} = \psi\,|_{\theta=\pi} = 0$

2)   orientation distribution is a periodic function of ϕ with a period of 2π, $\psi(\phi) = \psi(\phi + 2\pi)$

and a normalization condition:

$$\int_0^\pi \int_0^{2\pi} \psi(\phi,\theta)\,\mathrm{d}\phi\,\mathrm{d}\theta = 1$$

In order to solve the partial differential equation, Folgar converted the periodic boundary condition into a symmetry or zero-flux condition. However, the periodicity condition is not necessarily a symmetry condition. The forced symmetry condition causes erroneous results, leading to negative probabilities.

Mutel and Kamal [3,4] have recently solved Eq (12.22) using a finite difference scheme which can incorporate periodic boundary condition. For high-fiber concentration the diffusion coefficient was not taken to be a linear function of the shear rate. The bulk stresses due to the fibers were calculated using Eq (12.4) with the second and fourth moments for the orientation function calculated from the explicit solution of the Fokker–Planck equation. A shear thinning viscosity function was predicted in simple shear flow with asymptotic plateaux at low and high shear rates, as shown in Fig. 12.3. Large stress overshoots for both transient shear and normal stresses were predicted and observed experimentally. The magnitude of stress overshoot was a function of the initial orientation distribution of fibers.

Non–Newtonian behavior of fiber suspensions is not restricted to the observation of a shear rate dependent viscosity function. Normal stresses [17–19], large increases in extensional viscosities [33,34], and entrance pressure drops [24,35,36] have been observed with the addition of fibers to Newtonian fluids. All non–Newtonian effects are enhanced by an increase in p, ϕ_V, or flexibility. At high shear rates the first normal stress differences have a power-law dependence on shear rate. Deviations from power-law behavior are observed only either for rigid fibers at low shear rates and high concentrations or for flexible particles. The effect of a $\Delta\phi_V$ increase in concentration is smaller at higher concentrations, probably due to the detrimental effect of crowding on orientation distribution.

Extensional viscosities measured for fiber suspensions in Newtonian fluids are one or two orders of magnitude higher than that of the suspending fluid [33,34] for $\phi_V < 0.01$. Fiber concentration, although apparently small, is too large for the dilute suspension theories to be valid, $nL^3 \ll 1$ being the criterion. However, Batchelor [32] has extended the dilute suspension theory to concentrated suspensions for which the lateral spacing between the particles, b, satisfies the condition $d \ll b \ll L$, d and L being the diameter and length of the particles, respectively. The particle contribution to the extensional viscosity is given by [32]:

$$\sigma_t = 4\,\phi_v\,p^2/9\ln\,(\pi/\phi_v) \qquad (12.23)$$

where σ_t refers to the elongational stresses caused by the particles. Mewis and Metzner [33] found good agreement between the predictions of Eq (12.23) and their experimental results. Kizior and Seyer [34] reported qualitative agreement in terms of concentration and aspect-ratio dependence, but the theoretical predictions were lower than the results based on thrust measurements.

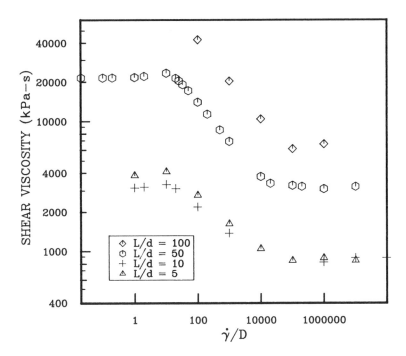

Figure 12.3 *Steady-state shear viscosities for concentrated fiber suspensions calculated using Kamal and Mutel model [4]. L/d is the aspect ratio of fibers.*

12.2.4 Suspension of Fibers in Non-Newtonian Fluids

Suspensions in non-Newtonian fluids constitute a considerable part of widely used commercial suspensions, which also include fiber-reinforced polymer melts. The interest in the rheology of non-Newtonian suspensions has been spurred by the increasing share of filled polymeric materials in the plastics industry. There is a need for a better understanding of the flow behavior and structure development during flow of these materials, in order to utilize the full potential of enhancement in material properties by the inclusion of particles. However, the theoretical treatment of the problem becomes complicated due to the additional nonlinearity brought into the flow equations. Mathematical tools developed for linear Stokes equations cannot be utilized for these systems. The effect of non-Newtonian properties of the suspending fluid enters both in the evaluation of the stress contribution from a particle with a specific configuration and in the kinematics of orientation distribution.

Some of the observed differences between suspensions in Newtonian and non-Newtonian fluids result from instantaneous, weak, but cumulative non-Newtonian contributions. In such a case, the flow analysis can be restricted to the slowly varying, weak flow asymptote and the relevant constitutive relation becomes the Rivlin-Ericksen nth-order fluid. However, analysis of O(1) non-Newtonian effects necessitates the use of a full nonlinear constitutive model for the suspending fluid. In most cases, the solution of the equations of motion can only be achieved by numerical methods and the success of the theoretical predictions is determined by the choice of the constitutive model. On the other hand, microscale flow is nonviscometric, and the lack of information on the utility of different constitutive models in nonviscometric flows offers little guidance for the choice of suitable constitutive models.

Theoretical studies on suspension mechanics in non-Newtonian fluids have been scarce. Particle motions in non-Newtonian fluids [37] and near-Newtonian fluids have been the main areas of theoretical interest. Most of the work on these materials has emphasized experimental observations in an effort to develop a qualitative understanding of the significant phenomena. Quantitative analyses have been limited to correlations using semi-empirical and empirical expressions proposed for Newtonian fluids. Utracki [38,39] and Kamal and Mutel [40] have reviewed relevant work in non-Newtonian suspension rheology.

12.3   EXPERIMENTAL OBSERVATIONS

12.3.1 Steady-Shear Viscosity

Steady-state viscosities are increased by the addition of fibers. The effect of fibers on the shear viscosity is most pronounced at low $\dot{\gamma}$ and the relative effect of fibers diminishes with increasing $\dot{\gamma}$. The viscosity function at low $\dot{\gamma}$, either approaches an asymptotic plateau [3,17,41-48] or exhibits an unbound increase with decreasing $\dot{\gamma}$ [3,41,49], depending on the aspect ratio and the concentration of fibers. Typical viscosity functions obtained for glass-fiber-filled systems are shown in Fig. 12.4. Yield stresses were calculated from modified Casson plots in cases of unbound viscosity increase at low $\dot{\gamma}$, especially for long, flexible fibers at high concentrations [50]. However, the pronounced effect of fibers at low $\dot{\gamma}$ and enhanced shear-thinning properties are reflections of changes in the structure formed by fibers under shear couples. It is possible that calculated yield stresses are apparent and that they may decrease at very low $\dot{\gamma}$, similar to observations with spherical particles [51]. Fiber-orientation distribution with respect to shear planes is shown in Fig. 12.5 for 10 wt% glass-fiber filled polypropylene at different $\dot{\gamma}$ [3,52]. The fibers are aligned closer to shear planes with increasing $\dot{\gamma}$, leading to lower viscous dissipation and shear thinning properties at low $\dot{\gamma}$. At high $\dot{\gamma}$, the structural changes are exhausted with the majority of fibers aligned in the direction of flow and parallel to shear planes.

The structural dynamics at low $\dot{\gamma}$ are strongly affected by the concentration and p of fibers. Fibers with small p or at low concentrations usually approach an asymptotic plateau, while fibers with large p or high fiber concentrations lead to a shear thinning viscosity function even at the lowest $\dot{\gamma}$ employed. The effect of p on the shear viscosity function is shown in Fig. 12.6 for polyamide-6 melts filled with glass fibers of different p [49]. Shear viscosities increase with increasing p at all $\dot{\gamma}$. The effect of p on the viscosity function is most pronounced at low $\dot{\gamma}$. The asymptotic plateau at low $\dot{\gamma}$ disappears for p > 10.

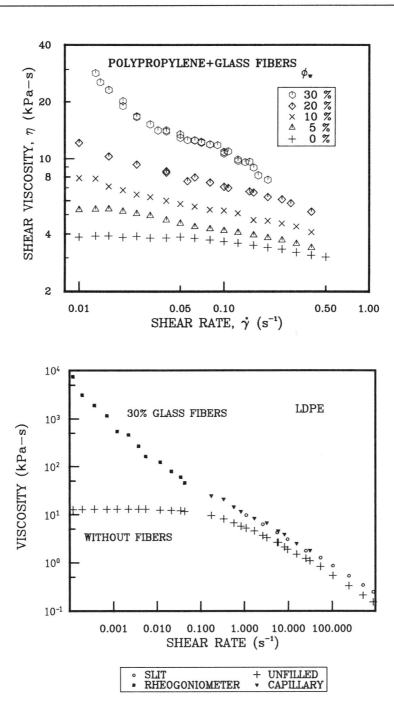

Figure 12.4 *Steady-state shear viscosities measured for glass-fiber-filled (a) polypropylene [3] and (b) polyethylene melts [49].*

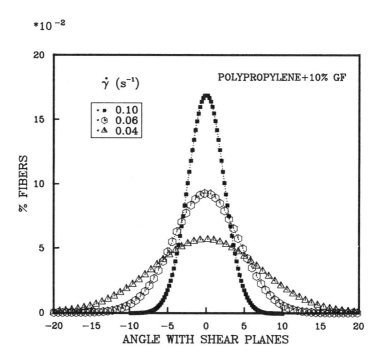

Figure 12.5 *Steady-state fiber orientation with respect to shear planes for glass-fiber-filled polypropylene melts [3].*

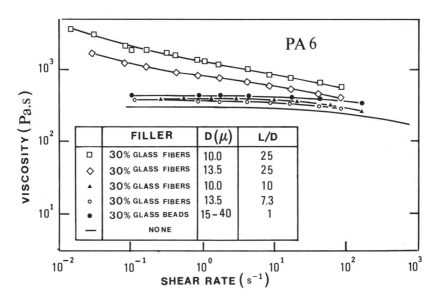

Figure 12.6 *The effect of aspect ratio on steady-state shear viscosity for glass-fiber-filled polyamide-6 melts [49].*

The effect of fiber addition on the viscosity function is best represented by the relative viscosities defined at the same $\dot{\gamma}$ or shear stress. As shown in Fig. 12.7 two regions are usually distinguished in the relative-viscosity-shear-rate curve: a decrease at low $\dot{\gamma}$ due to structural changes (changes in orientation distribution) and an asymptotic plateau at high $\dot{\gamma}$, where all the structural changes are exhausted [3,49,53]. In some cases, relative viscosities defined at the same $\dot{\gamma}$ fail to approach an asymptotic plateau at high $\dot{\gamma}$. However, relative viscosities defined at the same stress exhibit asymptotic plateaux at large stresses [18,41,46,48,50].

In the absence of structural changes it is possible to reduce the viscosities at different concentrations to a master curve, by vertical ($f_1 \cdot \eta_r$) and horizontal ($f_2 \cdot \dot{\gamma}$) shifting [3,18,44,54-57]. In most cases, $f_2 = 1/f_1$ and is equal to the asymptotic relative viscosities ($\eta_{r,\infty}$), similar to a time-temperature superposition. The reduced viscosities are shown in Fig. 12.8 for glass-fiber-filled polypropylene melts [3]. It is not possible to obtain a master curve at low $\dot{\gamma}$, where structural changes occur due to shearing. It was suggested that the particles cause a shift in the relaxation spectra of the fluid to longer times, similar to the effect of a reduction in temperature [3,43,58,59].

Structural changes during flow are irreversible, which may lead to shear history-dependent steady-state properties [3,53]. If the composite melt is previously sheared, leading to an irreversible preorientation of fibers near the shear planes and flow direction, the steady-state properties may not necessarily be the same, as compared to a sample with initially random orientation of fibers. The effect of shear history on the steady-state viscosities is shown in Fig. 12.9, where individually measured viscosities at each $\dot{\gamma}$ are compared with the viscosities measured for presheared samples at the highest $\dot{\gamma}$ employed [3]. Presheared samples exhibit lower viscosities at low $\dot{\gamma}$, where the shear-thinning properties are due to structural changes.

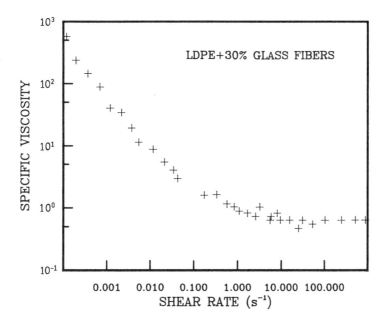

Figure 12.7 *Specific viscosities for glass-fiber-filled polyethylene melts as a function of shear rate [49].*

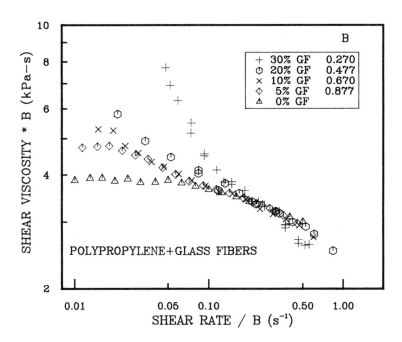

Figure 12.8 *Master curve for the viscosities of glass-fiber-filled polypropylene melts at different concentrations. B is the reciprocal of specific viscosity at high shear rate plateau [3].*

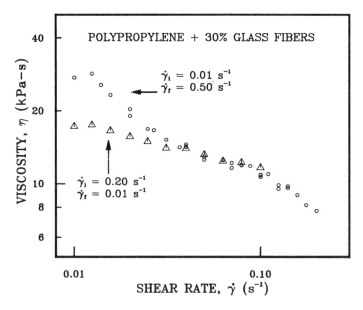

Figure 12.9 *Effect of shear history on the steady-state viscosities for glass-fiber-filled polypropylene melts [3]. $\dot{\gamma}_i$, initial shear rate and $\dot{\gamma}_f$, final shear rate.*

12.3.2 Transient Behavior under Shear

Thixotropy and rheopexy are common phenomena in suspensions and are associated with time-dependent structural changes. Fiber-filled melts exhibit complex transient behavior during flow due to superimposed viscoelastic behavior of the fluid and time-dependent changes in fiber-orientation distribution.

Large stress overshoots in shear flow and long time scales to achieve steady-state stresses are observed with fiber-filled melts compared to unfilled melts [3,49]. The peak stress is usually observed at 3-10 strain units, while steady state is achieved at 50-100 strain units. As shown in Fig. 12.10 for glass-fiber-filled polypropylene melts [3], both the peak stress and the time (or strain) to achieve steady state are increased by increasing concentration. These phenomena are associated with slow changes in fiber-orientation distribution during flow. Time scales associated with structural changes are the same as those for steady state of rheological properties. The effect of fiber orientation on the shear flow properties was illustrated in shear creep experiments with preoriented samples [49], shown in Fig. 12.11. Fibers oriented perpendicular to the shear planes exhibit the largest transient stresses, while the effect of orientation relative to flow direction is minimal if the fibers are oriented close to the shear planes.

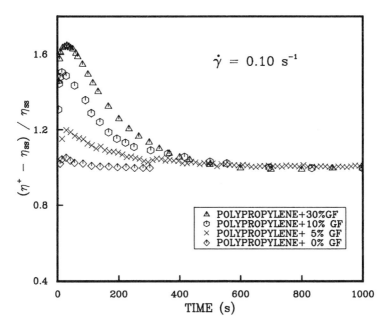

Figure 12.10 *Effect of fiber concentration on the transient shear viscosity function for glass-fiber-filled polypropylene melts [3].* η^+ *and* η_{SS} *are transient and equilibrium viscosities, respectively.*

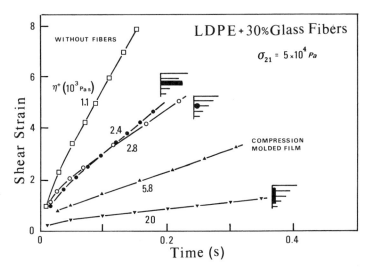

Figure 12.11 *Effect of initial fiber orientation on the shear creep of glass-fiber-filled polyethylene melts [49].*

For filled polymer melts, transient first normal stress differences also exhibit large stress overshoots, unlike unfilled polymer melts. The stress overshoot is again a reflection of changes in the anisotropic fiber structure; it disappears if the fibers are preoriented by preshearing until the structural changes are exhausted (shown in Fig. 12.12) [52].

The aspect ratio of fibers is another important parameter in the formation of structure and is expected to have a strong effect on transient properties. Stress overshoots and the time scales associated with steady state are both increased with increasing aspect ratio as shown in Fig. 12.13 for glass-fiber-filled polyamide-6 melts [49].

Transient properties may depend on the flow geometry due to the topological constraint imposed on the maximum orientation angle, as a result of comparable flow dimensions and fiber length. The effect of flow geometry on transient behavior is shown in Fig. 12.14 for glass-fiber-filled polypropylene melts with different geometries [3,52,53]. Steady-state properties are not as amenable to topological constraints due to the orientation of fibers close to the shear planes (parallel to the geometrical constraint). The steady-state properties are the same with different flow geometries.

Structural changes during flow are irreversible due to negligible restoring forces for the particle considered sizes. Interrupted shear results for glass-fiber-filled polypropylene melts are shown in Fig. 12.15 [3]. The sample was first subjected to shear at $\dot{\gamma} = 0.2$ s^{-1} until equilibrium was reached. The same sample was subsequently sheared at the same rate after rest periods of 15 and 30 min. Stress overshoot due to structural effects was not recovered even after 30 min. Similar results were obtained by Laun [49] for glass-fiber-filled polyamide-6 melts, as shown in Fig. 12.13. After a total shear strain of 7, the shearing was stopped until the shear stress was relaxed completely. The shear stresses attained the same value within 0.5 strain units. This shows that the orientation of the fibers was not changed during the rest period.

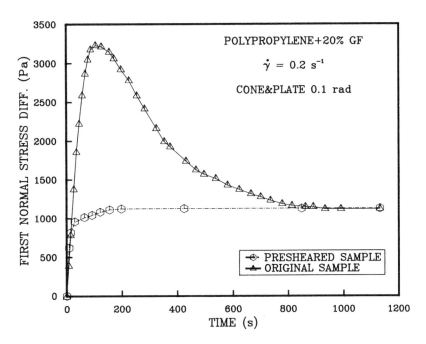

Figure 12.12 *Effect of preshear on the transient first normal stress difference for glass-fiber-filled polyprolene melts [53].*

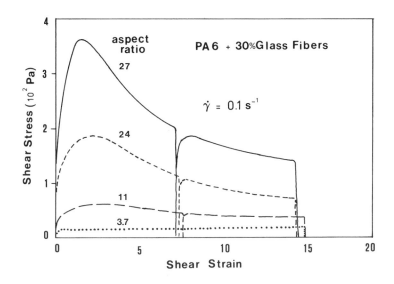

Figure 12.13 *Effect of aspect ratio on transient shear behavior of fiber-filled polyamide-6 melts [49].*

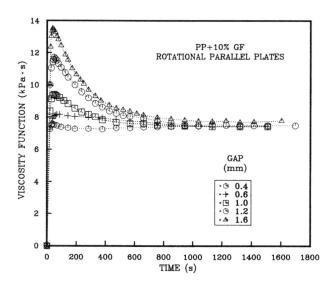

Figure 12.14 *Effect of gap size on the transient viscosity function for glass-fiber-filled polypropylene melts in rotational parallel plate geometry [3].*

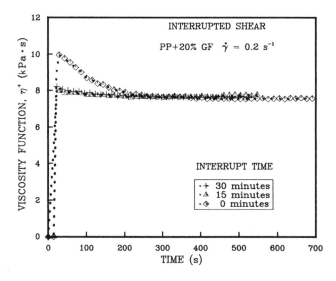

Figure 12.15 *Interrupted shear results for glass-fiber-filled polypropylene melts in cone-and-plate geometry [3].*

12.3.3 Normal Stresses

First normal stress difference, N_1, increases by the addition of fibers when compared at the same shear rate or shear stress [3,42,44,46,47], as shown in Fig. 12.16. This indicates that the effect of fibers on N_1 is larger than their effect on the shear stresses:

$$N_1 = f(\phi_v, p)\, \sigma^m \qquad (12.24)$$

where $m > 0$, $df/d\phi_v > 0$ and $df/dp > 0$. If the fibers are not rigid, a deviation from power-law behavior is observed.

The effect of fibers on the first normal stress difference decreases with increasing shear rate (or stress). In some cases, an approach to an asymptotic plateau was observed at low shear rates [3].

12.3.4 Small-Amplitude Oscillatory Shear

Small-amplitude oscillatory shear measurements yield information regarding the structural changes in suspensions. The stress response measured for suspensions of small particles with strong interparticle forces is highly dependent on the strain amplitude. It is also anharmonic,

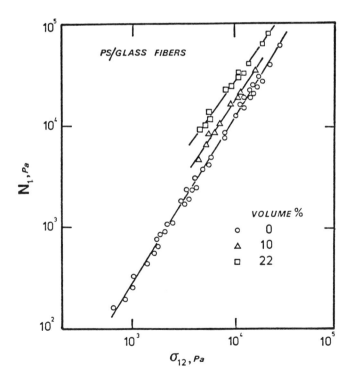

Figure 12.16 First normal stress difference for glass-fiber-filled polystyrene melts [44].

except at very small strain amplitudes, $\gamma_0 < 0.005$-0.010 [60]. The effect of γ_0 on the dynamic properties of fiber suspensions in viscoelastic fluids was not considered in the few available studies [41,61-63]. The tests were carried out with an arbitrarily chosen small strain amplitude ($< 4\%$). Mutel and Kamal [53], however, observed a strong strain dependence of the dynamic properties for fiber-filled melts, especially at low $\gamma_0 < 10\%$. However, the stresses measured were harmonic except at frequencies, $\omega < 0.1$ rad/s [3,53]. Anharmonicity is usually associated with structural changes for which the relaxation times are on the order of the characteristic time scale for the flow. Transient shear behavior of fiber-filled melts indicates long relaxation times associated with structural changes. Hence, the structural changes are arrested at relatively short time scales for the flow, leading to a harmonic stress response. Anharmonicity is observed at low frequencies where the time scales for the flow and the structural changes are comparable. Harmonic stresses allow the definition of material response by two parameters, which are functions of frequency and strain amplitude, e.g., $G'(\omega, \gamma_0)$ and $G''(\omega,\gamma_0)$.

Loss and storage moduli usually increase with the addition of fibers, indicating that the fibers enhance both the viscous dissipation and the elastic energy storage for the system. Second plateaux for G' were observed at high concentrations and with long fibers [41,62]. Dynamic viscosities measured for fiber-filled systems were not comparable with the steady shear viscosities at the same shear rate (or frequency), although dynamic and steady shear viscosities for the base resins were found to be similar. The structural changes occurring during steady shear did not occur in dynamic due to short characteristic time for the flow. Hence, the measured properties reflected the properties associated with different structures. Consequently, the dynamic measurements can be sensitive to the shear history of the material in cases where no effect of shear history is expected under steady shear [3].

12.3.5 Extensional Flows

There are few studies on the elongational flow behavior of fiber-filled polymer melts [3,47,49,64]. Polymer melts in elongational flows exhibit extension-rate-independent transient viscosities at small strains. The transient viscosities usually deviate from this envelope at large strains. Fiber-filled polymer melts exhibit strain rate dependent transient behavior even at very small strains [47,64]. Equilibrium viscosities are achieved with fiber-filled polymer melts at lower strain units compared to unfilled melts. Equilibrium viscosities were observed for fiber-filled melts at strain rates at which no equilibrium value was achieved for unfilled melts within the maximum strain range attainable [46,61]. However, the strain at sample failure is much lower for fiber-filled melts due to increased imperfections caused by the inclusions.

Equilibrium elongational viscosities are substantially increased by the addition of fibers, even at low fiber concentrations [47,49,64]. The effect has been predicted by the suspension theories [32,65].

An order of magnitude increase in elongational viscosities (ten to fifty times) has been reported for fiber concentrations of commercial interest (20-40 wt%) [64], as shown in Fig. 12.17. The substantial increase in elongational stresses is due to the orientation of fibers in the principal direction of extension. The equilibrium viscosities decrease with increasing strain rate, even at the lowest rates employed, where unfilled polymer melt viscosities are independent of the applied strain rate. Strain rate thinning with filled melts is the result of microscale shear flows between fibers.

Goddard [65-67] extended Batchelor's [32] analysis for the particle stress in extensional flows of fiber suspensions in Newtonian fluids to the case of fiber suspensions in a

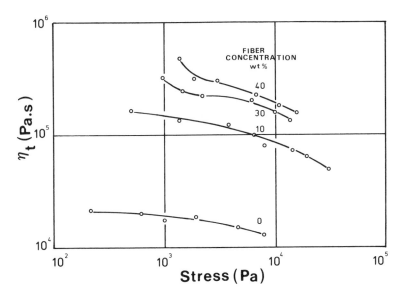

Figure 12.17 *Equilibrium elongational viscosities for glass-fiber-filled polypropylene melts [64].*

non-Newtonian fluid whose properties are described by the power-law model. The particle contribution to tensile stresses in a power-law fluid is greatly diminished compared to the Newtonian case. The intrinsic particle stress for dilute suspensions is given by [67]:

$$[\sigma_t] = \frac{\sigma - \sigma_0}{\sigma_0 \; \phi_v} = (2/9) \, p^{1+1/m} \, (m-1)^{1/m} \, B \, [1 + 0(p^{1/m-1} \, \ln p)] \tag{12.25}$$

and for concentrated suspensions:

$$[\sigma_t] = (2/9) \, p^{1+1/m} \left[\frac{m-1}{1-\beta^{1-m}} \right] B \tag{12.26}$$

with $B = [9/(2 + 1/m)] \, [\tau_0(\dot{\epsilon})/\sigma_0(\dot{\epsilon})]$, where σ is the tensile stress for the suspension at strain rate $\dot{\epsilon}$, σ_0 is the tensile stress in the suspending fluid with no fibers at $\dot{\epsilon}$, m is the reciprocal of the power-law index for the suspending fluid, and β is a spacing parameter. Comparison with Chan et al. [47] results gave theoretical/experimental value ratios of ≈ 1 to 4. White and Czarnecki [68] suggested the use of a different definition for the calculation of average aspect ratio which improved the agreement between theoretical predictions and experimental results.

12.3.6 Extrudate Swell and Entrance Effects

Extrudate swell, which is usually attributed to the elastic properties of viscoelastic materials, decreases with the addition of fibers to polymer melts. The swell ratio decreases with increasing concentration and aspect ratio [56,57,69-72]. The origin of the apparent contradiction between the effect of particles on the normal stresses and extrudate swell behavior is not clear.

There are few experimental data available on entrance-pressure losses for fiber-filled polymers in capillary or slit flow, although Poiseuille flow has been extensively utilized in the characterization of filled systems. In general, the entrance-pressure drop increases by the addition of fibers [36,56,65,70-74]. The effect of fibers is increased with an increase in the aspect ratio [56].

12.4   CONCLUSIONS

Bulk rheological behavior of fiber-filled systems is governed by the dynamics of internal structure created by the particles. The transient and steady-state rheological behavior of filled systems is a manifestation of structural changes with time or deformation rates, like large stress overshoots or enhanced shear thinning. The evolution of the structure is specific to the kinematics of the flow considered; it is irreversible and is associated with long time scales. Fibers, in general, increase the viscous dissipation and elastic properties and the effect of fibers is more pronounced in elongational than in shear flows. The effect of fibers is enhanced with increasing concentration, aspect ratio, and flexibility of the fibers and is most pronounced at low shear rates.

Available suspension theories address the strong relation between the structural dynamics and the bulk rheological properties. However, the theories so far developed are limited to either very dilute suspensions or concentrated suspensions in specific flows. Furthermore, the solutions are limited to Newtonian fluids. Consequently, semi-empirical relations for concentrated suspensions have been used to correlate the experimental results. Recently, the interaction effects have been included in terms of effective parameters or as randomizing processes for suspensions in Newtonian fluids.

12.5   OUTLOOK

The above discussion suggests that substantial progress has been made toward understanding the rheological and flow behavior of fiber-reinforced polymer melts. This understanding should be helpful in the design of composite products and relevant processing equipment and in the optimization of processing conditions. However, much remains to be done for obtaining a more clear and concise understanding of the phenomena encountered during the flow of fiber reinforced polymer melts. Some of the important issues in need of immediate attention are outlined below:

1)   The key to understanding the rheological behavior of composite melts relates to obtaining experimental data under well-controlled and documented conditions. Information must be obtained with regard to the distributions of fibers, their orientations, and their lengths as functions of time and space. Ideally, on-line monitoring techniques should be used for this purpose. However, it is more likely that the successful approach in the foreseeable future will be based on off-line techniques, thus requiring laborious and extensive experiments.

2) Recent studies have indicated a strong dependence of transient rheological behavior on flow dimensions, which are frequently of the same order of magnitude as fiber dimensions. This is especially true for long, discontinuous fibers, which are gaining commercial importance. Studies relating to the effect of flow dimensions on transient rheological and flow behavior would be invaluable.

3) The effect of strain amplitude on dynamic properties under oscillatory shear has not been extensively studied. Dynamic properties may lead to useful information on the state and dynamics of internal structure.

4) There is a need for rheological models that address the relationship between the internal structure and the bulk behavior explicitly. The incorporation of wall effects and the effect of non-Newtonian properties into the model is needed for more realistic predictions.

NOMENCLATURE

C	Orbit constant
D	Diffusion coefficient
E	Strain rate tensor
G'	Storage shear modulus
G''	Loss shear modulus
h	Lateral distance between particles
L	Fiber length
K	nth-order tensor for the description of microstructure
$\dot{K}$	Time derivative of K
N_1	First normal stress difference
n	Number of fibers
p	Aspect ratio
p_e	Equivalent aspect ratio
t	Time
U	Velocity vector
V	Volume
γ^0	Cauchy strain tensor
γ_0	Strain amplitude
$\dot{\gamma}$	Shear rate
$\dot{\varepsilon}$	Strain rate in uniaxial extension
η^∞	High shear rate asymptote for shear viscosity
η^0	Low shear rate asymptote for shear viscosity
η	Shear viscosity
η_0	Shear viscosity of the suspending fluid
$[\eta]$	Intrinsic shear viscosity, $(\eta - \eta_0)/\eta_0 \, \phi_v$ at infinite dilution
η_r	Relative shear viscosity, η/η_0
η_{sp}	Specific viscosity, $\eta_r - 1$
θ	Spherical angle describing the orientation of a fiber
ξ	Friction coefficient
σ	Stress tensor

σ_0	Stress contribution of the suspending fluid
σ_p	Stress contribution of the particles
σ_t	Elongational stress
τ	Relaxation time
ϕ	Spherical angle describing the orientation of a fiber
ϕ_v	Volume fraction of particles
ψ	Orientation distribution function
ω	Frequency
Ω	Vorticity tensor
ℓ	Unit vector describing the orientation of a particle
$\wp$	Probability distribution function

REFERENCES

1.   Dinh, S. M., Armstrong R. C., *J. Rheol.*, **28**, 593 (1984).
2.   Folgar, P. F., *PhD thesis*, University of Illinois at Urbana Champaign (1983).
3.   Mutel A. T., *PhD thesis*, McGill University (1989).
4.   Kamal, M. R., Mutel, A. T., *Polym. Compos.*, **10**, 337 (1989).
5.   Leal, L.G., *Proc. VIIth Inter. Congress Rheology*, p. 392, Gothenburg, Sweden, Aug. (1976).
6.   Chung, B., Cohen, C., *J. Non−Newtonian Fluid Mech.*, **25**, 289 (1987).
7.   Cohen, C., Chung, B., Stasiak, W., *Rheol. Acta*, **26**, 217 (1987).
8.   Batchelor, G. K., *J. Fluid Mech.*, **41**, 545 (1970).
9.   Batchelor, G. K., *Ann. Rev. Fluid Mech.*, **6**, 227 (1974).
10.   Jeffery, G. B., *Proc. Roy. Soc. (London)*, **A102**, 161 (1922).
11.   Goldsmith, H. L., Mason, S. G., *Rheology Theory and Applications*, Vol. 4., Eirich, F. D., Ed., Academic Press, New York (1967).
12.   Eisenschitz, R., *Z. Physik. Chem.*, **A158**, 85 (1971).
13.   Burgers, J. M., *The Second Report on Viscosity and Plasticity*, Nordemann, New York (1938).
14.   Guth, E., *Phys. Rev.*, **53**, 926A (1938).
15.   Nawab, M. A., Mason, S. G., *J. Phys. Chem.*, **62**, 1248 (1958).
16.   Blakeney, W. R., *J. Coll. Interf. Sci.*, **22**, 324 (1966).
17.   Goto, S., Nagazono, H., Kato, H., *Rheol. Acta*, **25**, 119 (1986).
18.   Schroder, R., *Rheol. Acta*, **25**, 130 (1986).
19.   Kitano, T., Kataoka, T., *Rheol. Acta*, **20**, 390 (1981).
20.   Simha, R., *J. Res. Natl. Bur. Std.*, **42**, 409 (1949).
21.   Brodnyan, J. G., *Trans. Soc. Rheol.*, **3**, 61 (1959).
22.   Simha, R., *J. Phys. Chem.*, **44**, 25 (1940).
23.   Yang, J. T., *J. Am. Chem. Soc.*, **80**, 1783 (1958).
24.   Maschmeyer, R. O., Hill, C. T., *Adv. Chem. Ser.*, **134**, 95 (1974).
25.   Maschmeyer, R. O., Hill, C. T., *Trans. Soc. Rheol.*, **21**, 195 (1977).
26.   Ziegel, K. D., *J. Colloid Interface Sci.*, **34**, 185 (1970).
27.   Quemada, D., *Rheol. Acta*, **17**, 632 (1978).
28.   Quemada, D., *Rheol. Acta*, **17**, 643 (1978).
29.   Cross, M. M., *J. Colloid Sci.*, **20**, 417 (1965).
30.   Dinh, S. M., Armstrong, R. C., *Proc. 2nd World Congress of Chemical Engineers*, **6**, 297 (1981).

31. Dinh, S. M., Armstrong, R. C., *J. Rheol.*, **28**, 593 (1974).
32. Batchelor, G. K., *J. Fluid Mech.*, **46**, 813 (1971).
33. Mewis, J., Metzner, A. B., *J. Fluid Mech.*, **62**, 593 (1974).
34. Kizior, T. E., Seyer, F. A., *Trans. Soc. Rheol.*, **18**, 271 (1974).
35. Miles, J. N., Murty, N. K., Modlen, G. F., *Polym. Eng. Sci.*, **21**, 1171 (1981).
36. Goto, S., Nagazono, H., Kato, H., *Rheol. Acta*, **25**, 246 (1986).
37. Leal, L. G., *J. Non−Newtonian Fluid Mech.*, **5**, 33 (1979).
38. Utracki, L. A., Fisa, B., *Polym. Compos.*, **3**, 193 (1982).
39. Utracki, L. A., *Rubber Chem. Tech.*, **57**, 507 (1984).
40. Kamal, M. R., Mutel, A. T., *J. Polym. Eng.*, **5**, 293 (1985).
41. Kitano, T., Funabashi, M., Klason, C., Kubat, J., *Intl. Polym. Proces.*, **3**, 67 (1988).
42. White, J. L., Czarnecki, L., Tanaka, H., *Rubb. Chem. Technol.*, **53**, 823 (1980).
43. Nicodemo, L., Nicolais, L., *Polymer*, **15**, 589 (1974).
44. Czarnecki, L., White, J. L., *J. Appl. Polym. Sci.*, **25**, 1217 (1980).
45. Kitano, T., Kataoka, T., *Rheol. Acta*, **19**, 753 (1980).
46. Kitano, T., Kataoka, T., Nagatsuka, Y., *Rheol. Acta*, **23**, 20 (1984).
47. Chan, Y., White, J. L., Oyanagi, Y., *J. Rheol.*, **22**, 507 (1978).
48. Hinkelmann, B., Mennig, G., *Chem. Eng. Commun.*, **36**, 211 (1985).
49. Laun, H. M., *Colloid Polym. Sci.*, **262**, 257 (1984).
50. Kitano T., Kataoka, T., *Rheol. Acta*, **20**, 403 (1981).
51. Matsumoto, T., Hitomi, C., Onogi, S., *Trans. Soc. Rheol.*, **19**, 541 (1975).
52. Mutel, A. T., Kamal, M. R., *SPE Techn. Pap.*, **33**, 732 (1987).
53. Mutel, A. T., Kamal, M. R., *SPE Techn. Pap.*, **32**, 679 (1986).
54. Kitano, T., Kataoka, T., Nishimura, A. T., Sakai, T., *Rheol. Acta*, **19**, 541 (1980).
55. Nicodemo L., Nicolais, L., *J. Appl. Polym. Sci.*, **18**, 2809 (1974).
56. Oyanagi Y., Yamaguchi, Y., *J. Soc. Rheol. Japan*, **3**, 64 (1975).
57. Knutsson, B. A., White, J. L., *J. Appl. Polym. Sci.*, **26**, 2347 (1981).
58. Masi, P., Nicodemo, L., Nicolais, L., Taplialatela, G., *Rheol. Acta*, **21**, 598 (1982).
59. Nicodemo L., Nicolais, L., *J. Appl. Polym. Sci.*, **18**, 2809 (1974).
60. Onogi, S., Matsumoto, T. T., *Polym. Eng. Rev.*, **1**, 46 (1981).
61. Onogi, S., Mikami, Y., Matsumoto, T., *Polym. Eng. Sci.*, **17**, 1 (1977).
62. Kitano, T., Kataoka, T., Nagatsuka, Y., *Rheol. Acta*, **23**, 408 (1984).
63. Mills, N. J., *J. Appl. Polym. Sci.*, **15**, 2791 (1971).
64. Kamal, M. R., Mutel, A. T., Utracki, L. A., *Polym. Compos.*, **5**, 289 (1984).
65. Goddard, J. D., *J. Non−Newtonian Fluid Mech.*, **1**, 1 (1976).
66. Goddard, J. D., *J. Fluid Mech.*, **78**, 177 (1976).
67. Goddard, J. D., *J. Rheol.*, **22**, 615 (1978).
68. White, J. L., Czarnecki, L., *J. Rheol.*, **24**, 501 (1980).
69. Münsted, H., *Proc. VIIth Inter. Congress Rheology*, Gothenburg, Sweden, Aug. (1976).
70. Yarlykov, B., *Inter. Polym. Sci. Tech.*, **4**, T/7 (1977).
71. Crowson R. J., Folkes, M. J., *Polym. Eng. Sci.*, **20**, 934 (1980).
72. Wu, S., *Polym. Eng. Sci.*, **19**, 638 (1979).
73. Chan, W. W., Charrier, J. M., Vadnas, P., *Polym. Compos.*, **4**, 9 (1983).
74. Han, C. D., *Polym. Eng. Rev.*, **1**, 363 (1981).

CHAPTER 13

THERMOPLASTIC MATRIX-SHEET COMPOSITES BY SLURRY DEPOSITION

by D. F. Hiscock and D. M. Bigg

BATTELLE
505 King Avenue
Columbus, OH 43201-2693
U.S.A.

The use of papermaking technology for producing long-fiber-reinforced thermoplastic matrix composites is described. This process consists of filtering a mixture of a polymeric powder and reinforcing fibers from an aqueous slurry, and then consolidating the resulting mat by the application of heat and pressure. An important factor in the implementation of this process is the means to insure homogeneity of the mixture and the ability to handle the dried sheet prior to consolidation. Several techniques have been developed to provide composite homogeneity and sheet integrity, including the use of binders, foam stabilizers, and the addition of pulp fibers to the slurry. In order to maximize the mechanical properties of the composite the sheet must be thoroughly dried prior to consolidation. The tensile properties and the degree of reinforcement efficiency of composites produced by the slurry deposition process have been shown to be comparable to those produced by melt-impregnation techniques.

13.1   INTRODUCTION

The need for strong lightweight structures has led to the considerable growth in the development and application of polymer–matrix composites over the past 20 years. This trend is expected to continue well into the next century. Polymer–matrix composites are commonly divided into two major classifications: conventional and advanced. Conventional composites are considered to be those utilizing low-temperature polymers that have been reinforced with glass fibers, usually of an E-glass composition. Advanced composites consist of higher-temperature polymers and are most commonly reinforced with carbon fibers. A frequently used temperature guideline separating the two types of composites is 150°C.

For both conventional and advanced composites the predominant matrix resins are thermosetting polymers. There are both advantages and disadvantages associated with the use of thermosetting polymers in composites. Prior to curing, thermosetting resins are usually low-viscosity liquids. This facilitates impregnation of the resin into the fiber bundles insuring adequate wetting and adhesion of the polymer to the fibers. The composite prepreg structure can be prepared either manually or by semiautomated processes at room temperature. This has proven to be the most economical method for low production runs of complex structures. The properties of thermoset matrix composites have been found to be quite adequate for most applications.

However, thermosetting polymers are not without problems. Since they are chemically reactive, the resins have a limited shelf life, the properties of the composite can vary throughout the molded structure, as well as from part to part, and the overall time required to layup and cure the composite is often longer than is economically desirable. Long cycle times reduce the economic competitiveness of composites in high-volume applications. There are also instances in which the properties of thermoset matrix composites are inadequate. This is particularly true when impact strength is of particular concern.

Thermoplastic matrix composites have been shown to provide comparable mechanical properties to thermoset matrix composites, with the addition of improved impact resistance [1,2]. Thermoplastic polymers offer the potential for producing composites with improved chemical resistance, closer part-to-part reproducibility, and more uniform homogeneity within a part than thermoset matrix composites. After thermoplastic composites have been brought to the prepreg stage they can be fabricated into parts at a much faster rate than thermoset composites. Scrap recovery, while still difficult, is much more feasible with thermoplastic polymers than thermosetting polymers.

The primary reason that thermoplastic matrix composites have not achieved widespread use is that the prepreg forming processes have proved to be more complex than those used to manufacture thermosetting matrix composites. Thermoplastic polymers must be processed at elevated temperatures ($> 200\,^{\circ}$C), and require high pressures (> 7 MPa) to pump the high viscosity molten polymer into the reinforcing fibers. These conditions require the use of automated process equipment except for the production of a few small parts. Large-scale automated equipment is economically justified only for high-volume applications.

There are several processes that have been investigated for producing long-fiber-reinforced thermoplastic matrix composite prepregs. These processes include solvent impregnation [3], film stacking [4], melt impregnation [5], fluidized bed impregnation of expanded fiber bundles [6], and the deposition of a polymeric powder/fiber mixture from a water-based slurry [7-9]. Solvent impregnation is undesirable because of the solvent handling problems. Film stacking is primarily a laboratory technique. The remainder of these processes are either in commercial use or at the developmental stage.

Of the commercial processes AZDEL^{TM} sheet, produced by the melt impregnation of a nonwoven glass-fiber mat with molten polypropylene, is the most advanced [5]. This process has been recently expanded to produce composite sheets with polymers other than polypropylene. Both glass-fiber- and carbon-fiber-reinforced polyphenylenesulfide composite sheets are commercially available from Phillips Petroleum Company by a melt impregnation process [10]. Carbon-fiber-reinforced polyetheretherketone prepregs for advanced composite applications are also produced by melt impregnation technology [11].

A fluidized bed process has been developed in which very-small-diameter polymer particles are suspended among the fibers of an expanded continuous fiber tow [6]. After the powder infiltrates the fibers, the tow is constricted, trapping the powder, and the impregnated tow encapsulated by an extruded coating of the polymer. The encapsulated tows can be used to produce woven or nonwoven prepregs which are then consolidated into sheets by the application of heat and pressure.

Several slurry-deposition processes have been developed that produce composite sheets of long discontinuous fiber-reinforced thermoplastic polymers. While the slurry-deposition processes have several advantages relative to the aforementioned processes, one in particular stands out high production rates at low processing costs. In this chapter the technology behind the development of this processing technique will be discussed, and the properties of the resulting composites compared with those made by conventional polymer-processing techniques.

13.2 THEORY

13.2.1 Tensile Strength

The tensile strength of a fiber-reinforced composite can be estimated with reasonable success from the modified rule of mixtures [12]:

$$\sigma_c = \sigma_p\, \phi_p + \sigma_f\, \phi_f\, e_I\, e_0 \tag{13.1}$$

where σ and ϕ are, respectively, tensile strength and volume fraction, e is efficiency factor, subscripts c, p, and f refer to the composite, polymer, and fiber, respectively, and subscripts o, I refer to orientation and matrix–fiber interaction.

If the composite consists of unidirectionally oriented fibers, e_0 is 1.0 in the direction of fiber orientation. Composites containing fibers that are randomly dispersed in the plane of the sheet have a value of e_0 equal to approximately 0.33. The value of e_I is more difficult to quantify because it is strongly affected by the adhesion between the polymer and fibers. The degree of adhesion between the polymer and fibers depends on a multitude of factors, including the effectiveness of coupling agents, polarity of the polymer, and length of the fibers. For continuous fibers e_I is 1.0. For discontinuous fibers e_I is related to the critical aspect ratio for the fiber–matrix combination. The critical aspect ratio occurs when the strength of the interfacial adhesion between the polymer and fiber equals the tensile strength of the fiber. The equation used to describe the relationship between the critical fiber aspect ratio and polymer–fiber adhesion is [12]:

$$(L/D)_{cr} = \sigma_f / 2\tau_I \tag{13.2}$$

where σ_f is the tensile strength of the fiber, and τ_I is the interfacial strength between the fiber and matrix.

The interfacial strength between a fiber and the matrix is strongly influenced by the use and effectiveness of coupling agents. The relationship between e_I and the length of discontinuous fibers in a composite has been presented in graphic form similar to that shown in Fig. 13.1. This is a generalized plot typical of those presented for many polymer–fiber combinations. Figure 13.1 shows that for randomly oriented fiber-reinforced composites the maximum degree of reinforcement is reached when the average fiber aspect ratio is at least ten times the critical aspect ratio. For many polymer–fiber combinations the critical aspect ratio is between 20 and 50 [12–14]. Therefore, maximum reinforcement in a discontinuous fiber-reinforced composite is achieved when the fibers have a minimum average aspect ratio of 500.

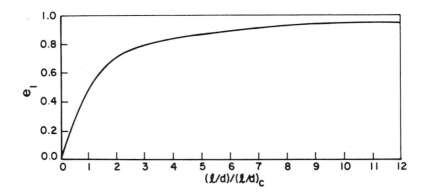

Figure 13.1 *Reinforcement efficiency versus fiber aspect ratio relative to the critical fiber aspect ratio.*

13.2.2 Modulus

The theoretical relationship used to estimate the modulus of a short-fiber-reinforced composite depends on whether or not the fibers can be considered long relative to the smallest dimension of the molded part. For such fillers as mineral fibers and reinforcing flakes, where the average long dimension of the particles is on the order of tens of micrometers the Halpin-Tsai, Kerner, and Nielsen equations can be used to estimate the modulus of the composite [15-17]. These models assume that the particulate phase is randomly dispersed in a three-dimensional manner within the matrix. This packing arrangement is possible only with particles that are small relative to the smallest dimension of the part.

For the case where the fibers are long relative to the thickness of the sheet, and as a result cannot be randomly oriented in three dimensions, they will assume a predominantly two-dimensional packing arrangement. In this instance a relationship similar to that used to describe the tensile strength of composites can be used to estimate the modulus. The modified rule of mixtures equation for modulus enhancement is [12]:

$$E_c = E_p \, \phi_p + E_f \, \phi_f \, e_E \tag{13.3}$$

where e_E is an efficiency factor that depends on many of the same factors as the efficiency factors used in the modified rule of mixtures equation for tensile strength. However, there are significant differences in the application of the rule of mixtures to the determination of modulus and tensile strength. Specifically, while the tensile strength of the composite depends on the degree of adhesion between the fibers and matrix for short fibers, this is not the case for modulus enhancement. The effect of interfacial adhesion on modulus enhancement is considerably less than it is on tensile strength. This is evident from the fact that nonreinforcing fillers will enhance the modulus, but not the strength of polymers. The orientation of the fibers has a very strong influence on tensile strength, but a lesser effect on tensile modulus. For example, if the reinforcement is in the form of a woven fabric, the weft-direction fibers do not contribute to the strength of the composite in the warp direction, but they do contribute to the modulus. Therefore, the efficiency factor in Eq (13.3) will have no numerical relationship to the efficiency factors in Eq (13.1).

13.2.3 Impact Strength

Theoretical developments capable of predicting the impact-strength improvement of composites from an engineering point of view are not as well developed as models predicting tensile strength and modulus. In essence, the impact strength of a material represents its ability to absorb energy prior to failure. The impact strength of composites and unreinforced polymers depends strongly on the testing procedure, rate of impact, shape of the impacting implement, degree and form of crystallinity (if any), and the existence of microdefects in the vicinity of the impact. One way of representing impact strength is the energy required to fracture a sample in a high-speed tensile test [18]. This energy is the area under the stress-strain curve. Mathematically, the area under a stress-strain curve can be represented by the following equation:

$$A = \int \sigma(\gamma) \, d\gamma \tag{13.4}$$

where A is the energy absorbed during impact, $\sigma(\gamma)$ is the stress as a function of strain, γ.

Qualitatively, the shape of a stress-strain curve changes as a polymer is reinforced. Many "tough" polymers have peak stresses that are relatively low compared with that of composites or metals, but an elongation to break that is quite high. This results in very high energy absorption because of the ability of the material to deform in response to an impact. As the polymer is reinforced its elongation to break drops sharply, but the peak stress it is capable of withstanding significantly increases. An increase in impact strength results if the increase in energy absorption associated with the increase in strength exceeds the reduction in energy absorption associated with the reduction in the elongation to break. For that reason many "tough" polymers lose some of their impact resistance when reinforced, because the increase in tensile peak strength does not offset the reduction in elongation to break. On the other hand many "brittle" polymers show an increase in impact strength when reinforced because, while the tensile strength increases sharply, the elongation to break is not significantly reduced.

The maximum amount of energy absorbed occurs when all of the fibers equal the critical fiber aspect ratio [12]. At this point the maximum amount of energy is expended in pulling the fibers from the matrix. Unfortunately, the tensile strength enhancement of the composite is far from maximized in this condition.

In addition to these factors, which are present in all impact tests, the effect of fiber orientation, fiber aspect ratio, fabric construction, and interfacial adhesion affect the impact strength of composites.

13.3   MATERIALS

13.3.1 Polymers

Thermoplastic polymers are conveniently divided into two classes; semi-crystalline and amorphous. Tables 13.1 and 13.2 contain pertinent data for many common polymers. The polymers listed in the two tables are organized according to increasing melting point (semicrystalline polymers) or increasing glass transition temperature (amorphous polymers). While any thermoplastic matrix polymer can be used to produce long-fiber-reinforced composites, only a few polymers have been investigated in such composites. These include polyvinylchloride, polypropylene, polybutyleneterephthalate, polyethyleneterephthalate, polyamide-6, polyphenylenesulfide, and polyetheretherketone. Polypropylene and polyvinylchloride are low-cost commodity polymers that, if suitably reinforced, can compete with structural materials where high temperatures are not experienced. Polybutyleneterephthalate, polyethyleneterephthalate, and polyamide-6 are modestly priced engineering resins which have higher-temperature utility than the lower-priced commodity thermoplastics. Polyphenylenesulfide and polyetheretherketone are more expensive polymers whose chemical and thermal resistance make them acceptable in demanding applications. All of these polymers are semicrystalline, although the degree of crystallinity of polyvinylchloride is low enough that it has a negligible effect on the properties of the polymer.

The polymers used as matrices in this investigation included two semicrystalline polymers; polypropylene (PP) and polybutyleneterephthalate (PBT). Both polymers were cryogenically ground to powders with average particle diameters of approximately 100 μm.

Table 13.1 Properties of Semi−crystalline Polymers

Polymer	Code	Density kg/m^3	Strength MPa	Modulus GPa	T_g °C	T_m °C	HDT °C
High density polyethylene	HDPE	960	34	1.23	-20	132	55
Polypropylene	PP	910	31	1.50	-10	176	60
Polyvinylidenefluoride	PVF$_2$	1780	50	2.90	-30	178	115
Polyoxymethylene	POM	1420	68	3.56	-65	180	125
Polybutyleneterephthalate	PBT	1320	52	2.50	40	228	85
Polyamide-6	PA-6	1140	79	1.20	65	228	85
Polyethyleneterephthalate	PET	1380	70	3.10	80	255	40
Polyamide-6,6	PA-6,6	1140	82	1.20	78	260	75
Polyphenylenesulfide	PPS	1340	83	3.28	93	288	135
Polyetheretherketone	PEEK	1360	95	3.85	143	338	160

Table 13.2 Properties of Amorphous Polymers

Polymer	Code	Density kg/m^3	Strenght MPa	Modulus GPa	T_g °C	HDT °C
Polyvinylchloride	PVC	1350	60	4.10	85	75
Polystyrene	PS	1040	80	4.10	108	95
Styrene-Acrylonitrile	SAN	1100	80	3.15	108	100
Acrylnitrile-Butadiene-Styrene	ABS	1060	48	2.40	108	105
Polyphenyleneether[a]	PPE	1060	66	2.60	145	125
Polycarbonate	PC	1200	65	2.40	150	130
Polyurethane	PU	1050	25	0.50	160	55
Polysulfone	PSO	1240	70	2.69	190	175
Polyetherimide	PEI	1330	105	2.95	217	200
Polyethersulfone	PES	1370	85	2.60	223	203
Polyimide[b]	PI	1310	102	3.76	250	232
Polyamide-imide	PAI	1400	63	4.60	275	270

[a] *General Electric Company's Noryl (referenced blend contains 50% polystyrene).*
[b] *DuPont's Avimid K−III.*

13.3.2 Fibers

The primary reinforcing material in long-fiber-reinforced composites is E-glass. Some applications utilize carbon fibers, but these are relatively low in usage. Chopped fibers are often bound together in bundles such that the individual fibers must be separated by a mixing process. A bundle usually consists of 800 individual filaments. The length of chopped fibers is usually between 3 and 12 mm. The diameter of the most common glass filament is 13.5 μm. Each filament is also sized to prevent damage to the fibers prior to incorporation into the polymer. Sizings are often coupling agents (usually silanes), starches, antistatic coatings, wetting agents, and lubricants.

Table 13.3 lists the dimensions of the glass fibers used in this investigation. The fibers were supplied by Owens Corning with surface treatments designed to improve their dispersability in water. In all cases the aspect ratios of the fibers exceeded 500.

Two types of pulp fibers were used. They were Pulpex^{TM}, a polypropylene pulp supplied by Hercules, and a Kevlar^{TM} pulp supplied by DuPont. The pulp fibers were approximately 2 mm long and had a branched fibrillar structure.

Table 13.3 *Characteristics of Reinforcing Glass Fibers Investigated*

Length mm	Diameter μm	L/D	Surface treatment
6.35	3.8	1671	OCF 636
12.70	9.0	1411	OCF S1451
4.76	9.0	529	OCF 416CB
12.70	13.5	694	OCF S1371
12.70	9.0	1411	OCF 691

13.4   THE WET-FORMING PROCESS

A schematic of the wet-forming process is shown in Fig. 13.2. The ingredients were well mixed in a dispersing tank to form a aqueous slurry. The basic ingredients were chopped fibers and a polymeric powder. Depending on the specific materials and processing scheme used, other minor components could be added to the slurry as well; some of the ingredients may require pretreatment prior to being fed to the mixing tank. After the ingredients have been well mixed they are fed to the headbox, where mixing has continued, due to the differences in specific gravity and size of the different components in the slurry. The screen moving through the headbox collected the solids from the slurry to form a wet mat. The screen then transported the wet mat to the drying zones. Initially, the wet mat drips loose water. Later the mat must be completely dried in a hot-air tunnel oven.

A number of researchers have investigated the wet-forming process for making composite-sheet materials. Biggs et al. developed a process for producing long-glass-fiber-reinforced polypropylene sheet on conventional papermaking equipment, such as described above [7]. In this process polypropylene powder and glass fibers were suspended in an aqueous foam instead of a slurry. This overcame the problem of differential settling due to the differences in specific gravity and size among the ingredients. The foam was filtered and dried to produce a sheet which was then consolidated under heat and pressure to form the prepreg sheet. Wesling et al. used a similar process to produce long-fiber-reinforced thermoplastic composite prepreg sheets, but without the use of the suspending foam [8]. The authors added a latex binder to the slurry to hold the dispersion of fibers and powder together for subsequent consolidation. Vallee and Cortinchi [9] also used a papermaking process to produce a thermoplastic matrix sheet. In this process the slurry consisted of the polymeric powder, reinforcing fibers, thermoplastic pulp fibers, latex binder, and flocculating agents. The pulp fibers, used in concentrations between 5 and 25 wt%, aided in the retention of polymeric powder in the sheet, and in improving the wet-sheet handling characteristics during processing on the papermaking machine. The latex binder held the constituents of the sheet together prior to consolidation. The flocculating agents agglomerated the small-diameter polymer particles together, which was claimed to improve the distribution of materials within the sheet.

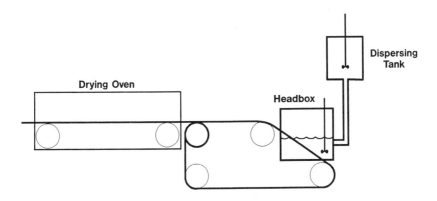

Figure 13.2 *Schematic of the slurry deposition process.*

These wet-forming processes take advantage of the low-cost, high-productivity characteristics of the basic papermaking process to produce thermoplastic matrix composite sheets. This type of process offers significant potential for producing a large volume of high-performance prepreg composites sheets at a competitive cost. In this chapter a modification of the wet-forming process is presented in which no binding or flocculating agents are used, and the concentration of pulp fibers is reduced to below 5 wt%.

An aqueous slurry of the polymeric powder, reinforcing fibers, and pulp fibers was prepared. The solids content for dispersing the materials was approximately 5 wt%, and the solids content for sheet forming ranged from 0.02 to 1 wt%. The solids content for sheet forming varied within this range as the fiber length, polymer content, and other process variables changed. Generally, longer glass fibers required more dilute suspensions than shorter ones. Hand sheets were formed from the fiber/polymer/pulp slurry on a Noble and Wood handsheet mold using a commercial polyester-forming fabric to filter the solids from the slurry. Handsheets were made with weights per unit area ranging from 50 to over 2000 g/m^2.

Nonpolar polymers, such as polypropylene, required a surface treatment to assist wetting and dispersion. The density of the polymer was found to have a significant effect on the dispersion and filtration steps. Because both the polymer and fibers have different densities, the slurries had to be constantly agitated to insure a uniform dispersion of ingredients. This was particularly true for PP, which would float on the water's surface unless strongly agitated. The rapid filtration of the papermaking process resulted in a uniform, random dispersion of polymer particles and fibers.

The wet sheets produced by this process were thoroughly dried in a vacuum oven in order to remove all traces of water. While vacuum and forced-air ovens were found acceptable, forced-air ovens dried the sheet more rapidly than the vacuum ovens. The dried sheets were stacked to build up a laminate to the desired final consolidated composite thickness. The number of plies required depended on the dry weight of the sheets produced in the filtration step. In order to achieve a consolidated sheet thickness of 2.5 to 3 mm the sheets were stacked to an areal weight of between 3000 and 4500 g/m^2. Up to this point the basic wet-forming operation was independent of the melting point of the polymer, and all of the steps were carried out at temperatures below 100°C. At this point, the dry laminated mats were heated above the melting point of the polymer and consolidated under pressure to form the composite sheet.

The sequence of heating and applying pressure was varied to determine the most effective method of forming void-free composite sheets. It was found that the highest quality sheets were produced when the mats were preheated under contact pressure, and then fully consolidated under pressures between 3.5 MPa and 10 MPa. For example, a PBT composite containing 15.6 vol% glass fibers had a tensile strength of 80.2 MPa when the pressure was applied before heating, and a tensile strength of 120.7 MPa when the sheet was preheated prior to compression. This trend was observed in every case investigated. The consolidation temperature used for PP was 190°C, while the consolidation temperature for PBT was 240°C. In most cases consolidation pressures between 3.5 and 10 MPa were used with a typical holding time of about 20 min. The consolidated sheets were cooled to room temperature under pressure.

The concentration of reinforcing fibers investigated ranged from 15 to 40 vol% of the final sheet. The concentration of pulp fibers ranged from 0 to 7 vol%. The fiber-length distribution was not determined qualitatively, but visual examination of the fibers in the consolidated sheet indicated that most of the original fiber length was retained during the process.

13.5   SAMPLE TESTING

The mechanical properties of the composites produced in this program were measured on an Instron Tensile Testing Unit at a jaw speed of 25 mm/min. The tensile strength obtained from these measurements was then used to calculate the appropriate reinforcement efficiency for comparison with other composite systems. An overall efficiency factor, $e = e_0 e_l$, was used to analyze the reinforcement effectiveness of the various polymer-fiber combinations on the tensile strength of the composites. The use of the combined efficiency factor was justified by the fact that all of the fibers had at least ten times the critical aspect ratio. This meant that in all cases e_l was greater than 0.95. The factor was, therefore, a measure of fiber orientation, degree of adhesion between the fibers and polymer, and fiber-fiber interactions.

13.6   RESULTS

The presence of the pulp fibers was extremely important to the preconsolidated sheet-forming operation. Without the presence of the pulp fibers the dried filtered mat had a low-density, open structure with very little integrity. The sheet could not be handled without falling apart. As little as 0.2 vol% of the pulp fibers produced a sheet with a significantly higher dry density, and sufficient sheet integrity that it could be easily transferred to the consolidation step. The concentration of pulp fibers did not appear to affect the sheet handling performance of the composite within the range of 0.1 to 5 vol%. Both the PP pulp and the Kevlar pulp produced similar results.

The type of glass fiber used had little effect on the performance of the sheet-forming process. The 3.8 μm diameter fibers were somewhat easier to disperse, but not to the extent that a significant difference in final sheet properties was observed. Fiber length had no effect on sheet-forming characteristics until they reached 18 mm where orientation effects became significant. The choice of the glass fibers listed in Table 13.3 produced no appreciable effect on the tensile strength obtained from measurements of the composite sheets.

Table 13.4 Effect of Fiber Concentration on the Mechanical Properties of Glass Fiber Reinforced Composites

Matrix	Volume fraction	Weight fraction	Strength MPa	Efficiency e
PBT	0.16	0.27	120.6	0.20
PP	0.19	0.40	120.0	0.20
PBT	0.26	0.41	99.0	0.10
PBT	0.38	0.55	146.6	0.13

Table 13.4 shows the effect of the volume fraction of reinforcing fibers on the tensile strength of several composites reinforced with the 3.8 μm diameter fibers. The reinforcing efficiency of the fibers began to fall when the fiber concentration exceeded 20 vol%. This was attributed to the inability to produce void-free composites at concentrations above this level. Examination of the cross section of a fractured tensile specimen containing 16 vol% fibers showed a homogeneous matrix phase without the presence of voids. Samples containing 26 vol% contained regions with an appreciable void content where the polymer was unable to fill. This lack of a continuous matrix phase was due to the packing arrangement of the fibers. For every packing arrangement there is a minimum level of polymer required to produce a continuous matrix phase. The packing arrangement was determined by the fiber characteristics and the manner in which the fibers were deposited on the screen during the filtering process. Longer fibers entangle and bridge more than short fiber, and therefore, to insure a continuous matrix phase, require a higher matrix-phase concentration than short fibers.

Table 13.5 summarizes the tensile-strength and reinforcement-efficiency data for selected composites produced in this program along with those of an amine-cured, randomly dispersed, long-glass-fiber-reinforced epoxy composite, sheet-molding compound (SMC) [19], AZDEL polypropylene-glass mat sheet [20], and injection-molded, short-glass-fiber-reinforced polypropylene [21]. The reinforcement efficiencies for these composites were calculated from published tensile-strength data. The reinforcement efficiencies ranged from 0.13 for the sheet-molding compound (which contains a significant concentration of inert fillers) to 0.22 for the epoxy composite. Not surprisingly, the reinforcement efficiency of the short-fiber-reinforced injection-molding composite, 0.17, was less than that of the long-fiber-reinforced composites, ~ 0.20 (except for the SMC for reasons explained above).

Table 13.5 *Mechanical Properties of Selected Composites*

Matrix	Volume fraction	Strength MPa	Modulus GPa	Efficiency e
Epoxy	0.43	250.4	--	0.22
PP[a]	0.19	103.4	6.2	0.17
PP[b]	0.19	120.0	5.0	0.20
SMC	0.35	158.6	15.0	0.13
PP	0.19	120.0	6.3	0.20
PBT	0.16	120.6	--	0.20

Notes: [a] *Injection molded,* [b] *AZDEL.*

The reinforcement efficiency, tensile strength, and tensile modulus of the wet-formed composites were comparable to those of other long-glass-fiber-reinforced composites at similar fiber loadings. A reinforcement efficiency of 0.20 was obtained with both the PP and PBT composites. This is the same level of reinforcement efficiency calculated for AZDEL, and very close to the level obtained with the epoxy-matrix composite. Consequently, the tensile strength of the wet-formed composites was very similar to that reported for AZDEL, 120 MPa, as was the tensile modulus, 6.3 GPa for the wet-formed polypropylene composite versus 5.0 GPa reported for AZDEL. Table 13.6 allows a comparison of the mechanical properties of the various wet-formed composites produced. The lack of binder in the sheets produced in this program resulted in stronger composites.

Table 13.6 *Mechanical Properties of Wet Formed Polypropylene Composites containing 40wt% Glass Fiber*

Tensile strength MPa	Flexural strength MPa	Flexural modulus GPa	Reference
110	140	6.3	This work
96	117	5.4	[22]
97	165	6.5	[7]
80	132	5.3	[23]

One of the forming techniques that is used to produce parts from sheet composites is stamping. This term encompasses fast compression molding (also called flow molding) and solid-state forming (thermoforming). Prior to actually forming the part by either of these methods the sheet must be heated to some forming temperature. It has been observed that when wet-formed composite sheets are heated above the melting points of the matrix polymer they expand to a thickness approximately twice their consolidated thickness. This phenomenon has been called "lofting" and was first reported by Yats and Edens [22]. It has since been verified that all wet-formed composites exhibit the same behavior. The foamed reinforced sheet can be formed into a rigid low-density product or reconsolidated by stamping.

13.7   CONCLUSIONS

It has been shown that high-strength thermoplastic matrix composite sheets can be produced by the deposition of a powdered polymer and long reinforcing fibers from an aqueous slurry. An important feature of this process was the addition of a small concentration of pulp fibers. These fibrillated fibers aided in the distribution of the fibers and polymer, as well as in densifying the dried sheet so that it could be handled prior to the melt-consolidation step. Melt consolidation was found to be most effective when the sheet was heated to a temperature above the melting point of the polymer before the application of pressure.

The reinforcement efficiency of consolidated sheets was similar to that obtained in other long-fiber-reinforced composite systems. The maximum fiber concentration depended on the initial aspect ratio of the reinforcing fibers, with 20 vol% easily obtainable with fibers having an aspect ratio of 1600. A PP composite containing 20 vol% glass fibers was produced that had a tensile strength of 120 MPa and a tensile modulus of 6.3 GPa.

The wet-forming process imposes less thermal processing on the polymer than conventional thermoplastic composite sheet-forming operations. Only in the consolidation step are high temperatures required. This reduces molecular-weight losses during processing, allows thermally sensitive polymers to be used, and reduces the energy requirements needed to process polymers with high melting points.

NOTATION

A	Area under stress-strain curve
D	Fiber diameter
E	Modulus
HDT	Heat-distortion temperature
I	Subscript denoting interface property
L	Fiber length
PBT	Polybutyleneterephthalate
PP	Polypropylene
SMC	Sheet-molding compound
T_g	Glass transition temperature
T_m	Melting point
e_E	Efficiency factor related to modulus
e	Efficiency factor related to tensile strength
c	Subscript denoting composite property
p	Subscript denoting polymer property

f	Subscript denoting fiber property
o	Subscript denoting property associated with fiber orientation
cr	Subscript denoting critical value
γ	Strain
ϕ	Volume fraction
σ	Strength
$\sigma(\gamma)$	Stress as a function of strain
τ	Shear stress

REFERENCES

1. Bigg, D. M., Hiscock, D. F., Preston, J. R., Bradbury, E. J., *Advanced Composites III: Expanding the Technology*, ASM International, p. 253 (1987).
2. Seiler, E., Welz, M., Wald, H. H., *Kunststoffe*, **72**, 341 (1982).
3. Hill, G. G., House, E. E., Hoggatt, J. T., *Advanced Thermoplastic Composite Development*, NTIS ADA081978 (1979).
4. Hartness, J. T., *Proc. SAMPE, 25th Tech. Conf.*, **25**, 376 (1980).
5. Temple, C. S., Matthews, J. R., *U.S. Pat.* 3,684,645 (Mar. 25, 1969).
6. Anonymous, *Thermoplastic Polymers in Powders for Composites*, Atochem product brochure (1984).
7. Biggs, I. S., Radvan, B., Willis, A. J., *Plastics on the Road '86*, The Plastics and Rubber Institute, London, 4/1 (1986).
8. Wessling, R. A., Yats, L. D., Tolbert, D. K., *U.S. Pat.* 4,426,470 (Jan. 17, 1984).
9. Vallee, A., Cortinchi, H., *U.S. Pat.* 4,645,565 (Feb. 24, 1987).
10. Anonymous, *Advanced Thermoplastic Composites*, Phillips Petroleum Company product brochure (1984).
11. Cogswell, F. N., Leach, D. C., *Plast. Rubb. Proc. Appl.*, **4**, 271 (1984).
12. Folkes, M. J., *Short Fibre Reinforced Thermoplastics*, Wiley, Letchworth, UK (1982).
13. Bigg, D. M., *Polym. Compos.*, **6**, 20 (1985).
14. Weiss, R. A., *Polym. Compos.*, **2**, 89 (1981).
15. Halpin, J. C., Tsai, S. W., *AFML Tech. Rep.*, TR67-423 (1967).
16. Kerner, E. H., *Proc. Phys. Soc.* (London), **69B**, 802 (1956).
17. Nielsen, L. E., *Ind. Eng. Chem. Fundam.*, **13**, 17 (1974).
18. Lee, L. H., *SPE Techn. Pap.*, **14**, 1 (1968).
19. Lubin, G., *Handbook of Composites*, John Wiley & Sons, New York (1982).
20. Anonymous, *AZDEL Reinforced Thermoplastic Moldable Sheet* product brochure, PPG Industries (1981).
21. Anonymous, *Modern Plastics Encyclopedia*, McGraw-Hill, New York (1986-87).
22. Yats, L., Edens, M., *Proc. 42nd Annual Conf. SPI*, 19-F (1987).
23. Anonymous, *ARJOMIX*^{TM}, *The New Generation of Glass Reinforced Thermoplastics*, Arjomari Prioux S. A., product brochure (1986).

CHAPTER 14

DEVELOPMENT OF FINE MORPHOLOGY IN POLYPROPYLENE COMPOSITES

by S. F. Xavier

Research Centre
Indian Petrochemicals Corporation Ltd.
Baroda 391 346, Gujarat
INDIA

Polypropylene develops a transcrystalline interfacial morphology in the vicinity of an incorporated fiber or a flake. In certain cases it gives rise to shish-kebab structures and sheets of chain-extended material. Besides developing such fine morphological changes, a reinforcement is also found to interfere with the overall crystallization kinetics of the matrix. Different surface treatments to a reinforcement, while modifying the interfacial morphology, alter the skin-core thicknesses of injection molded bars. All these morphological developments cause significant changes in composite properties. A considerable effort has been made over the past fifteen years to understand and control the nature of such interfacial morphologies. Certain fundamental issues are not yet clarified.

14.1   INTRODUCTION

Incorporation of reinforcement or fillers into thermoplastics is a common practice in polymer technology. These materials improve the physical properties of neat polymers, help reduce the cost of the molded products, and aid in tailoring the materials to suit the requirements [1].

During processing of a reinforced polymer which exhibits spherulitic morphology, the polymer melt is often cooled in contact with the reinforcement surface that may act as a nucleating agent. The morphology of the polymer, in such a situation, is considerably altered in the vicinity of the nucleating reinforcement surface [2]. The subtle nature of the interlayer region between polymer and fiber, or the interphase, is crucial in determining the properties of polymer composites. Apart from such special morphological features which result from the presence of the reinforcement, certain other microstructural changes can also occur [3] depending on the melt flow prior to the polymer crystallization. Several attempts have been made to understand the nature and origin of such morphological features. Although many basic questions are not yet completely answered, significant advances have been made during the past fifteen years. In this chapter an attempt is made to provide a comprehensive understanding of the morphological developments in reinforced polymers with special emphasis on polypropylene composites.

14.2   CRYSTALLIZATION AND NUCLEATION IN POLYPROPYLENE

14.2.1   Principal Aspects of Crystallization

Thermoplastics, in general, can be classified as either crystalline or amorphous polymers. The crystallization of polymers has been extensively studied [4-8]. In this section, we are generally concerned with the semicrystalline nature of polymers and of polypropylene (PP) in particular.

The crystalline phase of polymers consists of thin plates or ribbons with the chains oriented along the thin dimension [9]. The plates stack one above the other, and are separated by layers of amorphous material as sketched in Fig. 14.1. The amorphous region consists of chain-end cilia, totally occluded chains, and chains which are incorporated into two or more crystals (tie chains). The tie chains determine the mechanical continuity of the system and thereby control the elastic modulus as well as yield strength. When crystallization is initiated

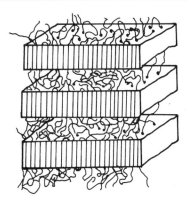

Figure 14.1 *Sketch of the basic crystallite microstructure of semicrystalline polymers [10]. The layers separating crystallites are composed of chain ends, tie-chains, and fully occluded chains.*

the nucleation simultaneously begins at various sites in the melt. The transformation proceeds spherically from these centers, through the radial growth of ribbonlike crystals. The spherulites grow until they impinge on each other, forming a polygonal array in the solid phase. The dimension of spherulites ranges from 0.1 to 1000 μm. As a spherulite grows, all less-crystallizable matter is pushed ahead of the crystallization front. This matter is collected at the interface separating two spherulites [10].

The nature of the nuclei initiating such microstructurally complex systems is still a matter of debate. Nevertheless, a large number of nucleating agents for polyolefins have been discovered by Wijga [11], Wales [12], Binsbergen [13], and others [14,15]. Binsbergen [16] tried 2000 substances for their possible nucleating effects on crystallization of PP. The effect of different shapes of crystalline surfaces, physical state, and chemical nature of the nucleating agents were also studied in detail. In general, it is believed that the spherulitic nucleation takes place at heterogeneities in the melt such as catalyst residues, dust particles, low-molecular-weight fractions, branch points, stereo-irregular molecules, fillers, or specific nucleating agents. The spherulite size is primarily dictated by the volumetric nucleation rate which depends on thermal history and nature of the surfaces. The nucleation rate increases with degree of undercooling defined as $\Delta T = T_m - T_c$ where T_m and T_c are, respectively, the melting point and the crystallization temperature. The final spherulite size decreases with increasing undercooling and the cooling rate. These two are the most frequently used for controlling the spherulite size.

Polypropylene exhibits isotactic, syndiotactic, and atactic forms, defined by spatial disposition of the methyl groups. The main-chain carbon backbone has a helical conformation in three dimensions. When a melt of these helical chains is allowed to cool, it may either organize into a rather loose, noncrystalline structure or into a highly ordered crystal. Isotactic polypropylene (PP) exhibits spherulitic morphology (Fig. 14.2). Crystal nucleation in PP is more complicated as it was observed that not only solid substances, but also liquids and gases act as nucleating agents. Furthermore, the same substance does not exert the same nucleation effect on various commercial PP types [17,18]. Boon [19] proved that heterogeneous nucleation of isotactic polystyrene was prompted by catalyst residues containing Ti and Al. On the other hand, Binsbergen [20] doubted the decisive role of the catalyst residues left in the polymer, because the number of these residues is several orders of magnitude larger than the number of observed spherulites. Turnbull [21] showed that in pores, crevices, and surface steps of solid heterogeneities the crystal fragments may survive far above the bulk polymer melting temperature and serve as nuclei on subsequent cooling. Rybnikar investigated the isothermal crystallization of isotactic polypropylene and its composites with talc [22]. He observed two types of heterogeneous nuclei affecting the crystallization process, dependent on the melting conditions. These are: 1) metastable nuclei represented by the remnants of the polymer crystalline phase, which are effective at low melting temperatures or at short melting periods; and 2) stable crystallization nuclei associated with solid heterogeneities such as catalyst residues.

Figure 14.2 Spherulites in polypropylene as viewed under optical microscope.

Spherulite size and spherulite boundaries significantly influence yield and failure behavior of polymers. These effects are manifested in macroscopic measurements and in the microscopic nature of the fracture path. Kargin et al. [23] as well as Lovinger and Williams [24] clearly demonstrated that the fine spherulitic microstructure produced by adding nucleating agents gives rise to greatly enhanced ductility, yield strength, and impact strength. In general, refinement of the spherulite size by either homogeneous or heterogeneous nucleation or by reducing the crystallization temperature results in improved ductility and strength. It is clear that large spherulites exhibit cracks at spherulite boundaries, whereas systems with smaller spherulites draw smoothly. It has been suggested [25,26] that the spherulite boundaries are effective fracture sites because of the presence of foreign matter.

14.2.2 The Origins of Nucleation

Heterogeneous nucleation is the commonly observed mechanism of initiation of polymer crystallization from the melt [5,6,27]. This type of nucleation is promoted by surfaces of foreign bodies, known as nucleating agents. The necessary qualifications for a material to act as a nucleating agent for a polymer and the mechanism of the process are yet to be clarified, though several attempts have been made in this direction. It appears that certainly there is no universally strong nucleating agent. Many of the hypotheses available in the literature are based on nucleation at different fiber surfaces. It is almost certain that the same nucleation phenomenon holds true for nucleation by particles or flakes. Turnbull and Vonnegut [28] postulated that nucleating efficiency should increase with increasing closeness of match between lattice parameters of the substrate and the forming crystal. Binsbergen [16] observed that, in the case of polyolefins, the nucleating agents are mostly insoluble hydrocarbons whose activity strongly depends on the degree and the method of dispersion. He also reported that the orientation of the nucleating particles in the polymer melt causes oriented crystallization of the polymer. In contrast to this, Campbell and Qayyum [29] observed that continuous, uniaxially oriented polyamide, PA, or polyethyleneterephthalate, PET, fibers, which differ considerably in their unit-cell parameters from those of PP, develop preferential nucleation on their surfaces when the latter is crystallized in their presence.

Chatterjee et al. [30] considered that chemical similarity between the crystallizing polymer and the substrate was necessary. This postulate was also found not to be valid since glass is amorphous and it does not possess any chemical similarity with PP but causes preferential nucleation, leading to transcrystalline structure [31]. Furthermore, nonpolar PP crystallizes on polar material such as PA, PET, or aromatic PA. In the case of polyethylene (PE), the nucleating ability of alkali single crystals did not show significant variation with considerable changes in surface energies [32]. Measurements of isothermal crystallization kinetics showed [33] that the number of active crystallization centers markedly depends on the degree of supercooling. The nucleation is almost completely instantaneous and the temperature of melting has no influence on the nucleation density.

Binsbergen [20] critically studied various models of heterogeneous nucleation. He concluded that in the case of polyolefins, the nucleating activity is neither based on: high surface free energy of the substrate, epitaxy, a minimum particle size, nor persistence of crystalline material in holes of the nucleating particles. The author proposed a theory and explained the mechanism of heterogeneous nucleation of polyolefins, according to which the process takes place at steps of limited length on the surface of the nucleating particles. The nucleation is enhanced by a high degree of accommodation of polymer molecules at the surface of the substrate through the presence of ditches. In the latter case, the polymer chains are prealigned thereby facilitating crystallization.

So far, the reviewed literature leads to the conclusion that the exact origins of nucleation sites of polymers remain uncertain. The effect of introducing a foreign material (which acts as a nucleating agent) during the crystallization of PP is addressed in the subsequent section.

14.3   TRANSCRYSTALLINITY

When a polymer melt is cooled in contact with a foreign surface which can nucleate crystalline growth, the proximity of many nucleation sites on the surface inhibits lateral growth of the resultant spherulites; thus the crystallization develops only in a direction normal to the foreign surface. The development of such a layer has been termed transcrystallization [2]. These transcrystalline regions have been observed in several polymers, such as polyurethanes [2], polyamides [2,34,35], polyethylenes [36,37], polypropylene [38-40], and polyetheretherketone [41]. In the case of PP the transcrystalline morphology was observed when it was crystallized on graphite [42], thermoplastic [43], cellulosic [44], and glass fibers [45]. Since transcrystallization at the surface of such reinforcing fibers could improve the interfacial bonding between the fibers and the polymer and, hence, it should improve the resultant composite strength [46], this phenomenon is of general interest.

14.3.1 Transcrystallinity in Polypropylene

Several workers have investigated the nature of transcrystallinity and various factors influencing it. In the transcrystalline region, the columnar growth of embryonic spherulites is constrained to one dimension. Keller [47] reported that the transcrystalline structure is otherwise identical to that of normal spherulites. However, others [48,49] found that the transcrystalline surface layers have mechanical properties different from those of the bulk phase. These layers are usually 10 to 100 μm thick and therefore are effectively macroscopic. Kantz and Corneliussen [43] observed that in the case of PP reinforced with PA fibers, the transcrystalline region extends two to four fiber diameters into the bulk prior to its termination by conventional spherulites. The authors also observed that, in the absence of impingement effects from the neighboring fibers, the transcrystalline regions are usually cylindrically symmetrical about the fiber axis.

The suggested factors influencing transcrystallization are temperature gradients at the surface [2], the wettability and, hence, surface energy of the foreign surface [46], the chemical composition of the surface [16,50,51], the evolution of volatile products from the nucleant [17], and the crystalline morphology of the nucleating surface [42,52]. However, Fitchmun and Newman [40] observed that temperature gradients are not necessary for the occurrence of transcrystalline growth, although they may influence the spherulitic nucleation. Gray [44], in his studies on PP transcrystallization at the surface of cellulose fibers, observed that differences in surface energy between the cellulose fibers do not justify the difference in transcrystallization. Considering the surface chemistry of the cellulosic fibers, Gray pointed out that the regenerated cellulose fibers possess extensive surface carbonyl groups which were reported to be effective nucleating agents [40]. However, these fibers failed to produce transcrystallization. Folkes and Hardwick [53] had shown that in the case of PP, the transcrystalline growth depends on the molecular weight of the polymer. On the other hand, Burton and Folkes [54] observed that PA induced vigorous transcrystallinity over a wide range of molecular weights.

Transcrystallization of PP at glass surface was not reported in literature until 1974, although nucleation along cracks in a cover glass, during optical microscopic studies had been observed [55]. It was Gray [31] who reported transcrystallization across glass fibers,GF, in PP. Apart from the factors discussed above, he observed that slight mechanical stress at the PP/GF and PP-air-bubble interface lead to transcrystalline morphology. He also found that preferential nucleation of PP takes place at slightly deformed vapor bubbles which certainly presents a low-energy surface to the melt [56]. This is, again, in contrast to the suggestion that transcrystalline morphology develops more readily on the high-energy surface [57].

In support of the findings by Gray [31], Misra et al. [58] observed transcrystalline regions develop in PP when glass fibers were intentionally pulled under the hot-stage polarizing microscope. PP was melted on a glass slide preheated at 220°C with GF on top of it, then more PP was added in order to embed the GF into molten PP, ensuring that the fibers protruded out of the glass slide. A covering glass slide was put on top, and the system was allowed to acquire thermal equilibrium for about 4-5 min. Then the glass slide was transferred to a hot stage (maintained at 136°C) attached to a polarizing optical microscope, and immediately a few of the protruding GF were pulled with the help of tweezers. The occurrence of transcrystallization could be seen (Fig. 14.3) after about 8 min. From the Figure it is clear that the transcrystalline regions developed only around those fibers that were pulled while no such phenomenon occurred along the undisturbed (uppermost) identical fiber. It is obvious that the shear stresses at the polymer-fiber interface initiated the nucleation. Moreover, there is a gradual change of temperature from the edge to the interior of the glass slide which resulted in the transcrystalline growth variation, although without apparent change in nucleation density. Burton et al. [59] also reported variation in transcrystalline growth due to temperature gradient along high-modulus carbon fiber in PP. The undisturbed fiber did not develop any preferential nucleation even though there was some thermal gradient across the fiber length.

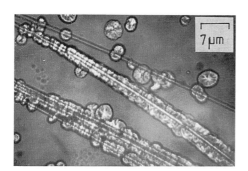

Figure 14.3 *Transcrystallization close to identical glass fibers [58]. Except the fiber at the top, all others were gently pulled. The polarizers were kept at (a) 50° and (b) 80° apart, respectively.*

Campbell and Qayyum [60] suggested that the differences in nucleation density arise from heterogeneities in the bulk which the fibers would attract onto their surface. They visualized some type of attractive, possibly polar, forces between the surface of the fibers and the heterogeneous nucleating sites in the region of a fiber at higher crystallization temperatures. With increasing fiber volume fraction, the overall strength of attraction could increase resulting in building up nucleating sites in the region adjacent to the fibers. The precise nature of these heterogeneous nucleating sites and their rates of diffusion in the melt are unknown. However, the nucleating sites (if they exist at all) should have been attracted both by the undisturbed and the pulled fibers.

Campbell and White [61] indicated that observations of the stress-induced transcrystallinity [58] agree with the earlier proposed mechanism of diffusion of heterogeneities [60]. The authors [61] reconciled the reported results [58,60] by invoking the effects of molecular orientation. Molecular orientation, produced by flow during a molding operation or by shear stresses induced by the fiber motion under the hot-stage microscope, increase chances of nucleation both at the nucleating sites attached to fibers and at those existing independently within the melt. Thus, they argued that it may be possible to produce crystallization at or near to moving fibers while stationary fibers surrounded by unoriented PP melt show little nucleation within a critical temperature range near 136°C. This is quite independent of the nature and distribution of heterogeneous nucleation sites.

The results of Gray [31] and Misra et al. [58] are hardly surprising in view of the well-recognized relationship between stress and crystallization. Nevertheless, they suggest that care must be exercised in interpreting heterogeneous nucleation at the polymer-fiber interface based solely on the properties of the surfaces. Mechanical stresses and shearing forces, with or without thermal gradient, appear almost unavoidable when polymer composites are compression or injection molded. The shear stress produces mechanical orientation along the direction in which the fiber is pulled, especially in the case of short-fiber-reinforced polymer composites, and thus causes a reduction in entropy which enhances the melting temperature and the degree of supercooling. Change in supercooling, however, cannot fully explain the effects of stress-induced crystallization [62]. The spherulitic growth rate has also been found to be higher for stress-induced crystallization than for quiescent melts [63].

Huson and McGill [64] discounted some of the above-mentioned factors such as surface energy, chemical composition, crystal morphology, and moisture content of the substrate. They concluded that surface topography of the substrate plays a major role. Holes, channels, or cracks present in the surface of crystalline polymers, such as PET, are of suitable size and allow a degree of prealignment of polymer chains leading to preferential nucleation at the surface. Similar prealignment of chains can also occur in microcracks in the glass slide or in suitable crevices between asbestos fibers. The authors also observed that when the substrate is very active (ideal topography) transcrystallinity occurs regardless of the rate of cooling while for less active substrates rapid cooling is essential for the transcrystalline growth [64].

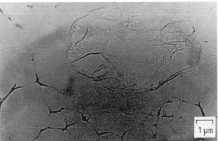

Figure 14.4 *Transcrystalline morphology of polypropylene in the presence of mica flakes [69]; (a) untreated mica; (b) location of the mica flakes (uncrossed polarizers); and (c) untreated mica in polypropylene after repetition of the crystallization process.*

The morphology of PP is also affected by the presence of mica flakes [65]. When used as a reinforcing agent in plastics, mica offers many advantages including planar reinforcement, reduced creep, and reduced wear of processing equipment [1]. PP reinforced with mica has received particular attention because of its high stiffness, low cost, good dimensional stability, and adequate temperature performance [65,66]. Garton et al. [67,68] observed that high-aspect-ratio mica flakes (Suzorite H200, phlogopite variety) nucleate PP. However, when they treated mica flakes with a silane coupling agent (Dow Corning 6032) or added a small amount of chlorinated hydrocarbon to the resin, they observed a reduction in the nucleation ability of mica. Xavier and Sharma [69] also observed that untreated mica (muscovite variety) as well as mica coated with isopropyltriisostearoyltitanate (TTS) acted as a nucleating agent for PP. The micrograph in Fig. 14.4a was taken after melting PP in the presence of untreated mica and then letting the system cool down to room temperature. The micrograph in Fig. 14.4b was taken to show the location of the mica flakes dispersed in crystallized polymer held between the glasses. It is clear from Fig. 14.4a that PP crystallized preferentially on the surface of the mica flakes. These results are in agreement with the observations made by Garton [68]. The polymer/mica system was reheated to 210 °C in order to reduce frozen-in stresses then allowed to cool to room temperature. The transcrystallization persisted, although some changes in the positioning of the spherulites were noticed at the interface (see Fig. 14.4c). These changes may be due to the changes in the nucleation sites on the mica surface. The density of nucleation, however, apparently remained the same. These results reveal that transcrystallinity takes place at the PP/mica (uncoated) interface which can persist on reheating and cooling. The TTS-treated mica also acts as a PP nucleating agent (Fig. 14.5).

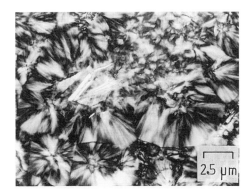 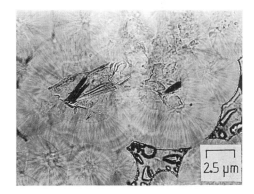

Figure 14.5 *Transcrystalline morphology of polypropylene in the presence of titanate coated mica flakes (a) and location of the mica flakes (uncrossed polarizers) (b).*

14.3.2 Transcrystallinity in Other Polymers

It has already been mentioned that transcrystallinity is a general phenomenon. A number of polymers that develop transcrystallinity were mentioned in the previous section. Adding to the list are polycarbonate and polyphenyleneether [70,71].

Two-zone transcrystallinity was observed by Tianbai and Porter [72] in HDPE matrix reinforced with gel-spun, high-modulus PE fiber. The authors concluded that structural similarity of the matrix and the fiber was responsible for the observed orientation of crystallites in planes perpendicular to PE fiber axis in the inner transcrystalline zone.

Lee and Porter [41] reported that at equivalent thermal histories, PEEK with carbon fiber exhibited higher nucleation density than PEEK itself. They argued that the surface of carbon fibers and the nuclei in PEEK matrix compete for crystallization growth. As the holding time in the melt was increased, the number of matrix spherulites decreased, developing a more pronounced transcrystalline region. Velisaris and Seferis [73] made similar observations and proposed a dual mechanism for the crystallization of PEEK. They observed both patterns typical of spherulitic crystalline growth as well as of epitaxial transcrystalline growth emerging from the fiber surfaces in carbon-fiber reinforced PEEK composites. Further studies by Tung and Dynes [74] on the morphological characterization of these composites revealed that spherulite size as well as the degree of transcrystallinity at the graphite-fiber surface are both reduced with increased quench rates.

Bessel et al. [75,76] had developed an alternative method of producing transcrystallinity, by in-situ polymerization of PA-6 matrix around glass and carbon fibers. The method has the disadvantage that the degree of polymerization near the fibers is unknown. Frayer and Lando [77] crystallized hexamethylene diammonium adipate (HMDA) on graphite fibers then polymerized it in the presence of an electric field. The authors reported that the specific crystallographic direction was preferentially oriented along the fiber axis. HMDA crystallized in this manner undergoes an oriented (topotactic) reaction to the cyclic dimer of PA-6,6 under conditions which, in the absence of substrate, would yield oriented PA-6,6.

14.4   MICROSTRUCTURE OF MOLDED POLYPROPYLENE COMPOSITES

The influence of various fibers and fillers on the nucleating effects of PP were presented in Section 14.3.1. The process of molding either by injection or by compression further influences the polymer morphology which in turn affects performance of the molded article. The fact that in a molded specimen the mechanical properties vary considerably from one location or direction to another was reported by Spencer [78]. The problem of polymer melt flow into injection molds has been extensively studied [79-84].

14.4.1 Skin-Core Morphology in Injection-Molded Fiber/Flake Composites

Injection-molded PP has been reported [85] to exhibit three morphological zones: the skin with no perceptible crystalline development, the shear zone with nearly identical spherulites bunched tightly together in thin layers, and the core with larger but randomly oriented spherulites.

Figure 14.6 shows the shear zone of PP composite (with 10 wt% glass fiber) observed under an optical microscope [86]. The spherulites are elongated towards the core (bottom) and their size is reduced remarkably towards the skin (top). Introduction GF apparently did not affect the orientation in the shear zone. The row nucleation is still present and the spherulites are elongated towards the core zones. This middle zone was observed clearly only in the case of PP and the composite with 10 wt% GF. The optical microscopic study of the microtomed sections of PP composites revealed that the characteristic three-zone morphological structure of PP gradually disappears with increasing fiber concentration [87]. With the addition of GF, the melt viscosity of the composite considerably increases and the polymer chains are increasingly obstructed in getting oriented in the flow direction. Lamellar formation is seriously affected and row-nucleation was destroyed when GF concentration exceeds 20 wt%.

Consequently, for the composites with 35 wt% GF, the skin-core morphology vanished completely (Fig. 14.7) [88].

Figure 14.6 *Polarizing photomicrograph of thin microtomed section near the edge of polypropylene with 10 wt% glass fiber [86].*

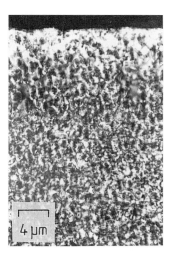

Figure 14.7 *Polarizing photomicrograph of thin microtomed section near the edge for polypropylene composite with 35 wt% glass fiber [86].*

It was also noted [88] that for composites with constant molecular weight, molecular weight distribution, degree of isotacticity, and fiber concentration, the injection-molding parameters, such as injection pressure, rate of injection, screw speed, and melt temperature, exert profound influence on morphology of the polypropylene/short-glass-fiber composites. It has to be borne in mind that the influence of each of these molding parameters cannot be isolated since their effects are intermingled. A detailed discussion on the influence of each parameter on morphology of PP composites is beyond the scope of this chapter. However, a brief discussion on the influence of one parameter, injection speed, on the core morphology of short-glass-fiber-reinforced PP composites [88] ought to be given.

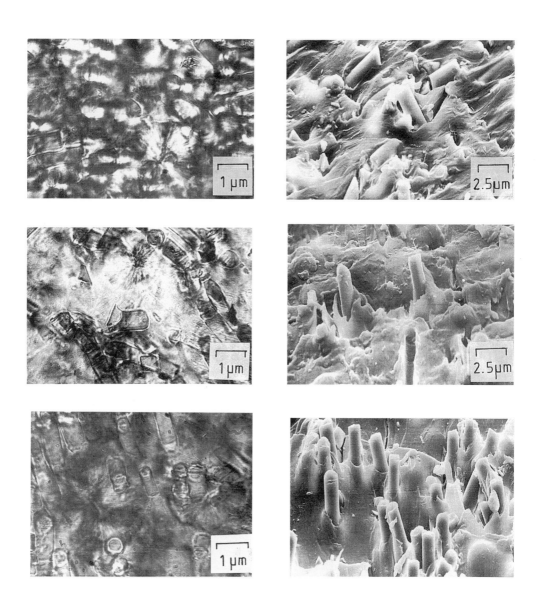

Figure 14.8 *Thin microtomed sections of polypropylene composites with 28 wt% glass fiber prepared at injection rates of 10, 20, and 30 mm/s viewed under polarizing microscope (a, b, and c) and scanning electron microscope (d, e, and f), respectively [88].*

Figure 14.8 presents the polarizing and scanning electron microscopy, SEM, results obtained on PP/28 wt% GF composites, prepared under three different injection speeds. As the injection speed increases from 10 to 30 mm/s there is considerable change in the polymer morphology and the GF orientation. An increase in the injection speed influences the polymer melt flow during the mold-filling process, and thereby influences the molecular orientation not only in the skin and shear zones, but also in the core (see Chapter 11). As the injection rate is increased, the shear rate is also increased, resulting in a greater number of polymer chains orienting in the elongating-force direction. In these cases, spherulites are scarcely visible in the core zone. Increase of injection speed increased not only the orientation of the molecules but also of the GF that are dragged along by the polymer chains. In Fig. 14.8d,e,f, the sample with highest injection speed has shown fewer voids compared to the rest and the bonding between the fiber and the matrix materials was appreciably better. Menges and Wubken [81], also observed that an increase of injection speed increased shrinkage caused by the molecular orientation.

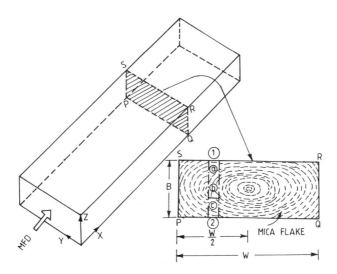

Figure 14.9 *Injection molded flexural bar with melt flow direction [92]. PQRS is the polished cross-sectional area where the microstructure was examined under scanning electron microscope: top side surface - (1), to bottom side surface - (2), within the area indicated by the dotted lines and subdivided as (a), (b), and (c) sections.*

Investigations of fiber-filled PET [3] and PEEK [89] also revealed that the microstructure of a composite is extremely important for understanding its mechanical properties. Recently, Malik and Prud'homme [90] made quantitative measurements of the orientation distribution function of mica flakes in compression-molded polymer composites. These composites are also becoming increasingly important because of several advantages mentioned earlier. It has been observed that in the case of the injection-molded PP/mica composites, the flake orientation in the skin zone is parallel to the surface of the specimen and is ellipsoidal in the core [91]. Figure 14.9 illustrates the section of the specimen subjected to metallographic polishing and later to microscopic observations [92]. A schematic illustration of the mica flakes on the polished surface and the area where the microscopic scanning was

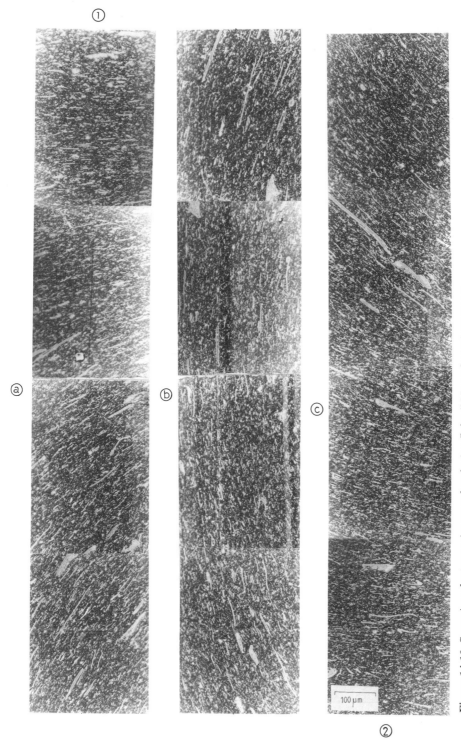

Figure 14.10 *Scanning electron micrographs of mica flake (325 mesh size) orientation from one mold surface (1) to another (2), subdivided as (a), (b), and (c) [92].*

undertaken (midway between the center and the edge, represented by the dotted lines) are also shown in Fig. 14.9. The orientation of mica flakes (325 mesh size) in the YZ plane is shown in Fig. 14.10. At the top the flakes are oriented parallel to the mold-wall surface. As one moves away from the mold surface to the core, the orientation angle of the flakes (α) gradually turns from 0 to 45° and to 90° at the core. Figure 14.11 shows α as a function of the thickness of the sample. The flake orientation is symmetrical with respect to the core. The limits of the core region can be defined as having mica flake orientations of ±45°. Thus, the central layer for the composite with mica of 325 mesh (denoted by G1 G2) is larger in comparison to that of the composite with mica of 1000 mesh. This indicates that with bigger mica flakes, the thickness of the core is increased and, simultaneously, the skin thickness on either side is correspondingly reduced. These changes in flake orientation and the corresponding changes in skin-core thicknesses are known to affect the mechanical performance of the composites [93].

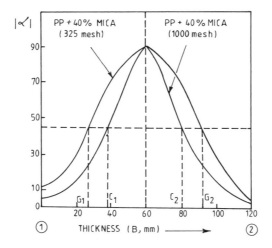

Figure 14.11 Variation of mica flake orientation (α) as a function of the thickness of the sample for polypropylene composites with mica of 325 and 1000 mesh sizes [92].

The changes in mica flake orientation angle with respect to the mold wall, for PP composites containing 20, 40, and 60 wt% mica, were measured from the scanning electron micrographs taken across the thickness of the sample, as discussed above. The results are presented in Fig. 14.12 [94]. Confining discussion to the same definition of the skin-core regions, it is evident that increasing mica concentration considerably increases the thickness of the core regions. Similar results were observed for short-glass-fiber reinforced PET [3]. In addition, with increasing mica concentration α in the skin increases, i.e., a disorientation of the flakes from the plane parallel to the mold wall gradually develops.

Remillard and Fisa [95] observed that during processing of mica-reinforced PP the air from small, intragranular pores in the molten composite can escape only with great difficulty. The problem is enhanced by the disklike shape of the mica flakes, which block the airflow. The pressure from this entrapped air, which builds up during the injection-molding process, may be sufficient to disorient the flakes in the skin. Because of the distortions, the molded product loses its surface gloss.

The influence of mica-surface treatments on the microstructure of PP composites was studied by Xavier et al. [96]. Untreated mica, mica treated with TTS, and mica treated with 3-aminopropyltriethoxysilane were used. Figure 14.13 presents the variation in mica-flake orientation angle with depth below the surface of the injection-molded bars. It is evident that different surface treatments resulted in different skin-core thicknesses. The titanate coupling agent reduced the core thickness, while the silane coupling agent enhanced it. The extent to which the coupling agents act through matrix plasticization or through the actual coupling of filler to matrix during the flow is not known. However, it is likely that coupling, itself, is important, since when the silane coupling agent (an effective coupler) is used, the flake orientation increases, resulting in enhanced core thickness. In fact, the core thickness has increased to approximately three times that of the composite with TTS-treated mica.

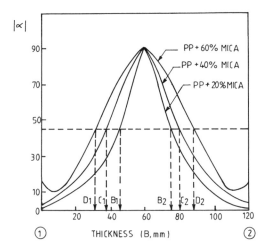

Figure 14.12 *Variation of mica flake orientation (α) as a function of thickness of the polypropylene composite with 20, 40, and 60 wt% mica [94].*

14.4.2 Transcrystallinity in Molded Composites

Very little work on the microstructure of matrix around individual fibers or flakes in a molded composite has been reported. This is partly because of the practical difficulty of producing thin sections of the composite needed for such an investigation. Nevertheless, the transcrystallinity was directly observed in injection-molded GF-reinforced PP [58,88]. Figure 14.14 shows the variation of transcrystallinity with increasing fiber volume content in composites molded under identical conditions. An interesting aspect illustrated in Fig. 14.14a,b is that the transcrystalline growth is not present on all the fibers. The obvious reason for this is nonavailability of sufficient internal stresses at all the fiber-polymer interfaces that promote the molecular orientation (responsible for the transcrystallization). With a rise in the fiber content to 35 wt%, the transcrystalline regions are found (Fig. 14.14c,d) across almost all the fibers. From the microscopic observations it was found [88] that the regions of

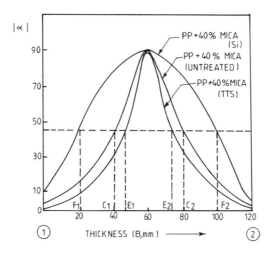

Figure 14.13 *Variation of mica-flake orientation (α) as a function of thickness of the sample for polypropylene composites with untreated, TTS-treated, and silane-treated mica [96].*

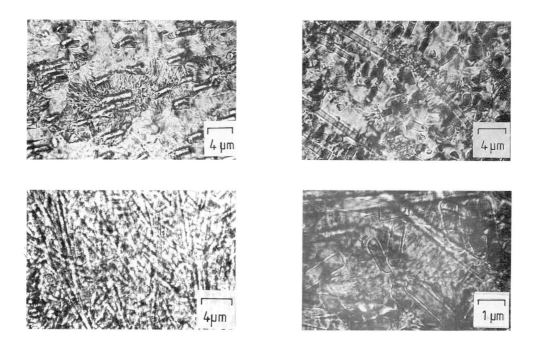

Figure 14.14 *Transcrystallinity variation across fiber surfaces in injection-molded polypropylene composites containing glass fibers 10 wt% (a), 28 wt% (b), and 35 wt% (c,d), respectively [58].*

transcrystallinity were greater in the case of samples molded at low injection pressure. It was reported that low pressure leads to high internal stresses (giving rise to more molecular orientations) and vice versa [97]. Besides the inherent internal stresses that give rise to transcrystallinity, in an injection-molded composite, Burton et al. [54,59] have observed that the irrotational component of the local flow field around high-modulus carbon fibers would produce shish-kebab morphology as well as sheets of chain-extended material in PP composites.

Transcrystalline morphology is also observed in the case of continuous carbon-fiber-reinforced unidirectional composites of PP prepared by the method of film stacking [98,99]. In this method continuous carbon-fiber tows, unidirectionally hand-laid and gripped at the ends, are interleaved with polypropylene films. The whole system was consolidated under heat and pressure using a compression-molding machine. The consolidated laminate was microtomed perpendicular to the unidirectionally laid carbon fibers. The thin sections of the composite with 25 wt% of carbon fiber, when viewed under optical microscope, revealed prevalent transcrystalline interphase (Fig. 14.15).

The interfacial morphology in carbon, Kevlar, and glass-fiber-reinforced PA-6,6 composites was studied by Burton and Folkes [35]. Although under hot-stage optical microscope, the authors observed transcrystalline morphology with Kevlar and high-modulus carbon fibers, their injection-molded components showed no evidence of transcrystallinity. Remelting the sections at 300°C resulted in the development of transcrystalline growth across the high-modulus carbon fibers. The relative importance of the shear and extensional components of the local flow around reinforcing fibers is yet to be investigated and their consequent effects on mechanical properties have yet to be established.

 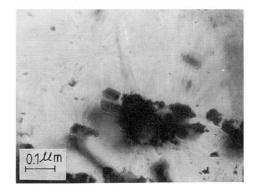

Figure 14.15 *Transcrystallinity in microtomed sections of unidirectional carbon-fiber reinforced compression-molded polypropylene composite: (a) crossed polars and (b) uncrossed polars [99].*

14.5 INFLUENCE OF TRANSCRYSTALLINITY ON COMPOSITE PROPERTIES

The interface between polymer and reinforcement is crucial in determining the properties of a composite since this is where the stress is transferred from the polymer to the reinforcement. In the case in which transcrystallinity occurs around a reinforcing fiber, the matrix properties in the transcrystalline region are observed to be different. Kwei et al. [48] reported that the dynamic modulus of the transcrystalline region in PP and PE is higher than that of the bulk phase. Furthermore, with increasing glass-fiber concentration, the PP transcrystalline regions are increased [58]. Also shish-kebab and chain-extended structures are formed in the case of high-modulus carbon-fiber-reinforced PP [59]. These observations indicate that the reinforcing fibers not only modify the fiber-matrix interface, but also the entire matrix material may be affected. This is a significant factor to be kept in mind while processing thermoplastic composites.

It has been argued, particularly by Schonhorn and his co-workers [57,100], that transcrystalline morphology results in better adhesion between polymer and reinforcement. Obviously, any such method of modifying a polymer so as to avoid the need for a costly coupling agent would be a great advantage to the composite industry. Campbell and Qayyum [29] observed an increase in tensile strength and in modulus along with a marked increase in elongation to break when they introduced continuous uniaxially orientated PA and PET fibers in PP. The significant rise in elongation was attributed both to the increased ductility of the matrix arising from the decrease in spherulite size and to restrained necking of the fibers due to the transcrystalline growth around the fibers.

However, later work in this direction led to negative results. Ritchie and Cherry [101] observed no correlation between the joint strength and the presence or absence of a transcrystalline region. They arrived at this conclusion from the lap shear tests using glass blocks as substrates and HDPE as the adhesive. Their observations are in agreement with those of Tordella [102] and Nakao [103].

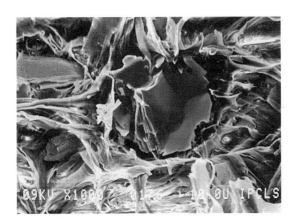

Figure 14.16 *Fracture surface of polypropylene and mica (TTS-treated) composite as viewed under SEM [69].*

Xavier and Sharma [69] investigated the influence of the transcrystallinity on interfacial bonding in PP/mica composites. It was already mentioned in Section 14.3.1 that untreated mica (muscovite mica) as well as mica treated with TTS acted as a nucleating agent giving rise to transcrystalline growth in PP. However, the fracture surfaces of the injection-molded specimens revealed no polymer adhering to the mica flakes, which implies that the interfacial bonding between mica (either coated or uncoated with TTS) and PP is poor. Figure 14.16 shows an SEM micrograph of the tensile-fracture surface of the composite containing 20 wt% of TTS-treated mica. The micrograph shows failure at the polymer-flake interface; there is no polymer sticking to the flakes. This clearly indicates that the interfacial bonding is poor. Similar results were observed [69] in the case of composite with 50 wt% TTS-treated mica. Thus, presence of the transcrystalline morphology does not seem to be related to interfacial bonding in the solid material.

Works of Boaira and Chaffey [104] and Garton [68] lend support to the above observations. From the fracture-surface studies the first authors found that a substantial amount of PP adhered to the mica flakes (phlogopite variety) coated with a silane coupling agent (Dow Corning 6032) and no polymer was adhering to uncoated mica flakes. However, Garton's investigations [68] revealed that the mica treated with the same silane coupling agent did not cause surface nucleation. However, the physical and chemical state of the polymer at the interface was reported to be very different from that of the bulk polymer. Thus, combining the reports by Boaira and Chaffey, with observations made by Garton, one notes that when transcrystalline morphology takes place, the interfacial bonding can be nonexistent and when mica was treated with silane coupling agent, the interfacial bonding was extremely good, but transcrystallinity disappeared. In addition to this, the fracture propagation studies [96,105] on injection-molded PP composites gave an indication that interfacial adhesion between PP and mica (treated with 3-aminopropyltriethoxysilane) exists at least at the flake edges (shown in Fig. 14.17). One can observe PP in the form of elongated fibrils adhering at the mica edges, and indication that adhesive strength here is apparently greater than the cohesive strength of PP. This small improvement in the interfacial adhesion is probably responsible for causing the above changes in microstructure presented in Fig. 14.13 as well as an increase of yield strength and Young's modulus [93].

Apart from the transcrystalline effects, the overall crystallinity of the PP matrix (determined with the differential scanning calorimeter) varied with mica surface treatment as well as mica concentration [91]. The recent investigations [92,94,96] on fracture propagation in particulate-filled PP composites revealed that incorporation of a filler, such as talc, $CaCO_3$, or mica, enhances the ductility of PP in the filler vicinity and thus the composite is enabled to undergo a stable crack propagation instead of a brittle failure. This behavior is likely associated with pinning of the crack front [106] by the particles and the consequent longer times available for plastic deformation of the matrix near the crack front. The crack-pinning

Figure 14.17 (a) SEM fractograph of composite with silane-treated mica, fractured at 25°C, (b) small portion of (a) magnified to resolve the polypropylene/mica interface [96].

mechanism is influenced by the nature of the filler, the filler concentration and also by the filler-surface treatment. It is worth noting that effects of the surface treatment of mica in composites of PP may generate transcrystallinity or not, depending on the temperatures [96]. In other words, the material test conditions somewhat mask the interphase effects.

Masuoka [106] measured the tensile bond strength of mild steel substrates bonded with PA-12. The author observed that the bond strength increased with degree of crystallinity of PA-12, rather than with the transcrystalline interphase.

In spite of all the above-mentioned observations, the phenomena of transcrystallinity still appears to be relevant. There are observations supporting the arguments of Schonhorn and his co-workers. Kardos [107] studied the microstructure and mechanical properties of glass- and carbon-fibers-reinforced polycarbonate composites. He concluded that formation of crystalline entities on the fiber surface was responsible for the significant enhancement of mechanical properties.

Unlike the earlier investigators [29,43], Huson and McGill [108] directly compared the effect of the same fiber in PP matrix with and without transcrystallinity. The authors observed that the transcrystalline layer provides a deformation mechanism for material adjacent to the fiber leading to an increase in elongation at yield.

Lee and Porter [41] studied the crystallization of PEEK in carbon-fiber composites. Their investigations revealed that the number of ordered regions or nuclei in PEEK decreased as the melt annealing time was increased. Reduction of the number of nuclei in the matrix was found to favor PEEK crystallization on the carbon fiber giving rise to transcrystalline interface, which made for a stronger interfacial bond as indicated by transverse tensile tests.

Folkes and Hardwick [109] made a direct study of the structure and properties of transcrystalline layers in PP. In agreement with Masouka [106] and Kwei et al. [48], they observed increased Young's modulus as well as shear and tensile-yield strengths in transcrystalline PP. Within the transcrystalline structure little interconnecting material was available thus aiding the propagation of cracks in these areas. This was reflected in the low strain to failure of the transcrystalline material.

14.6   CONCLUSIONS AND OUTLOOK

It was discussed in this chapter, that reinforcement significantly influences matrix morphology in its vicinity. The effect depends on the nature of reinforcement-matrix interactions, on processing, as well as on the test conditions. The fiber/flake orientation, the skin-core morphology, and the overall crystallization of the matrix undergo considerable change. However, certain issues such as the nature of the nucleation sites and the contribution of transcrystalline interfacial bonding to mechanical properties have yet to be clarified. To resolve these issues, an interdisciplinary approach making use of a variety of techniques is suggested. Since semicrystalline thermoplastic composites are being rapidly pushed to perform increasingly better, it would be desirable to achieve an understanding of all the factors and phenomena so that good correlation between processing parameters, morphological features, and mechanical properties would be established.

ACKNOWLEDGMENTS

I gratefully acknowledge the help and encouragement given by Dr. J. S. Anand and IPCL (R&D) Management.

NOTATION

GF	Glass fiber
HDPE	High-density polyethylene
PA	Polyamide
PE	Polyethylene
PEEK	Polyetheretherketone
PET	Polyethyleneterephthalate
PP	Polypropylene (isotactic)
SEM	Scanning electron microscopy
T_c	Crystallization temperature
T_m	Melting point
ΔT	Degree of undercooling
TTS	Isopropyl triisostearoyl titanate

REFERENCES

1. Woodhams, R. T., Xanthos, M., in *Handbook of Fillers and Reinforcements for Plastics*, Milewski, J. V., Katz, H. S., Eds., Van Nostrand Reinhold, New York (1978).
2. Jenckel, E., Teege, E., Hinrichs, W., *Kolloid–Z*, **129**, 19 (1952).
3. Friedrich, K., *Microstructure and Fracture of Fiber Reinforced Thermoplastic Polyethylene Terephthalate*, Center for Composite Materials, U. Delaware, Report CCM-80-17 (1980).
4. Geil, P. H., *Polymer Single Crystals*, Interscience Publishers, New York (1963).
5. Mandelkern, L., *Crystallization of Polymers*, McGraw-Hill, New York (1964).
6. Sharples, A., *Introduction to Polymer Crystallization*, St. Martin's Press, New York (1966).
7. Price, E. P., in *Encyclopaedia of Polymer Science and Technology*, Vol. 8, Mark, H. F., Gaylord, N. G., Bikales, N. M., Eds., John Wiley & Sons, New York (1968).
8. Price, E. P., in *Nucleation*, Zettlemoyer, A. C., Ed., Marcel Dekker, New York (1969).
9. Samuels, R. J., *Structured Polymer Properties*, John Wiley & Sons, New York (1974).
10. Schultz, J. M., *Polym. Eng. Sci.*, **24**, 770 (1984).
11. Wijga, P. W. O., *U.S. Pat.* 3,207,735; 3,207,736; 3,207,738 (1960).
12. Wales, M., *U.S. Pat.* 3,207,737; 3,207,739 (1961-1962).
13. Binsbergen, F. L., *U.S. Pat.* 3,326,880; 3,327,020; 3,327,021 (1963).
14. Kargin, V. A., Sogolova, T. I., Shaposhnikova, T. K., *Dokl. Akad. Nauk.*, **156**, 1156, 1406 (1964).
15. Vonk, G. C., *Kolloid Z.*, **206**, 121 (1965).
16. Binsbergen, F. L., *Polymer*, **11**, 253 (1970).
17. Rybnikar, F., *Polymer*, **10**, 747 (1969).
18. Rybnikar, F., *J. Appl. Polym. Sci.*, **13**, 827 (1969).
19. Boon, J., *J. Polym. Sci.*, Part A-2, **6**, 1835 (1968).
20. Binsbergen, F. L., *J. Polym. Sci.*, *Polym. Phys. Ed.*, **11**, 117 (1973).
21. Turnbull, T., *J. Chem. Phys.*, **18**, 198 (1950).
22. Rybnikar, F., *J. Appl. Polym. Sci.*, **27**, 1479 (1982).
23. Kargin, V. A., Sogolova, T. I., Rubshtein, V. M., *Polym. Sci. USSR*, **8**, 707 (1966).
24. Lovinger, A. J., Williams, M. L., *J. Appl. Polym. Sci.*, **25**, 1703 (1980).

25. Friedrich, K., Karsch, U. A., *J. Mat. Sci.*, **16**, 2167 (1981).
26. Sandt, A., *PhD thesis*, Ruhr-Universität, Bochum (1981).
27. Mandelkern, L., *Chem. Rev.*, **56**, 903 (1956).
28. Turnbull, T., Vonnegut, B., *Ind. Eng. Chem.*, **44**, 1292 (1952).
29. Campbell, D., Qayyum, M. M., *J. Mat. Sci.*, **12**, 2427 (1977).
30. Chatterjee, A. M., Price, F. P., Newman, S., *J. Polym. Sci., Polym. Phys. Ed.*, **13**, 2369 (1975).
31. Gray, D. G., *J. Polym. Sci., Polym. Lett. Ed.*, **12**, 645 (1974).
32. Kontsky, J. A., Walton, A. G., Bear, E., *J. Polym. Sci.*, Part B, **5**, 185 (1967).
33. Binsbergen, F. L., de Lange, B. G. M., *Polymer*, **11**, 309 (1970).
34. Barriault, R. J., Gronkolz, L. F., *Polym. Sci.*, **18**, 393 (1955).
35. Burton, R. H., Folkes, M. J., *Plast. Rubber Process. Appl.*, **3**, 129 (1983).
36. Eby, R. K., *J. Appl. Phys.*, **35**, 2720 (1964).
37. Schornhorn, H., *J. Polym. Sci.*, **B2**, 165 (1964).
38. Fitchmun, D. R., Newman, S., *J. Polym. Sci., Polym. Lett. Ed.*, **7**, 301 (1969).
39. Fitchmun, D. R., Newman, S., Wiggle, R., *J. Appl. Polym. Sci.*, **14**, 2457 (1970).
40. Fitchmun, D. R., Newman, S., *J. Polym. Sci.*, A-2, **8**, 1545 (1970).
41. Lee, Y., Porter, R. S., *Polym. Eng. Sci.*, **26**, 633 (1986).
42. Hobbs, S. Y., *Nature* (London), *Phys. Sci.*, **234**, 12 (1971).
43. Kantz, M. R., Corneliussen, R. D., *J. Polym. Sci., Polym. Lett. Ed.*, **11**, 279 (1973).
44. Gray, D. G., *J. Polym. Sci., Polym. Lett. Ed.*, **12**, 509 (1974).
45. Thomas, K., Meyer, D. E., *Plast. Rubber Mater. Appl.*, **1**, 35 (1976).
46. Schornhorn, H., Ryan, F. W., *J. Polym. Sci.*, Part A-2, **6**, 231 (1968).
47. Keller, A., *J. Polym. Sci.*, **15**, 31 (1955).
48. Kwei, T. K., Schornhorn, H., Frisch, H. L., *J. Appl. Phys.*, **38**, 2512 (1967).
49. Frisch, H. L., Schornhorn, H., Kwei, T. K., *J. Elastoplast.*, **3**, 214 (1971).
50. Beck, H. N., Ledbetter, H. D., *J. Appl. Polym. Sci.*, **9**, 2131 (1965).
51. Beck, H. N., *J. Appl. Polym. Sci.*, **11**, 673 (1967).
52. Hobbs, S. Y., *Nature*, **239**, 28 (1972).
53. Folkes, M. J., Hardwick, S. J., *J. Mat. Sci., Lett.*, **3**, 1071 (1984).
54. Burton, R. H., Folkes, M. J., in *Mechanical Properties of Reinforced Thermoplastics*, Clegg, D. W., Collyer, A. A., Eds., Elsevier Appl. Sci. Publ., London (1986).
55. Shaner, J. R., Corneliussen, R. D., *J. Polym. Sci.*, A-2, **10**, 1611 (1972).
56. Rybnikar, F., *Polymer*, **10**, 767 (1969).
57. Schonhorn, H., *Macromolecules*, **1**, 145 (1968).
58. Misra, A., Deopura, B. L., Xavier, S. F., Hartley, F. D., Peters, R. H., *Angew. Makromol. Chem.*, **113**, 113 (1983).
59. Burton, R. H., Day, T. M., Folkes, M. J., *Polym. Commun.*, **25**, 361 (1984).
60. Campbell, D., Qayyum, M. M., *J. Polym. Sci., Polym. Phys. Ed.*, **18**, 83 (1970).
61. Campbell, D., White, J. R., *Angew. Makromol. Chem.*, **122**, 61 (1984).
62. Mitchell, J. C., Meier, D. J., *J. Polym. Sci.*, **6**, 1689 (1968).
63. Haas, T. W., Maxwell, B., *Polym. Eng. Sci.*, **9**, 1225 (1969).
64. Huson, M. G., McGill, W. J., *J. Polym. Sci., Polym. Chem. Ed.*, **22**, 3571 (1984).
65. Garton, A., Kim, S. W., Wiles, D. M., *J. Appl. Polym. Sci.*, **27**, 4179 (1982).
66. Rexer, J., Anderson, E., *Polym. Eng. Sci.*, **19**, 1 (1979).
67. Garton, A., Kim, S. W., Wiles, D. M., *J. Polym. Sci., Polym. Lett. Ed.*, **20**, 273 (1982).
68. Garton, A., *Polym. Compos.*, **3**, 189 (1982).
69. Xavier, S. F., Sharma, Y. N., *Angew. Makromol. Chem.*, **127**, 145 (1984).

70.   Kardos, J. L., Cheng, F. S., Tolbert, T. L., *SPE Techn. Pap.*, **18**, 154 (1972).
71.   Kardos, J. L., in *Molecular Characterization of Composite Interfaces*, Ishida, H., Kumar, G., Eds., Plenum Press, New York (1985).
72.   Tianbai, H., Porter, R. S., *J. Appl. Polym. Sci.*, **35**, 1945 (1988).
73.   Velisaris, C. N., Seferis, J. C., *Polym. Eng. Sci.*, **26**, 1574 (1986).
74.   Tung, C. M., Dynes, P. J., *J. Appl. Polym. Sci.*, **33**, 505 (1987).
75.   Bessel, T., Hull, D., Shortall, J. B., *Faraday Spec. Discuss. Chem. Soc.*, **2**, 137 (1972).
76.   Bessel, T., Shortall, J. B., *J. Mat. Sci.*, **10**, 2035 (1975).
77.   Frayer, P. D., Lando, J. B., *J. Polym. Sci., Polym. Lett. Ed.*, **10**, 29 (1972).
78.   Spencer, R. S., *Modern Plast.*, **28**(12), 97 (1950).
79.   Harry, D. H., Parrot, R. G., *Polym. Eng. Sci.*, **10**, 209 (1970).
80.   Kamal, M. R., Kenig, S., *Polym. Eng. Sci.*, **12**, 294 (1972).
81.   Menges, G., Wubken, G., *SPE Techn. Pap.*, **19**, 519 (1973).
82.   Berger, J. L., Gogos, C. G., *Polym. Eng. Sci.*, **13**, 102 (1973).
83.   Tadmor, Z., *J. Appl. Polym. Sci.*, **18**, 1753 (1974).
84.   Clark, E. S., *Plast. Eng.*, **30**(3), 73 (1974).
85.   Kantz, M. R., Newman, H. D., Jr., Stigale, F. H., *J. Appl. Polym. Sci.*, **16**, 1249 (1972).
86.   Xavier, S. F., Misra, A., *Polym. Compos.*, **6**, 93 (1985).
87.   Tyagi, D., Misra, A., *53rd Annual Meeting of the Society of Rheology*, October (1981).
88.   Xavier, S. F., Tyagi, D., Misra, A., *Polym. Compos.*, **3**, 88 (1982).
89.   Friedrich, K., Walter, R., Voss, H., Karger-Kocsis, J., *Compos.*, **17**, 205 (1986).
90.   Malik, T. M., Prudhomme, R. H., *Polym. Compos.*, **7**, 315 (1986).
91.   Xavier, S. F., Sharma, Y. N., *Polym. Compos.*, **7**, 42 (1986).
92.   Xavier, S. F., Schultz, J. M., Friedrich, K., *J. Mat. Sci.*, **25**, 2411 (1990).
93.   Xavier, S. F., *Fracture Mechanics in Particulate Filled Polypropylene Composites*, UNDP/DST Report (1986).
94.   Xavier, S. F., Schultz, J. M., Friedrich, K., *J. Mat. Sci.*, **25**, 2421 (1990).
95.   Remillard, B., Fisa, B., *J. Polym. Eng.*, **6**, 135 (1984).
96.   Xavier, S. F., Schultz, J. M., Friedrich, K., *J. Mat. Sci.*, **25**, 2428 (1990).
97.   Kubat, J., Rigdahl, M., *Polymer*, **16**, 925 (1975).
98.   Cattanach, J. B., Guff, G., Cogswell, F. N., *J. Polym. Eng.*, **6**, 345 (1986).
99.   Xavier, S. F., Mukhopadhyay, D., *unpublished results*.
100. Schonhorn, H., *J. Polym. Sci., Part B*, **5**, 919 (1967).
101. Ritchie, P. J. A., Cherry, B. W., in *Adhesion 1*, Allen, K. W., Ed., Elsevier Applied Science Publishers, London (1977).
102. Tordella, J. P., *J. Appl. Polym. Sci.*, **14**, 1627 (1970).
103. Nakao, K., *J. Adhesion*, **4**, 95 (1972).
104. Boaira, M. S., Chaffey, C. E., *Polym. Eng. Sci.*, **17**, 715 (1977).
105. Newaz, G. M., *J. Mat. Sci., Lett.*, **5**, 71 (1986).
106. Masuoka, M., *Int. J. Adhesion Adhesives*, **1**, 256 (1981).
107. Kardos, J. L., *J. Adhesion*, **5**, 119 (1973).
108. Huson, M. G., McGill, W. J., *J. Polym. Sci., Polym. Phys. Ed.*, **23**, 121 (1985).
109. Folkes, M. J., Hardwick, S. T., *J. Mat. Sci., Lett.*, **6**, 656 (1987).

15. SUBJECT INDEX

D

16. AUTHOR INDEX

A

Abate, G., 223, 224, 230, *237*
Ablazova, T. I., 177, *184*
Ackhurst, S. R., 123, *135*
Acrivos, A., 172, 173, *184*
Adams, W. W., 215, *236*
Addonizio, M. L., 216, *236*
Advani, S. G., 280, *303*
Agassant, J. F., 32, *41*, 279, 280, 282, 293-299, *301—303*
Agur, E. E., 81, *90*
Ahmadi, A. A., 102-104, 109-116, 119-123, *135*
Ahnemiller, J., 138, *162*
Ahroni, S. M., 225, *238*
Akay, G., 279, 280, 282, 300, *302, 303*
Albers, A., 34, 35, *42*
Alfrey, T., 272, *275*
Alglave, H., 279, 301, *302*
Allan, P. S., 30, 31, *41*
Allen, R. C., 215, *236*
Allen, W. F., 15, *39*
Allison, G. R., 80, *90*
Altomare, R. E., 80, *90*
Alzner, B. G., 15, *39*
Amano, T., 216, *236*
Anczurowski, E., 293, *303*
Anderman, H., 32, *41*
Anders, D., 77, 78, 83, *90*
Anderson, D. P., 215, *236*
Anderson, E., 354, *369*
Andrews, E. H., 222, *237*
Andrianova, Z. S., 216, *236*
Anelich, M., 80, *90*
Angeli, S. R., 225, *238*
Annis, Jr., R. E., 138, *162*
Aral, B., 80, *90*
Archambault, P., 215, *235*
Aref-Azar, A., 230, *239*
Arman, J., 195, *211*
Armstrong, R. C., 23, *40*, 58, *67*, 141, *163*, 300, 301, *303*, 307, 313, *330, 331*
Ashby, M. F., 119, *135*
Astarita, G., 83, *90*
Atanassov, A. M., 215, *236*
Avrami, M., 215, *236*

B

H

M